Annual Review of Network Managment and Security

IEC
Chicago, Illinois

About the International Engineering Consortium

The International Engineering Consortium (IEC) is a non-profit organization dedicated to catalyzing technology and business progress worldwide in a range of high technology industries and their university communities. Since 1944, the IEC has provided high-quality educational opportunities for industry professionals, academics, and students. In conjunction with industry-leading companies, the IEC has developed an extensive, free on-line educational program. The IEC conducts industry-university programs that have substantial impact on curricula. It also conducts research and develops publications, conferences, and technological exhibits that address major opportunities and challenges of the information age. More than 70 leading high-technology universities are IEC affiliates, and the IEC handles the affairs of the Electrical and Computer Engineering Department Heads Association and Eta Kappa Nu, the honor society for electrical and computer engineers. The IEC also manages the activities of the Enterprise Communications Consortium.

Other Quality Publications from the International Engineering Consortium

- *Achieving the Triple Play: Technologies and Business Models for Success*
- *Business Models and Drivers for Next-Generation IMS Services*
- *Delivering the Promise of IPTV*
- *Evolving the Access Network*
- *The Basics of IPTV*
- *The Basics of Satellite Communications, Second Edition*
- *The Basics of Telecommunications, Fifth Edition*

For more information on any of these titles, please contact the IEC publications department at +1-312-559-3730 (phone), +1-312-559-4111 (fax), *publications@iec.org*, or via our Web site (http://www.iec.org).

ISBN: 978-1-931695-72-5

International Engineering Consortium
300 West Adams Street, Suite 1210
Chicago, Illinois 60606-5114 USA
+1-312-559-3730 phone
+1-312-559-4111 fax

Contents

Executive Perspectives

Network Management

Next-Generation Services

Security

Telecommunications

Contents by Author

University Program Sponsors

The IEC's University Program, which provides grants for full-time faculty members and their students to attend IEC Forums, is made possible through the generous contributions of its Corporate Members. For more information on Corporate Membership or the University Program, please call +1-312-559-4625 or send an e-mail to *cmp@iec.org*.

Based on knowledge gained at IEC Forums, professors create and update university courses and improve laboratories. Students directly benefit from these advances in university curricula. Since its inception in 1984, the University Program has enhanced the education of more than 500,000 students worldwide.

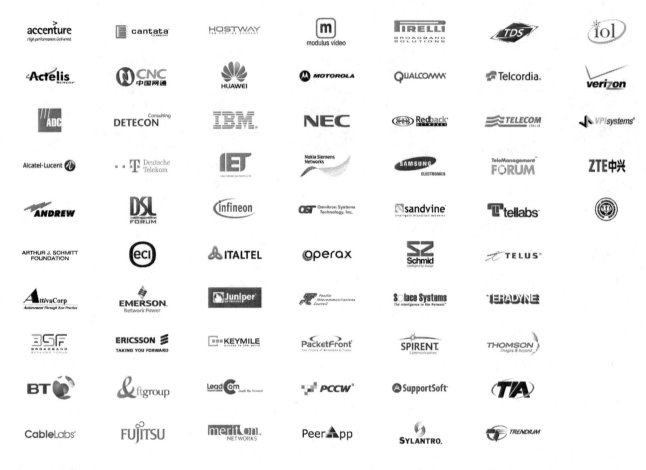

IEC–Affiliated Universities

The University of Arizona
Arizona State University
Auburn University
University of California at Berkeley
University of California, Davis
University of California, Santa Barbara
Carnegie Mellon University
Case Western Reserve University
Clemson University
University of Colorado at Boulder
Columbia University
Cornell University
Drexel University
École Nationale Supérieure des Télécommunications de Bretagne
École Nationale Supérieure des Télécommunications de Paris
École Supérieure d'Électricité
University of Edinburgh
University of Florida
Georgia Institute of Technology

University of Glasgow
Howard University
Illinois Institute of Technology
University of Illinois at Chicago
University of Illinois at Urbana-Champaign
Imperial College of Science, Technology and Medicine
Institut National Polytechnique de Grenoble
Instituto Tecnológico y de Estudios Superiores de Monterrey
Iowa State University
KAIST
The University of Kansas
University of Kentucky
Lehigh University
University College London
Marquette University
University of Maryland at College Park
Massachusetts Institute of Technology
University of Massachusetts

McGill University
Michigan State University
The University of Michigan
University of Minnesota
Mississippi State University
The University of Mississippi
University of Missouri-Columbia
University of Missouri-Rolla
Technische Universität München
Universidad Nacional Autónoma de México
North Carolina State University at Raleigh
Northwestern University
University of Notre Dame
The Ohio State University
Oklahoma State University
The University of Oklahoma
Oregon State University
Université d'Ottawa
The Pennsylvania State University

University of Pennsylvania
University of Pittsburgh
Polytechnic University
Purdue University
The Queen's University of Belfast
Rensselaer Polytechnic Institute
University of Southampton
University of Southern California
Stanford University
Syracuse University
University of Tennessee, Knoxville
Texas A&M University
The University of Texas at Austin
University of Toronto
VA Polytechnic Institute and State University
University of Virginia
University of Washington
University of Wisconsin-Madison
Worcester Polytechnic Institute

Executive
Perspectives

The RBOCs Are Going to Win the High-Speed Access Wars

Clifford R. Holliday
President
B&C Consulting Services

The cable companies have enjoyed a lead in high-speed access for many years, indeed from the start of the high-speed market. In the early 2000s, that lead was as much as 2-to-1. *Figure 1* shows how cable modems (CMs) have led so far this decade.

As an example point, in late 2000, there were about 2 million digital subscriber lines (xDSLs) and about 5 million CMs, and in late 2004, the cable companies had more than 13 million CMs, while the telcos had fewer than 7 million xDSLs. Now the lead is only about 27 million to 22 million.

This lead has been evaporating in recent quarters. This trend has changed so much that we are now forecasting that the telcos will overtake the cable companies in late 2006. Before looking at the graphs showing our new forecast, the qualitative reasons for this market change need to be considered.

Differences in Marketing Approaches: Why the Telcos Are Winning

In the past two to three years, the telcos have moved in a decidedly different direction. Before then, they were marketing high-speed access based largely on the speed of the product, along with the fact that it was essentially "always on" and therefore did not require a long (and sometimes tedious) setup period. Of course, this strategy kept the telcos behind the cable companies, because, while these marketing arguments worked well for marketing against dial-up, they did not differentiate from cable. Cable already had the lead (based as much as anything on a more aggressive initial attack in the market), and the telcos really did nothing to change that. In fact, the approach of the telcos probably helped the cable companies, because they made the case to convert to high-speed access but did not differentiate as to which high-speed access to use. Since cable already had a big lead, this advertising probably helped them more than it did the telcos.

The new direction of the telcos' marketing has been to emphasize a lower price and a communications bundle. The price difference has become dramatic. In fact, a price war seems to have started for high-speed access. SBC started the ball rolling with an offer of $14.95 per month for its introductory high-speed access service. Verizon has responded with the same-priced offer for a slower service but bundled with Yahoo Premium. Since SBC and Verizon really do not directly compete against each other (yet), this was an interesting response. BellSouth has also cut its xDSL offering to $24.95.

These prices are all substantially below what had been the normal rates of around $45 per month most of the cable companies were charging in early 2005.

The other new marketing direction has been to offer bundles. The telcos are offering bundles that include a regular voice landline, a high-speed access line, and some form of video (often satellite) entertainment service. This is the "triple play" that all of the telecommunications companies are seeking to offer. In addition to these services, many are also bundling long-distance and/or cellular service. These bundled offers bring lower prices (even below the discounted high-speed access rates mentioned above) and they offer a single bill each month instead of the flurry of paper that comes from separate subscriptions to these services.

As a result of these new marketing policies, the regional Bell operating companies (RBOCs) are doing much better in high-speed access. *Figure 2* graphs this improvement.

This chart shows the actual in-service high-speed lines and our previous forecast (this forecast was made in 2002) for CMs and xDSLs. Note that the CM actual line is falling more and more below our forecast. It is doing no better than following a straight-line growth path. On the other hand, the xDSL line is slightly above our parabolic forecast line. Even our old forecast had predicted that the telcos would eventually catch up, but we had been suggesting parity in 2007–2008, as the reader can see from the above graph.

In addition to the noted changes, the RBOCs are now making a strong attack on video offerings. Verizon opened its triple-play assault with its first video service on its fiber-to-the-premises (FTTP) architecture, FiOS. Verizon debuted video in late September in Keller, Texas, the same city that saw the original debut of Verizon's FTTP service last year. However, this is not the IPTV service that Verizon has said

FIGURE 1

DSL as a Percent of Total High-Speed Access Lines

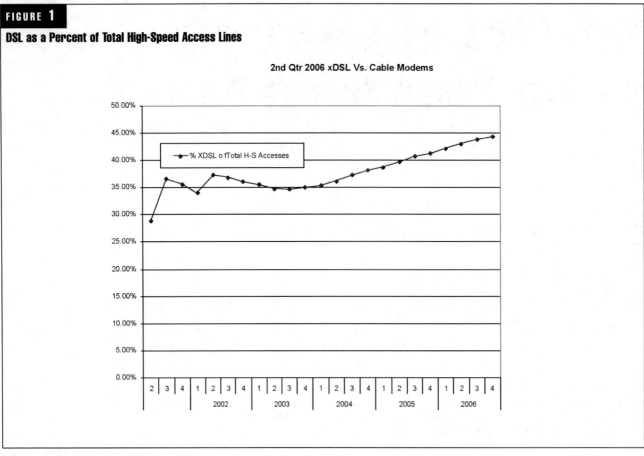

2nd Qtr 2006 xDSL Vs. Cable Modems

FIGURE 2

DSL as a Percent of Total High-Speed Access Lines

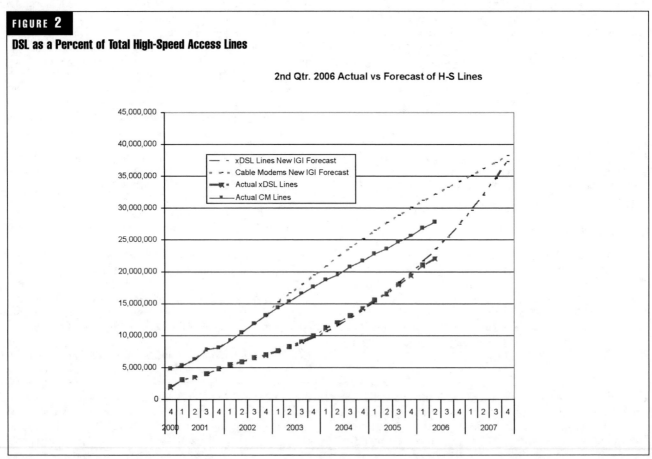

2nd Qtr. 2006 Actual vs Forecast of H-S Lines

it will eventually make available on FiOS. It appears that IPTV may not be quite ready for prime time. Verizon offers more than 180 channels on one of its more popular video packages.

Verizon followed up its Keller introduction by announcing that FTTP video would be available in 21 more Texas cities by the end of 2006. Texas has a statewide franchising law making it relatively easy to introduce video on the FTTP infrastructure by avoiding the need to negotiate franchises with each city. Verizon hopes to make similar announcements in other locations throughout the United States.

In November 2005, SBC was granted a state video franchise for Texas. SBC states that it plans to offer video via its Project Lightspeed infrastructure first in San Antonio in a pilot market (where its Lightspeed architecture has been on field trial) in a few months and expand to 20 nearby cities in 2006. The Lightspeed video offering has been planned to be IPTV. It will be interesting to see if SBC has an IPTV offering ready for commercial rollout.

These video ventures make the telcos' bundles even more attractive and promise to be a new and powerful competitor for the cable companies.

To further make this point, *Figure 3* illustrates the results of the major cable companies versus the RBOCs.

Figure 3 illustrates why the telcos are rapidly catching up with cable in high-speed access. The telco trend line is posi-tive, while the cable trend line is flat. We were, perhaps, the first analysts to forecast this turn in events, back when the cable companies were still going strong and had a 2-to-1 lead over the telcos. That cable lead has shrunk to only about 55 percent now.

Given the current trend in the data and what seems to be an accelerating advantage for the telcos, we are revising our forecast to predict that the telcos will overtake the cable companies by mid- to late 2006! We are including both xDSL– and FTTP–derived circuits in the telcos' forecast. *Figure 4* shows how we now think the market is going to develop.

The Cable Companies Strike Back, but Is It Enough?

The first reaction of the cable companies to this change in their fortunes was to attack on a feature basis. The cable companies appear to have been trying to compete with fea-ture differentiation—mainly by adding speed. Many are now offering speeds that approach even the FTTP speeds (at least the lower-level offerings of FTTP). Before, high-speed access was a niche market, with early adopters being the main market target. However, with more than 40 percent of the nation's households now equipped with high-speed access, this has turned into a commodity market. In a com-modity market, features really do not count. Those who would buy for features (speed) did so early in the growth curve. Those who are buying now will buy as one selects any commodity (assuming availability of options)—by price

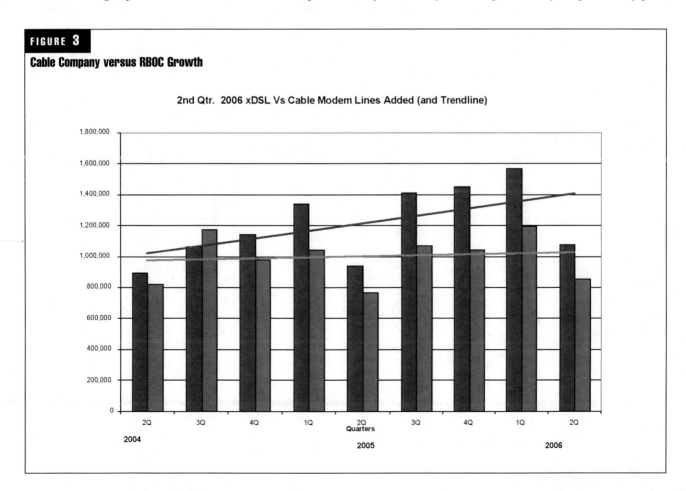

FIGURE 3

Cable Company versus RBOC Growth

2nd Qtr. 2006 xDSL Vs Cable Modem Lines Added (and Trendline)

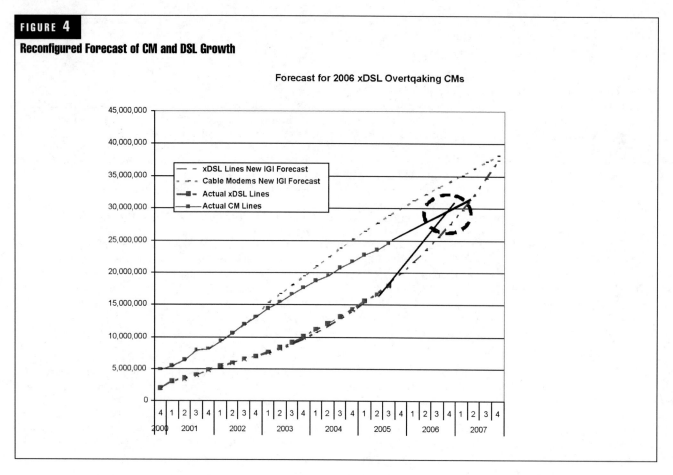

FIGURE 4

Reconfigured Forecast of CM and DSL Growth

and packaging. It appears that this rush to add speed may be a misdirected effort by the cable companies.

In late 2005 and early 2006, the cable companies responded strongly with price and packaging changes. Cox and Rogers now have cable modem services at $19.95 per month. Comcast and Charter are offering service for $19.99 per month, while Time-Warner is lagging behind with services at $29.95 per month.

In addition, many of the cable companies are offering triple plays of their own, bundling cable TV, high-speed access, and voice over IP (VoIP) phone service. This seems to be the right response, but there still is a question as to how con-

sumers will view the cable companies as providers of voice services. Many opinion surveys suggest that the cable companies are not viewed as desirable to entrust with what many consider to be a necessary service.

High-speed access has become of great importance to the cable companies, with Comcast reporting—for the first time for any company—more than $1 billion in quarterly revenue for high-speed access. This importance means that the cable companies will not lightly give up their lead in this market. This is a drama that will play out for the next year or two. It is, however, imperative for the cable companies to quickly develop an effective strategy before they lose the source of substantial revenue growth.

GPON: The Next Big Thing in Optical Access Networks

A Comparison of EPON, APON, and the Emerging GPON Technology

Oren Marmur

Chief Technology Officer
FlexLight Networks

Abstract

This paper is a comprehensive review of the various passive optical network (PON) technologies in the marketplace today, namely asynchronous transfer mode (ATM) PON (APON), Ethernet PON (EPON), and gigabit PON (GPON), and draws an in-depth comparison between them.

The emerging International Telecommunication Union Telecommunication Standardization Sector (ITU–T) G.984 series GPON technology will be examined in detail after a review of the history of the various PON flavors and of the service requirements set forth by service providers. System performance among the various protocols is compared using efficiency and scalability factors, and conclusions are drawn regarding the overall efficiency of and cost influence on the solution.

GPON carries a twofold promise of higher bit rates and higher efficiency when carrying multiple services over the PON. It offers a scalable framing structure from 622 Mbps to 2.5 Gbps, as well as support for asymmetric bit rates, exceptionally high bandwidth utilization for any type of service, and GPON encapsulation method (GEM) of any type of service (both time division multiplex [TDM] and packet) onto a synchronous transport protocol. It is shown that in the worst-case scenario, based upon the most conservative assumptions regarding traffic distribution, GPON is substantially more efficient, with an overall efficiency of 93 percent, compared to 71 percent with APON and 49 percent with EPON.

Using a more detailed analysis based upon a traffic model provided by the service providers in the Full-Service Access Network (FSAN) Group, it is shown in quantitative terms that GPON offers exceptionally higher bandwidth for the entire range of applications when compared to both APON and, especially, EPON, resulting in substantially lower cost per bit and a much faster payback period.

The History of PON

To fully understand the various PON flavors and the merits of GPON as the next leading technology, it is important to first cover the history of PON development over the past few years.

The basic principle of a PON is to share the central office (CO) equipment (optical line terminal [OLT]) and the feeder fiber among as many end units (optical network termination [ONT]) as possible within physical and bandwidth constraints. Since this solution requires less fiber layout to cover a specific area and less costly optical interfaces at the CO (one optical interface serves the entire network), the solution offered enables high-speed optical connections for businesses or residential units in scenarios that could not be served in an economical manner using traditional point-to-point or ring architectures.

The first PON activity was initiated in the mid-1990s when a group of major network operators established the FSAN Group. The group's goal was to define a common standard for PON equipment so vendors and operators could come together in a competitive market for PON equipment. The result of this first effort was the 155 Mbps PON system specified in the ITU–T G.983 series of standards. This system has become known as the BPON system, and it uses ATM as its bearer protocol (known as the APON protocol). The acronym BPON was introduced because APON by itself led people to assume that only ATM services could be provided to end users. Changing the name to BPON reflected BPON systems' ability to offer broadband services, including Ethernet access, video distribution, and high-speed leased line services. However, the most common and well-known name for the first generation of FSAN systems is APON.

The APON standards were later enhanced to support 622 Mbps bit rates as well as additional features in the form of protection such as dynamic bandwidth allocation (DBA).

On a parallel track, in early 2001, the Institute of Electrical and Electronics Engineers (IEEE) established the Ethernet in the First Mile (EFM) Task Force, realizing the enormous prospect that lies ahead in the optical access market. The group works under the auspices of the IEEE 802.3 group, which also developed the Ethernet standards, and as such is restricted in architecture and compliance to the existing 802.3 media access control (MAC) layer. The EFM's work is concentrated on standardizing a 1.25 Gbps symmetrical system for Ethernet transport only.

In 2001, the FSAN Group initiated a new effort for standardizing PON networks operating at bit rates above 1 Gbps. Apart from the need to support higher bit rates, the overall protocol has been opened for reconsideration, and the solution should be the most optimal and efficient in terms of support for multiple services and operation, administration, maintenance, and provisioning (OAM&P) functionality and scalability.

As a result of this latest FSAN effort, a new solution has emerged in the optical access market place, namely Gigabit PON (GPON), which offers unprecedented high bit-rate support while enabling the transport of multiple services, specifically data and TDM, in native formats and with extremely high efficiency.

Figure 1 depicts the general architecture and topology of PON networks in general and specific features related to

GPON, and will serve as a reference for the discussed network application. FlexLight's GPON architecture is used in *Figure 1* as an illustration.

Gigabit Service Requirements

An important attribute of FSAN is that it is a customer-driven group in which the telecommunications service providers set forth the service requirements. These in turn lay the basic foundations for the proposed solution. This is in strong contrast to the EFM effort, which is more vendor-driven toward a simple and Ethernet-compliant solution, rather than toward remedies to existing service and network demands.

As part of the GPON effort, a gigabit service requirement (GSR) document has been implemented based upon the collective requirements of all member service providers, representing the leading regional Bell operating companies (RBOCs) and incumbent local-exchange carriers (ILECs) of the world. The document was also recently submitted as an official recommendation to the ITU–T (recommendation G.984.1).

The main requirements resulting from the GSR document can be summarized as follows:

- Full service support, including voice (TDM, synchronous optical network [SONET], and synchronous dig-

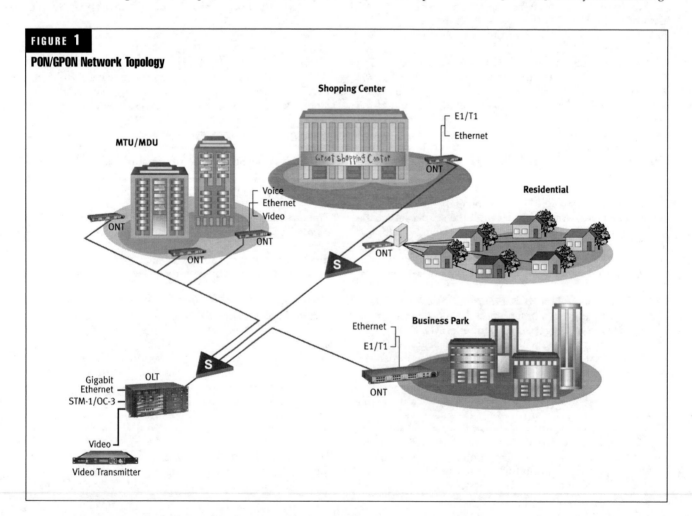

FIGURE 1

PON/GPON Network Topology

ital hierarchy [SDH]), Ethernet (10/100 BaseT-10 or 100 Mbps running on a twisted pair), ATM, leased lines, and others

- Physical reach of at least 20 km with a logical reach support within the protocol of 60 km
- Support for various bit-rate options using the same protocol, including symmetrical 622 Mbps, symmetrical 1.25 Gbps, 2.5 Gbps downstream, and 1.25 Gbps upstream
- Strong OAM&P capabilities offering end-to-end service management
- Security at the protocol level for downstream traffic because of the multicast nature of PON

Consequently, GPON systems that are standardized and developed nowadays are based on the entire set of requirements laid out by the GSR document, and thus represent a pervasive solution to service providers' needs.

Various Flavors of PON

Apart from GPON, two alternative technologies exist for PON networks: APON (representing the incumbency of lower bit-rate systems) and EPON (representing the emerging standards within the IEEE EFM group).

APON: ATM–Based PONs

APON systems are based upon ATM as the bearer protocol. Downstream transmission is a continuous ATM stream at a bit rate of 155.52 Mbps or 622.08 Mbps with dedicated physical layer operations, administration, and maintenance (PLOAM) cells inserted into the data stream. Upstream transmission is in the form of bursts of ATM cells, with a 3-byte physical overhead appended to each 53-byte cell to allow for burst transmission and reception.

The transmission protocol is based upon a downstream frame of 56 ATM cells (53 bytes each) for the basic rate of 155 Mbps, scaling up with bit rate to 224 cells for 622 Mbps. The upstream frame format is 53 cells of 56 bytes each (53 bytes of ATM cell plus 3 bytes overhead) for the basic 155 Mbps rate.

The downstream frame is constructed from two PLOAM cells, with one at the beginning of the frame and one in the middle, and 54 data ATM cells. Each PLOAM cell contains grants for upstream transmission relating to specific cells within the upstream frame (53 grants for the 53 upstream frame cells are mapped into the PLOAM cells) as well as OAM&P messages.

Upstream transmission consists of either a data cell, containing ATM data in the form of virtual paths and virtual circuits (VPs/VCs), or a PLOAM cell when granted a PLOAM opportunity from the central OLT.

APON provides a very rich and exhaustive set of operations, administration, and maintenance (OA&M) features, including bit-error rate (BER) monitoring, alarms and defects, auto-discovery and automatic ranging, and churning as a security mechanism for downstream traffic encryption. However, APON systems suffer from two substantial drawbacks: low overall efficiency for data transport and the complexity of adapting and provisioning services over an ATM layer.

EPON: Ethernet-Based PONs

Ethernet for subscriber access networks, also referred to as EFM, combines a minimal set of extensions to the IEEE 802.3 MAC and MAC control sub-layers with a family of physical

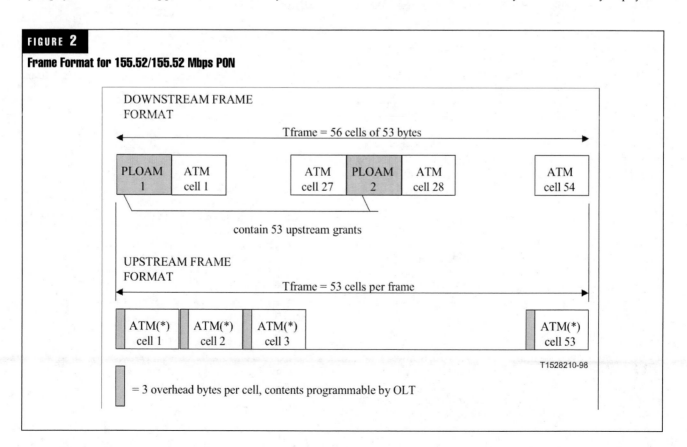

FIGURE 2

Frame Format for 155.52/155.52 Mbps PON

DOWNSTREAM FRAME FORMAT

Tframe = 56 cells of 53 bytes

| PLOAM 1 | ATM cell 1 | ... | ATM cell 27 | PLOAM 2 | ATM cell 28 | ... | ATM cell 54 |

contain 53 upstream grants

UPSTREAM FRAME FORMAT

Tframe = 53 cells per frame

| ATM(*) cell 1 | ATM(*) cell 2 | ATM(*) cell 3 | ... | ATM(*) cell 53 |

T1528210-98

= 3 overhead bytes per cell, contents programmable by OLT

layers (PHY). These PHYs include optical fiber and unshielded twisted pair (UTP) copper cable physical-medium dependent (PMD) sub-layers for point-to-point connections in subscriber access networks. EFM also introduces the concept of EPONs, in which a point-to-multipoint (PTMP) network topology is implemented with passive optical splitters, along with optical-fiber PMD sub-layers that support this topology. In addition, a mechanism for network OA&M is included to facilitate network operation and troubleshooting.

EPON is based on a mechanism called multipoint control protocol (MPCP), defined as a function within the MAC control sub-layer. MPCP uses messages, state machines, and timers to control access to a PTMP topology. Each optical network unit (ONU) in the PTMP topology contains an instance of the MPCP protocol, which communicates with an instance of MPCP in the OLT.

The basis of the EPON/MPCP protocol lies in the point-to-point (PTP) emulation sub-layer, which makes an underlying PTMP network appear as a collection of PTP links to the higher protocol layers (at and above the MAC client). It achieves this by attaching a logical link identification (LLID) to the beginning of each packet and replacing two octets of the preamble.

EPON—which, as a protocol, is still under work within the IEEE EFM Task Force—has many issues, including those relating to the protocol itself, OA&M functionality, and PHY specifications that still lack a detailed definition. EPON suffers from two substantial drawbacks, namely extremely low efficiency and the lack of ability to support any service but Ethernet over the PON, thus introducing quality of service (QoS) issues when dealing with voice/TDM services. While the efficiency issue will be discussed in more detail later on, it is important to mention here that EPON uses 8b/10b encoding as the line code. The line code ensures sufficient balance between 0s and 1s in the bit stream, which is crucial for the proper operation of any communication link.

The line code itself introduces a 20 percent bandwidth penalty, resulting in a starting point of 1 Gbps out of the line rate of 1.25 Gbps, before dealing with the protocol itself. APON and GPON systems, however, use scrambling as the line code, which is the same mechanism used for line coding in any SONET or SDH network, with the benefit of no bandwidth penalty as bits are only changed and not added.

GPON: The Native Mode PON

GPON carries a twofold promise of higher bit rates and higher efficiency when carrying multiple services over the PON. When initiated, the GPON was intended as a complete bottom-up reconsideration of PON applications and requirements and, as such, laid the foundation for new solutions that are not based upon the previous APON standard.

While much of the functionality that is not directly related to the PON is preserved, including OA&M messages and DBA, GPON is based upon a completely new transmission convergence (TC) layer. The FSAN has recently selected the proposal put forward by FlexLight and numerous additional vendors for a frame-based protocol using GEM for

service mapping, as the next GPON protocol. The standard was finalized as ITU–T Recommendation G.984.3 on February 2004. By now GPON is deployed by several carriers across the world and is gaining substantial traction by firm plans of several tier-1 operators to deploy GPON as their access solution.

Starting with the GPON work, the following objectives were put forward:

- Scalable framing structure for 622Mbps to 2.5Gbps, as well as asymmetric bit-rate support
- Exceptionally high bandwidth utilization/efficiency for any type of service
- GPON encapsulation method (GEM) of any type of service (both TDM and packet) into 125 msec periodic frames—high efficiency with no overhead transport of native TDM traffic required
- Dynamic allocation of upstream bandwidth via bandwidth maps (pointers) for each ONT

At this point, a few words on GPON GEM are required. GEM is based on the ITU GFP standard (ITU–T G.7041), with some minor modifications to make it optimized for PON topologies. GEM provides a generic mechanism to adapt traffic from higher-layer client signals over a transport network.

Since GEM provides a generic mechanism to transport different services in an efficient and simple manner over a synchronous transport network, it is ideally suited as the basis for the GPON TC layer. In addition, the fact that in using GEM, the GPON TC layer is synchronous in nature and uses the standard SONET 8 kHz (125 Ìsec) frame enables straightforward support for TDM services.

Summarizing the design choices for the GPON protocol, the following items can be mentioned:

- Frame-based, multiservice (ATM, TDM, data) transport over PON
- Upstream bandwidth allocation mechanism via slot assignments through pointers
- Support for asymmetric line rate operation—2.488 Gbps downstream and 1.244 Gbps upstream rates
- Line coding will be non-return-to-zero (NRZ) with scrambling
- Out-of-band control channel at PHY for OA&M functions using G983 PLOAMs
- Fragmentation and concatenation of data frames for bandwidth efficiency
- Upstream burst mode preamble, including clock and data recovery (CDR), will not be long
- DBA reports, security, and survivability overhead are integrated into the PHY
- Cyclic redundancy codes (CRCs) for framing header protection and bit-interleaved-parity (BIP) to support BER estimation
- QoS supported at the PHY

Figure 2 and Figure 3 depict the frame structure of the GPON protocol for upstream and downstream directions.

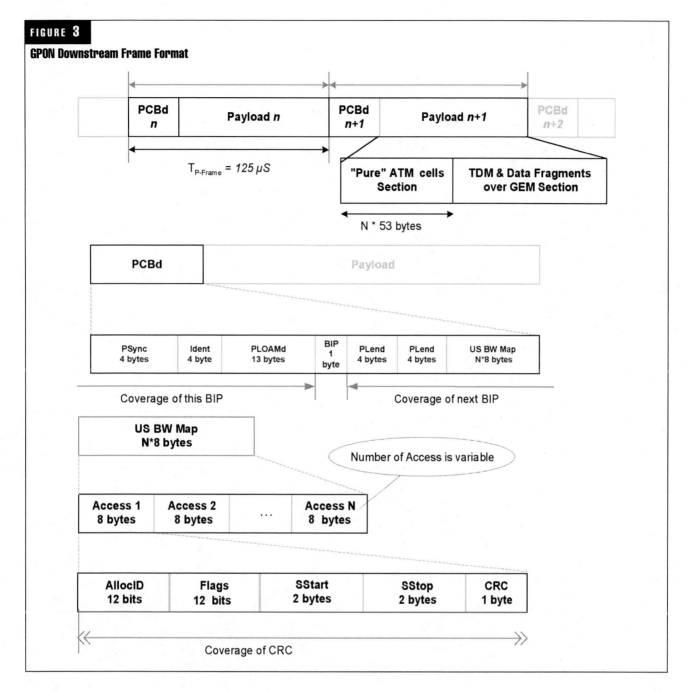

FIGURE 3

GPON Downstream Frame Format

Efficiency and System Performance

The most important factor in analyzing a solution's overall cost is efficiency, providing the overall bandwidth that can be sold as services over the system.

When comparing various PON systems such as APON, EPON, or GPON, and assuming a similar bit rate of 1.25 Gbps, it can be safely assumed that the system cost itself will be very similar. A substantial portion of the system cost originates from the optical interface, which is independent of the PON protocol, while the rest of the system components are expected to be of similar prices based upon application-specific integrated circuits (ASICs) and other standard components. This is not to say that the overall solution will be of the same cost as, for example, when using an EPON system

in which additional voice over Internet protocol (VoIP) equipment is required, which would add an additional cost factor.

Assuming similar cost figures for the system itself, efficiency is the single most dominant factor when determining the cost per bit or the amount of revenue bits that can be extracted from the network. A 100 percent efficient network will provide 1.25 Gbps of available throughout, while a 50 percent efficient network would provide only 622 Mbps of throughout and thus two systems would be required for the same network configuration or twice the cost of a system.

Efficiency Comparison
When comparing efficiencies of different PON protocols, four factors should be taken into consideration: line coding,

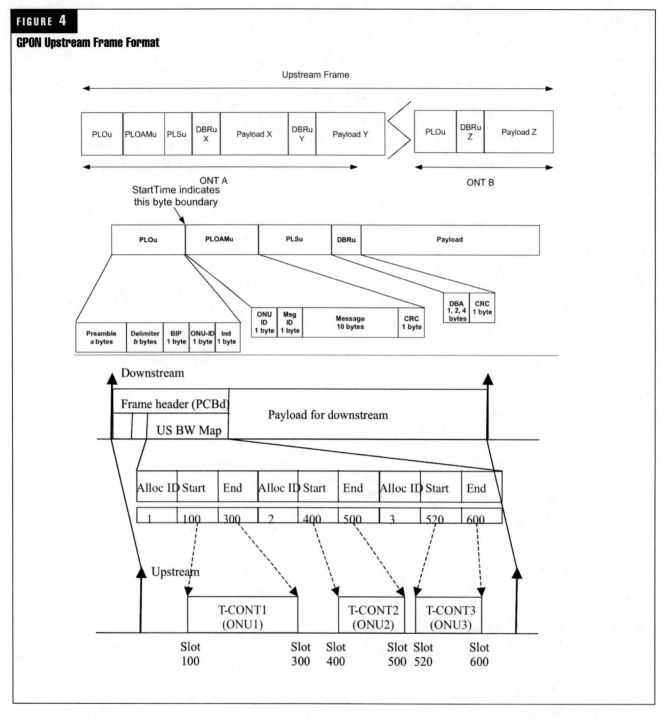

FIGURE 4

GPON Upstream Frame Format

PON TC or MAC layer efficiency, bearer protocol (ATM, Ethernet, or GEM) efficiency, and service adaptation efficiency. It should be noted that, when analyzing APON efficiency, the protocol assumptions for APON at 622 Mbps have been taken and extended for 1.25 Gbps, as the standard only reaches a 622 Mbps bit rate.

Table 1 summarizes the various factors in PON protocol efficiency analysis.

The Ethernet efficiency factor has been calculated assuming Ethernet packet size distribution according to collected data (56.3 percent 64-byte packets, 28.1 percent 512-byte packets,

and 15.6 percent 1,518-byte packets). Amazingly, EPON is the least efficient protocol, even when dealing solely with Ethernet services, because of the high tax as a result of the lack of fragmentation between consecutive frames.

Table 2 summarizes the overall PON efficiency for the different protocols, including all four factors for two scenarios of TDM/data distribution.

It should be stressed that the above results are absolute worst-case scenarios in terms of GPON, as the assumptions taken and the set of services including traffic distribution are all at their most conservative values.

TABLE 1

PON Protocols Efficiency Factors

	Line Coding	PON TC Layer Efficiency	Bearer Protocol Efficiency	Service Adaptation	
				T1	FE
APON	100%	96%	90%	98%	80%
EPON	80%	98%	97%	72%	63%
GPON	100%	99%	100%	96%	94%

TABLE 2

Overall PON Efficiency

	Overall Efficiency 10% TDM, 90% Data	Overall Efficiency 20% TDM, 80% Data
APON	71%	72%
EPON	49%	49%
GPON	93%	94%

System Performance

In order to analyze the overall effect of the PON protocol efficiency on system performance, operators within the FSAN group have issued a traffic model depicting several scenarios for PON deployment.

The traffic models cover all relevant parameters, including service distribution, latency requirements, and the number of ONTs over a PON, and relates to different applications such as fiber to the home (FTTH) and fiber to the business (FTTB).

Figure 5 presents the additional bandwidth provided by GPON, compared to APON and EPON, for the different applications according to the FSAN traffic model. As can clearly be seen from the figure, GPON offers exceptionally higher bandwidth for the entire range of applications when compared to both APON and, especially, EPON.

Coupled with the fact that an EPON system has no mechanism for supporting TDM traffic and thus requires external adaptation equipment, an EPON solution is expected to be substantially higher in both absolute cost per system and even more in price per bit of data.

Scalability in a Multiservice Environment

GPON not only provides substantially higher efficiency as a transport network, but also delivers simplicity and superb scalability for future expansion in supporting additional services.

As the access network is the closest layer within the network hierarchy to the end user, it is characterized by an abun-dance of protocols and services, starting with numerous TDM and data services today and expanding in the future to additional applications, including storage-area networks (SANs) and digital video.

GPON, through the GEM adaptation method, offers a clear migration path for adding these services onto the PON without disrupting existing equipment or altering the transport layer in any way. In contrast to APON and EPON, which require a specific adaptation method for each service and, furthermore, require development of new methods for emerging services, the core foundation of GPON is GEM, which already covers adaptation schemes for any possible service.

Summary

GPON is the most advanced PON protocol in the market-place today, offering multiple-service support with the rich-est possible set of OAM&P features. It offers far higher efficiency when compared to both APON and, especially, EPON. Additional bandwidth offered over the same system ranges from 40 percent to 160 percent, depending upon the specific application and supported services. GPON offers the lowest cost system for all modes of operation. Not only is the system cost itself expected to be lower, as no external adaptation is required, but the exceptionally higher efficiency also leads to much more revenue bits from the same system, for instance, as a much shorter payback period. GPON ensures simplicity and scalability when it comes to dealing with new and emerging services. A clear migration path is offered for emerging services without any disruption to existing GPON equipment or alterations to the transport layer.

Conceptual Barriers for Effective Network Design

George Mattathil

Chief Executive Officer

Strategic Advisory Group

Abstract

Communication networks have become a critical economic infrastructure, resulting in enhanced interdependency among technology, business, regulations, and financial markets. Factoring implications of technology in business and economic decision making has not kept up with the increased role of technology in the economy. Recent events have demonstrated that autonomous decision processes in these interdependent systems can result in large-scale negative economic impact, as when the choices are not on target. Unlike in the past, developing effective decision processes consistent across all interdependent economic systems are now essential for sustained network infrastructure enhancements. Despite the burst of the Internet bubble and the telecom meltdown, the enthusiasm for all-packet networks remains strong. However, the packet-centric network strategies overlook some technology-based fundamentals that cannot be overcome solely with enthusiasm. This report provides a historical perspective of the issue, discusses technical challenges for the packet-centric trends that are self-limiting, and offers an alternate model for evolutionary enhancement of the network infrastructure.

Historical Background

In 1961, Leonard Kleinrock provided the theoretical foundation for packet switching.[i] At the time, computers were an emerging new technology and the public switched telephone network (PSTN) was predominant, thus marking an accumulation of technology developed over a century ago since the first telephone call in 1876. The telephone network is based on circuit switching or time division multiplexing (TDM). Common carriers had established monopoly in network services by the 1920s and, compared to the emerging field of computers, telecommunication was mature with established disciplines.

In fact, two schools of thought emerged in regards to the use of computers and communications. The telecommunication designers viewed computers as the means for increasing the efficiency of switching systems. But J.C.R. Licklider, with his vision of galactic network, led the second school of thought, based on using computers as a way to enhance human communication.[ii] Computer-based human communication enhancement created the need to exchange information between computers. This data communication need is sporadic in nature and did not match the network services available. The common carrier usage-based business model was not cost-effective for computer networking. Further, the commercial interests of the common carriers prevented them from offering solutions that could have met the nascent computer networking needs. The result was to energize the early pioneers of computer networking to look for alternate methods to the circuit connections offered by carriers.[iii] The alternate network technology was packet switching, invented by Paul Baran and Donald Davies.[iv,v] This early divergence is the primary cause for the current polarization between packet-switching and circuit-switching camps. As consequence, fractious network design and divisive technology choices do not meet current network needs, which only resulted in wasted effort and resources.[vi] In fact, the current animosity of packet enthusiasts toward circuit switching is reminiscent of the reaction of telephony specialists toward packet-technology pioneers.[vii] Contrary to the popular myth that PSTN is horse-and-buggy technology, in fact, it is almost as modern as the packet technology. Major overhaul of the telephone network, the conversion of the old frequency multiplexing systems to time division multiplexing (TDM), and computerized switching took place in the 1960s; this took place during the same time as packet-switching developments.[viii] However, most TDM products have not been upgraded to take advantage of the recent developments in semiconductors and software. This technology-generation advantage of packet systems is often wrongly cited as benefits of packet switching. This overlooks the fact that TDM or hybrid systems could produce better results with the advances in semiconductor and software technologies.

Technical Background

An all-packet conceptual model (see *Figure 1*) serves as a guidepost for network design. The strength of the all-packet model is its simplicity, by mandating that all network service requirements be solved using packet switching, leading to the promotion of technologies beyond their technical merits, including voice over Internet protocol (VoIP), video over IP, and IPv6. This report presents an alternate reference model for network evolution, the Transfer Network Architecture (TNA). The TNA model (see *Figure 2*) closely represents the current state of networks, compared to the all-packet models in vogue. It introduces an evolutionary approach that makes realistic assumptions about the current state of the voice and data infrastructures while providing a framework for a unified next-generation network (NGN) infrastructure.

One of the assumptions challenged and corrected by the TNA model is that future networks will be packet-centric networks, displacing the existing circuit-switching networks.[ix] The nearly universal acceptance of this idea generated support for various technologies that could make packet-centric networks possible, including VoIP and asynchronous transfer mode (ATM). However, the pioneering initiatives for building fully capable packet-centric networks have failed.[x] This consequent failure was not because of technical feasibility, but rather the practicality of wholesale migration to packet-based systems when there is a huge installed infrastructure that is not packet-based. As the network industry looks to the future for planning, analyzing issues and developing solutions for network evolution, they must take into account practical considerations and be cognizant of the limitations of packet-centric strategies.

Current Trend

Current industry direction aims toward packet-only models for network evolution[xi] (see *Figure 1*). The underlying rationale is that the current network infrastructure, dominated by circuit switches for telephone services, cannot provide Internet (packet) services adequately. Therefore, if it were replaced by a packet infrastructure that supports any type of connection, including telephones, Internet, and video, then it would overcome the current mismatch between infrastructure capabilities and subscriber service needs.

This argument is simplistic because it does not take into account the full range of requirements that need to be considered when designing networks for human communication and instead focuses on a single metric (bit transfer). In addition, the voice network is already in place. So reimplementing the voice service capabilities using a packet network does not provide added value and creates problems[xii]. Another key factor overlooked is that the substitution cost of the current voice network with an equivalent packet network is cost-prohibitive, without any compensating benefits.

Another reason some favor packet-centric networks is the possibility of standardizing network operations management. Having to manage only one network (packet network)—rather than one for packet and another for voice as nowadays—would be simpler and could provide operational efficiency. This argument, however, is based on misplaced priorities, as there are major gaps between network service needs and deployed capabilities. Current network capabilities include full-featured rich-media applications, effective solutions for current network vulnerabilities, and containment of the nuisance of spam e-mails and phishing scams. Carrier operational efficiency will be a natural goal, especially once the customer service needs are met, at least

minimally. Further, the apparent operational efficiencies cited for packet-centric networks are more likely because of reduction in staffing with corresponding degradation in service level, improved management techniques, and basic technologies that are available for new implementations that were unavailable for older systems. It is possible that if older circuit-switched systems are upgraded to take advantage of new management techniques, the resultant operational efficiencies are likely to be comparable. There are no direct system management efficiencies that result from packet-based transport layer functions, but packet layer injects high overhead for real-time traffic.

Obviously, the PSTN operates inefficiently given that many technical problems were solved by using then–state-of-the-art means that are out-of-date by current standards. However, these inefficiencies do not diminish the technical and engineering reasons why the packet-only strategies will run into difficulties.[xiii] One of the major obstacles for building a packet-only network is that the investment in existing PSTNs runs into hundreds of billions or even trillions of dollars. What is often overlooked is that when there is a huge deployed infrastructure performing a useful telephone service, there is marginal value in replacing it with lower-grade VoIP service. The real problem is the usage-based pricing for phone services. The real solution lies in resolving the pricing discrepancies without dismantling a highly reliable and useful infrastructure.

The essence of the packet-centric solution is based on the assumption that surplus bandwidth can be taken for granted. Hence the net benefits derived by having to implement only one form of switching, namely packet switching, will produce better results than systems with more complex architecture. However, the availability of surplus bandwidth does not hold true at network edges, where efficient bandwidth allocation and utilization are critical requirements.

Changed Reality

The network landscape was totally different when key packet technology developments were contemplated in the 1960s, when bandwidth was at a premium. The proposed line speed for use in the ARPANET was 50 kbps.[xiv] Under those line speeds, maximizing bandwidth utilization was paramount, and statistical multiplexing (packet switching) provided higher bandwidth utilization.[xv] Assumptions based on these considerations were embedded in the packet-switching developments that followed. However, innovations in fiber optics and dense wave division multiplexing (DWDM) have made it no longer necessary to maximize bandwidth utilization.[xvi] With DWDM systems, a fiber that was built to carry 2.5 Gbps traffic can be enhanced to carry upward of 400 Gbps without changing the fiber. These developments have made the original rationale for packet-centric networking obsolete. In other words, packet technology is the natural choice when peak bandwidth demand is greater than available bandwidth. Yet, since bandwidth is no longer an absolute limitation, the case for packet switching is not axiomatic now as it was in the past.

The semiconductor industry has been a witness to a similar situation. Early integrated circuits (IC) had tens and later hundreds of transistors. Back then, optimizing the design to use every transistor in an IC was a worthwhile goal. Today, when chips contain transistors in tens and hundreds of millions, utilizing every transistor in an IC is not useful since the goal is to maximize the functionality of the overall design. The lack of such considerations regarding networks is the substantial resource allocations toward terabit networks, even though there are no practical applications that require such end-to-end network functionality.

Similar to the approach used with ICs, networks need to be designed for optimized functionality. For the most part, the current network products are designed for packet-centric or packet-only network infrastructure. This limits the scope of the issues being addressed within the capabilities of the network device. Current network devices are commonly categorized into two broad categories—routers and switches. However, this is a gross oversimplification of network service needs. Instead, if switching and routing were implemented alongside network service-level specialization, the resulting systems will provide higher performance and lower costs. The logical network service needs are as follows:

- Network access and connectivity
- Traffic aggregation, concentration, and multiplexing
- Transport and transmission
- Routing and switching
- Network security, protection, and privacy

If the network systems are designed to optimize these specialized network service needs, the resulting networks will match the requirements with better performance and functionality and lower costs. Expanding the design scope and considering the networks holistically as systems will allow for different perspectives. Through such an approach, the overall network functionality becomes the driver for the design considerations. Then, as in the case with ICs, the goal is not optimizing each of the network nodes, but the overall network functionality. This approach is admittedly more complex, but can help the development of products that will promote the growth of more capable networks, especially true broadband networks.

Technical Distinctions

There is another critical technical distinction between packet switching and circuit switching that is important for network design. In the case of packet switching, all incoming data needs to be received into local storage and regenerated for forward transmission with some level of processing at every network node. But with circuit switching, once the provisioning and connection setups are completed, the incoming traffic can be propagated without local storage or processing. Only the restoration of signal strength and fidelity are required for onward propagation.

This fundamental difference in packet switching and circuit switching manifests in the basic capabilities of a network node. This means that for the same amount of silicon (processing capability and memory), a circuit-switch device will be able to handle a significantly larger number of subscriber connections than a packet-switch device without sacrificing quality of service (QoS). Hence, a network node with reduced complexity will result when sufficient bandwidth is available in circuit switching. This will have significantly higher performance factors with full QoS than a comparable packet-switching node.

In essence, packet switching has intrinsic advantages for delay-tolerant traffic, and circuit switching has intrinsic advantages for real-time (delay-intolerant) traffic. Packet switching generates built-in inefficacies and additional overhead for real-time traffic (e.g., voice and video) at acceptable service levels.

If the fundamental differences and capabilities of packet and circuit switching are factored into network designs, the result will be systems that require less processing capacity and memory needs with superior performance. This point was articulated forcefully by Preston Marshall of the Defense Advanced Research Projects Agency (DARPA). Dr. Marshall said, "This (802.11 packet networks) is five orders of magnitude from what we can do. This is like selling cars that get 10 inches per gallon. We need to think as a community how we can get that efficiency up."[xvii]

Alternate Approach

When there is an existing infrastructure that provides useful functionality, there is no intrinsic value in replacing it with an alternate technology that does not provide any added benefits. Re-implementing voice telephony over the Internet is such a proposition. (Applications such as follow-me call routing are not a feature of packet switching, even though it is often cited as an advantage of packet telephony.) A better approach is to take the existing infrastructure components that are functional and to use them as the basis of the new solution. The packet-centric network products promoted for evolving the infrastructure obsoletes existing TDM technology investments, and hence they alone cannot provide a viable solution for near-term service needs. This is one of the major hindrances for the adoption of broadband. Before a viable solution for solving networking problems can be found, an alternate model for network evolution needs to be developed. The transfer network architecture (TNA) is such a model (see *Figure 2*).

The key advantage of the TNA model is that it closely represents the state of networks and therefore has more practical value than the packet-centric models being used. Underlying the new model is the assumption that a revolutionary approach of replacing existing systems with new ones is not viable. An evolutionary approach needs to be provided that can coexist with current systems and provide for gradual migration to superior solutions. TNA resembles, in some ways, the network models underlying the B–ISDN (broadband integrated services digital network) architecture. However, B–ISDN is circuit-centric, and the Internet had not emerged as the universal network of choice. TNA, in contrast, supports hybrid (intrinsic packet and circuit switching) transport for heterogeneous interconnection of networks with backward compatibility.

The PSTN consists of two networks—the voice-carrying data paths and signaling system 7 (SS7) network for control and management. Even though the SS7 network evolved as part of the PSTN, these networks are to be considered separate because the requirements on them are completely different. SS7 has to be fast, extremely reliable, and fail-safe. The voice-carrying data path segments have different use-characteristics. One of the current vulnerabilities of the Internet, such as distributed denial of service (DDOS) attacks, is the lack of out-of-band signaling such as the SS7 network.[xviii] Integrating the existing out-of-band signaling networks and building new signaling and control capabilities are necessary to achieve the full capabilities required for future networks.

The Internet is now evolving as an independent network infrastructure. Among other things, this infrastructure could include a separate network optimized for video services, including high-fidelity video, to take advantage of the available packet-based video technologies.

The essential step to effectively bridge PSTN with packet and other networks is a conceptual model that subsumes pre-existing infrastructure and incorporates future capabilities. This bridging component is the access network (see *Figure 2*), which consists of the local loop in the phone network, coax cable plants in the cable network, and wireless access networks. However, current products do not provide the capabilities required to support the TNA reference model, instead adopting a packet-centric or packet-oriented approach. At present, access networks represented in the TNA reference model are part of the PSTN. Many products that are offering local access solutions using various technologies such as VoIP and xDSL are implemented in a manner to make access networks part of the Internet and completely bypass the PSTN in the future. TNA is a reference model for designing networks with backward and forward compatibility with current networks and products, without making useful systems obsolete.

The backbone networks offer specialized capabilities for different types of traffic: voice; data; video; or special services such as utility meters, surveillance devices, alarm monitors, and appliance networks. The access networks transport and transfer network traffic between subscriber devices and backbone networks. Network devices in each of these networks provide specialized capabilities optimized for the networks they belong to.

The TNA model provides a natural path for phased network evolution by providing compatibility with existing network infrastructure and products while protecting current infrastructure, products, operation management systems, and human training investments. In addition, the TNA model allows flexible network design and deployment strategies with maximum flexibility without mandating a packet-centric or a circuit-centric approach. The design choices are left to be decided based on service, performance, and cost considerations.

Access Networks

The TNA model helps in making macro-level decisions about network systems and a framework for defining the functionality of network products based on its role in a sub-network. The primary barrier for network evolution is the lack of viable access networks as defined in the TNA model. Construction of viable access networks needs to overcome several major technical and investment barriers. Part of the

reason for the lack of viable solutions for building of effective access networks is the network industry's following of the packet-centric trend without considering its limitations and alternatives.

Macrodynamics

As a capital-intensive infrastructure capability, network technology adoption issues are not solely confined to the technical realm. Effective technology adoption requires self-reinforcing interaction of technology, market, and capital. The development of the Internet is an ideal example of these factors coming together fortuitously. Unfortunately, such examples are rare.[xix]

The recent dot-com bubble provided insights into the interplay of economical sectors during a technology adoption cycle.[xx] A successful technology adoption trend has four requirements: market demand/need, mature technology, institutional support, and promotion. These factors interacted adequately in the case of the railroad, aircraft, and telephone, but not the Internet. The popularity of the Internet created by the easy-to-use Netscape web browser provided the demand and resulted in the institutional support it received, thus creating the bubble.[xxi] The superseding factor of "newer is better" took over in favor of the Internet. However, the adoption cycle was not able to run its full course because all the success factors were not in place for the Internet. The Internet had three of the four factors—demand/need, institutional support, and promotion—but it lacked mature technology. The resulting collapse now makes it impossible for the Internet to supplant the PSTN[xxii] since the same level of enthusiasm for the Internet will not be possible in future, as it is no longer new.

Conclusion

There are technical and macroeconomic reasons why the current packet-centric network strategies are not viable. Taking a pragmatic approach will rather result in the earlier development of better networks for the future. The TNA model described in this report provides a model for such an approach. The diverged-packet and circuit-technology positions will not help meet network needs of today and tomorrow. Superior network capabilities will emerge by promoting technology and protocol-agnostic interoperating networks, with each network implemented by optimizing its design requirements derived from service needs, which were the original intentions for the Internet.[xxiii]

Acknowledgments

The author thanks Paul Cutt, Tom Minnis, Vlasta Pokladnikova, and Gary Travis for their feedback and suggestions for improving the draft.

Notes

1. Leonard Kleinrock, "Information Flow in Large Communication Nets," RLE Quarterly Progress Report 1961. www.lk.cs.ucla.edu/ LK/Bib/REPORT/PhD.

2. Joseph C.R. Licklider & Robert W. Taylor, "The Computer as a Communication Device," *Science & Technology*, 1968. gatekeeper.dec.com/pub/DEC/SRC/research-reports/abstracts/ src-rr-061.html.

3. "Putting it all together with Robert Kahn," *Ubiquity*, Vol. 4, Issue 3, March 11-17, 2003. www.acm.org/ubiquity/interviews/pf/r_kahn_1.html.

4. Paul Baran, "Introduction to Distributed Communications Networks," *RAND* Memorandum, RM-3420-PR, 1964. www.rand.org/publications/RM/RM3420.

5. Donald Davies, Roger Scantlebury, et al., "A Digital Communications Network for Computers," ACM Gaitlinberg Conference, 1967.

6. "Network King: CEO John Chambers talks about how Cisco's empire is facing new foes and new territories," *Info World*, Oct. 19, 1998.

7. "Technology History," Chapter 3, *IEEE Communication Society*. www.ieee.org/organizations/history_center/comsoc/chapter3.html.

8. "Bell System: A Brief History," *AT&T*. www.att.com/history/history3.html.

9. Andrzej Jajszczyk, "What is the future of Telecommunications Networking?" *IEEE Communications*, June 1999. www.comsoc.org/ci/public/1999/jun/cimess.html.

10. Margie Semilof, "Sprint ION Goes Down The Tubes," *InternetWeek*, Oct. 18, 2001. internetweek.cmp.com/showArticle.jhtml?articleID=6402508.

11. Andrzej Jajszczyk, "What is the future of Telecommunications Networking?" *IEEE Communications*, June 1999. www.comsoc.org/ci/public/1999/jun/cimess.html.

12. J. Nicholas Hoover, "Pitfalls and potholes can trip up your voice-over-IP implementations," *Information Week*, Nov. 14, 2005. internetweek.cmp.com/showArticle.jhtml?articleID=173602687.

13. Josef Lubacz, "The IP Syndrome," *IEEE Communications*, Feb. 2000. www.comsoc.org/ci/public/2000/feb/perspectives.html.

14. "A Brief History of the Internet," Internet Society. www.isoc.org/internet/history/brief.shtml.

15. Leonard Kleinrock, "Information Flow in Large Communication Nets," RLE Quarterly Progress Report, 1961. www.lk.cs.ucla.edu/LK/Bib/REPORT/PhD.

16. Jason K. Krause, "The Bandwith Glut: Bonanza and Bust," *The Standard*, Sep. 10, 1999.

17. Rick Merritt, "Darpa Looks Past Ethernet, IP Nets," *CommsDesign*, Apr. 26, 2004. www.commsdesign.com/story/showArticle.jhtml?articleID=19201035.

18. George Mattathil, "The Internet Could Learn from the Phone Network," *Silicon Valley Business Ink*, Nov. 8, 2002. strategygroup.net/svbiz_netattack.shtml.

19. "Putting it all together with Robert Kahn," *Ubiquity*, Vol. 4, Issue 3, March 11–17, 2003. www.acm.org/ubiquity/interviews/pf/r_kahn_1.html.

20. George Mattathil, "The Internet Not Like Any Other Technology," *Silicon Valley Business Ink*, June 14, 2002. strategygroup.net/svbiz_trend.shtml.

21. Charles P. Kindleberger, "Manias, Panics, and Crashes," Basic Books, Inc. New York, NY, 1978: 40.

22. "Network King: CEO John Chambers talks about how Cisco's empire is facing new foes and new territories," *Info World*, Oct. 19, 1998.

22. "A Brief History of the Internet," Internet Society.

Adaptive COTS OSSs: Introduction and Technologies

Mark H. Mortensen

President
Dynamic Network Consultants

Eric Heinzelmann

Vice President, Next Generation Systems
Telcordia Technologies, Inc.

Abstract

Commercial off-the-shelf (COTS) operations support system (OSS) software has been gaining popularity in the past 15 years in communications service providers (CSPs) of all sizes and complexity, due to their faster implementation, lower total cost of ownership, and lower implementation risk than custom developments. But COTS has not been without problems, mostly stemming from their inability to adapt easily to a CSP's specific business processes, and the well-known "integration tax" among systems. A new type of commercial software, dubbed "adaptive COTS," has been tackling these problems head-on. This article introduced the concepts of adaptive COTS and discusses some of the technological enablers that have allowed vendors to give CSPs unprecedented flexibility and adaptability in their COTS software products.

Why Adaptive COTS?

Traditionally, CSPs had the following basic strategies to support new services and technologies with their OSSs—and the best choice was usually obvious to each of the constituencies:

- Incrementally adding features to the existing legacy systems was obviously the best choice to the IT management—the hardware was already in place, the software was deployed, the support infrastructure was optimized, and the software developers were already familiar with the system. Plus, the legacy systems already had the interfaces to the other systems and were already optimized to the business rules in force.

- Building a new system, and perhaps even rebuilding some of the legacy functionality into it, using the latest software technology was obviously right to the software engineers. This would provide a much better platform for the future and had exactly the features and functions desired by the users. Plus, proprietary features in it could provide a competitive advantage to the enterprise.

- Buying COTS software was obviously the right choice to the chief financial officer (CFO) and procurement who read the vendor-provided literature, which showed that COTS systems have a much lower total cost of ownership (TCO) than custom-developed systems and can be deployed quickly. Plus, COTS systems have less risk—they already can be seen to be working at other enterprises, and with a reference call, the vendor claims can be checked out.

The arguments were always the same, and in each particular case, each service provider chose its strategy, but none were ever really satisfied. The legacy believers spent more money than they anticipated while shoehorning the new features in and found themselves with a system that increasingly looked like the clunker automobile that some of them had in their driveways—with rust in the wheel wells, constantly increasing maintenance expenses, and an eight-track player in the dashboard. The software engineer's shiny new systems—if they were finished at all, due to that last budget cut, constantly changing priorities, and feature churn from the users—were finished late, often delaying the introduction of the new services that marketing was counting on while greatly exceeding cost projections. Many CFOs were satisfied if the COTS product proved reliable and scalable enough, and the vendor was still in business after several years. But everyone agreed that adapting the software to the enterprise's business processes and policies (or adapting those policies to the software) was more difficult and time-consuming than the vendors' marketing literature led them to believe.

What is Adaptive COTS?

Adaptive COTS is not a shrink-wrapped "off the shelf" system that does what it does and that is all it does.

Adaptive COTS is not a "toolkit" or a "platform" or "middleware" sold by a vendor that can be used by an IT organization to build a working software system.

Nor is adaptive COTS something in between. It is another dimension, as shown in *Figure 1*. Its goal is to provide the

COTS benefits of low total cost of ownership (TCO) and quick and low-risk implementation while approaching the fit to current business practices of systems custom-developed for a CSP.

Adaptive COTS has the following attributes:

- It comes quickly "out of the box" and provides significant operational benefits with minimal configuration in usually 90 days or less.

- It has a wide area of applicability in a CSP's operation.

- It can be configured by the CSP's administrators to adapt to their particular environment and business operations over time.

- It can be extended by the vendor, the CSP's IT shop, or third parties to support new functions, services, and network technologies and elements.

- It can be easily integrated into the overall enterprise software structure by the CSP's IT shop or third parties.

Thus, it combines the best of COTS and custom, bespoke developments.

Attributes of Adaptive COTS

Adaptive COTS products are less than 10 years old in the marketplace, and the architectural and implementation details that have led to their flexibility were, for several years, closely guarded secrets by the vendors. But the general principles underlying them have recently begun to be revealed. In this section, we will talk discuss the general principles, without identifying any specific products. *Figure 2* provides a road map for these attributes.

Flexible Data Model

Many in the industry have been advocating a standard, common data model for many years, from the International Telecommunication Union (ITU) M.3100 Telecommunications Management Network (TMN) standards to the common information model (CIM) to the silence indicator description (SID) CIM of the TeleManagement Forum. In addition, many, many other "common" data models have been adopted in various technology arenas, the Internet Engineering Task Force (IETF), the Asynchronous Transfer Mode (ATM) Forum, the Optical Interconnect Forum, etc. Many vendors who seek to provide a software product for one of these areas often adopt one of these standards, "baking in" the data model deep within their database structure (for speed and scalability). But, a number of problems, including the following, have made this a very bad approach for a COTS system:

- The standards change with time, requiring changes in the system and to interfaces with other systems (O.K., that is just life, and life is hard).

- The product must interface with many other systems that do not adhere to the standards, requiring adapters that change with time (the classic "integration tax").

- The product must interface with other systems that attempt to adhere to the same standards but with somewhat different interpretations (a problem that ultimately caused the demise of many of the ITU's TMN standards, where even minor differences in the

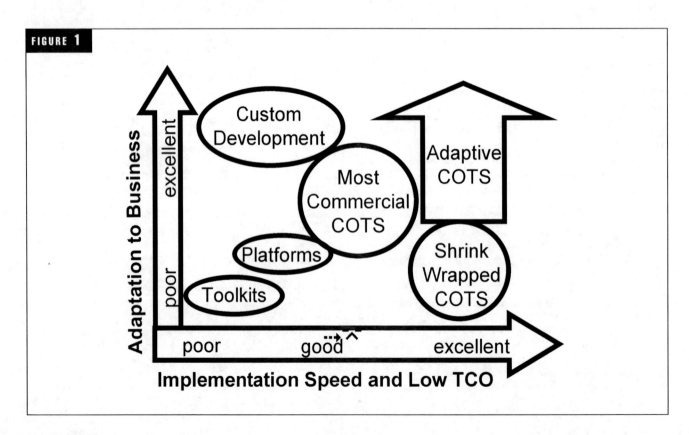

FIGURE 1

interpretation of the standards led to large differences in the final data model, through the power of inheritance in the object-oriented model).

- The standard often fell by the wayside, causing the demise of the product and often the vendor.

- The product could not easily be sold into other domains by the vendor, limiting their market and the company's growth.

- The implementation of the standard often was required by a customer to adhere to their particular interpretation of the standard, causing the COTS product to become more of a "one-off" project, rather than a true COTS product. This increased costs for the vendor and customer and hampered innovation, as the vendor had to often create a customer-specific code branch.

So what have the adaptive COTS vendors done that is different? They have adopted a simplified data model and provided built-in facilities for the customers to dynamically extend the data items and models themselves and to share the data with other systems, while making the data model used a configuration option of the system.

The simplified data model has few objects, uses recursive containment extensively, and is very abstract, often resembling the objects in a unified modeling language (UML). This model is implemented in a relational database, with a table structure chosen to "match" the object-oriented software to the relational database system. By delivery, a solution built on top of object-relational mapping techniques and supporting an open-ended model, users can model their application extensions with familiar objects from their application domain while simultaneously storing persistent application data in industrial strength relational databases that support the rich set of reporting, data integrity, and other infrastructure capabilities that CSP users require. This allows the vendor to implement any particular data model that a customer or engagement requires, without branching the source code—the data model so impressed becomes a configuration of the product, not a branch of the source code. The abstract model though, comes at a price—it could take an expert to do the modeling. Vendors mitigate against this by a combination of providing libraries of models of popular equipment, training for a few domain experts in the CSP's operations, and tools for these non-programmers to create their own support for network elements or services, implemented usually either using abstract models, rules engines, or templates.

The data model is also extensible, allowing the vendor and the CSP's administrator to add additional attributes to the data model for a particular service or network element (ideally, any object in the database) without the need for additional programming. Extensible data models have been a requirement since the very first database applications. However, new techniques using late binding approaches allow an entirely new level of capability. The following three levels of "data model extensibility" are required for adaptive COTS systems:

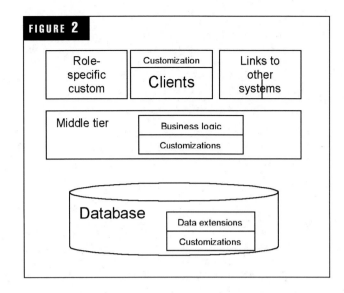

FIGURE 2

- *Attribute Extensions*: Allow the application to be customized by adding new attributes to existing objects. In relational terms, new columns are added to existing tables (although rarely implemented this way, the effect is the same).

- *Object Extensions*: Allow entirely new objects to be added and associated to the existing objects in the application's data model. In relational terms, new tables would be added.

- *External Object Projections*: Allow data values from external systems to be represented and manipulated within the adaptive COTS system as though it were a native object. This provides a degree of application integration that raises the bar for enterprise systems.

These extensions to the data model provide the CSP with a new level of flexibility for the system, but it must be implemented carefully, as performance can become a major issue.

Data Accessibility

Conventional COTS systems often effectively "locked up" the data in the system, even when there were transactional open interfaces provided, since they had to be queried individually. Adaptive COTS systems open up their data to other systems by providing open interfaces, usually implemented as publish/subscribe. These allow external systems to "subscribe" to changes to any data element they are interested in (though most implementations today do not provide the facility for on-line subscriptions, but require code added to the product, providing custom publication code for that interface). Other, bulk transactions, in and out of the product, are often provided for database loading and audit purposes, also.

For using data being shepherded by other systems, dynamic queries of the data are increasingly being implemented. The data is virtually stored within the product and when a transaction is performed requiring the data, instead of querying the database, a transaction is launched to another system and any required data transformation performed. Data caching can be done for performance purposes if latency is not a problem with these particular data.

Adaptable Business Logic

The business logic that used to be hard-coded into bespoke development systems, based on detailed user requirements derived from system analysts' detailed work, is another configuration attribute of the Adaptive COTS system. This is implemented in several ways in the adaptive COTS systems, including the following:

- User-accessible rules engines or proprietary or third-party workflow engines that give maximum flexibility to change the high-level workflow.

- Architectures and open interfaces to allow CSP IT shops to add their own code to the system. Approaches range from a limited set of interfaces (e.g., allowing a CSP to implement their own routing algorithms in an automated design and assign system) to a completely open system where the full business logic capabilities are available from a rich set of fully open APIs.

- Sets of "triggers" are provided in the product that bind in custom code, usually implemented as pre- and post-operation logic callouts to customer or third-party custom logic routines.

- Commercial workflow systems are integrated into the product.

Customizable, Extensible, and Role-Specific User Interfaces

Traditionally, the user interface for a COTS software system was static and extremely difficult to customize to any significant degree beyond minor look and feel. This has been the case whether the software was client/server-based or Web-based—especially with complex software such as OSSs. When an interface was finally decided upon, users had to be fully trained on it, even if their jobs require using only a small fraction of the functionality of the system.

CSPs were typically forced to rely on developers to improve usability and features. Changes were scheduled and released over time by the developers, with compromises made among the desires of the various types of users—experts, casual users, and administrators. This made users subject to the demands and cycles of the developers for both bespoke systems and COTS products.

But what else could you do? The user interface was tied into the guts of the functionality of the OSS software, often constituting over half of the source code, and it had to be maintained as a unit. One partial solution has been to offer two user interfaces, usually a limited read-only interface for casual users and a full-functionality, read-write interface for power users. But what about the user who only wants to do a simple job? Instead of the power-user interface, a simple input screen that makes one particular function immediately accessible to a technician with minimal training would suffice. So what is new about this? Have software developers not always been able to create specialized screens to input information and query a database?

Yes, but at the cost of creating and maintaining their own homegrown systems or contracting with a vendor to customize a COTS system just for them by creating their own special version, a complex and expensive proposition that recurs with each system upgrade and any upgrades to any systems connected to that system. What is new in adaptive COTS is the ease of creating and maintaining a specialized, user-defined, task-driven interface for the COTS system, giving users the best of both worlds—the complete customization of a homegrown system and the lower cost of ownership and feature speed of a COTS product. To do this, adaptive COTS vendors have done several things.

First, they have used a J2EE three-tiered software architecture, with a back-end database, a middle tier incorporating the business logic and data integrity functions, and a client tier providing the user interface. They have opened the interface between the middle tier and the client and maintained the integrity of that interface through subsequent versions of the software. Finally, they have provided a development framework for the IT shops, or even the users themselves, to create their own user interfaces to the open interface.

In effect, the specifics of the user interface have been decoupled from the functionality of the system. The purchasers of the adaptive COTS system can quickly create their own role-specific user interfaces—either Web- or Java-based—by using the business logic functions of the middle tier. This gives users the kind of interface they need to do their jobs—no more, no less—obviating the need for the specialized systems that provided these role-specific user interfaces, reducing the number of systems, lowering overall costs, and reducing the problems of multiple databases that must be synchronized.

Quick Network Element Adaptation

OSSs that communicate directly with network elements have a particular problem—the myriad vendors, multiplicity of boxes, and multiple versions of the software of each network element. Even in a small CSP, the combinatorics are impressive—with even only a half-dozen vendors, a dozen types of network elements from each vendor, and four software releases per year for three years would lead to almost 1,000 network element interfaces for an OSS. Traditionally, this would be handled by a subscription service from the vendor, combined with limitations on the list of network elements that the CSP directly connected to with the OSS.

Adaptive COTS vendors have tackled this problem, with varying degrees of success, with development toolkits for the network element interfaces that can be used by the vendor, the CSP's IT shop, or third parties. But this still remains a significant problem for the industry, limiting the growth of a vibrant set of COTS products in the activation and performance management spaces, where interfaces are complex and standards significantly lag network element capabilities.

Implementation Examples

Adaptive COTS has been gradually making its way into the marketplace for about the past 10 years. The best examples of these systems include Granite Systems' (now part of Telcordia Technologies) and Cramer Systems' inventory/resource management systems, Micromuse and SMARTS (now part of electromagnetic compatibility [EMC]) fault management systems, and Syndesis Networks' NetProvision activation and provisioning system. Other systems have been adding adaptive COTS features, including Telcordia Technologies' latest feature additions to trunk integrated record keeping system (TIRKS), which have radically opened the data architecture and the service order administration and control (SOAC) for order management that has added user-flexible business process features.

Technical Enablers to Adaptive COTS

Why has adaptive COTS only been available in the past five years? First, the OSS COTS market is still relatively immature, having only completed two or three three- to five-year cycles of major product innovations since the inception in the mid-1990s (it took PC office productivity software more than 20 major product iterations of six to 12 months each to mature). Secondly, successful business models for adaptive COTS, and its close cousin, open source software, were not well explored until recently—the business of software development is still taught, and practiced, based on a bespoke, waterfall or fast-prototyping development process, not on a COTS model. And third, although it was possible to create adaptive COTS with the software technology of the past, it is only recently that it has become much easier.

What are the major technical enablers, then, to adaptive COTS?

- Object-oriented programming, which has fostered better encapsulation of functions and less interdependency among subsystems in the product, allowing the product to evolve

- Java2 Enterprise Edition (J2EE) N-tiered architectures that bring with them high stability, excellent performance scalability, and excellent hardware independence

- Relational databases that also provide excellent reliability, scalability, hardware independence, and standard fault-resilient and fault-tolerant architectures, coupled with industry experience of how to best use these in an object-oriented system

- Late binding architectures that allow customization of the code at run time, allowing both the data model to be extended (as described above) and logical functionality to be extended through the application of custom logic insertion points within the framework of the adaptive COTS application

- Availability of major subsystems, either commercial or open source, that can be integrated into the system easily and have great flexibility, including reporting subsystems, trouble ticketing functions, display widgets, and user interface development toolkits

- Extensible markup language (XML) interfaces and tools for ease of specifying and implementing interfaces to other systems, and increasingly common models for describing the objects and attributes

- Service-oriented architectures (SOA) for reusable interface development

- Deep knowledge of the specific operations domains that have been encapsulated into the software, compiled and disseminated by the TeleManagement Forum and other industry groups

The Business and Service Models of a Vendor of Adaptive COTS Systems

But adaptive COTS requires not just architecture or a technical approach. The full range of business and service models of an adaptive COTS company must be tuned to the new approach that supports the customer in new ways.

First and foremost, an adaptive COTS company must make a deep and lasting commitment to a set of stable interfaces and capabilities that the CSP or other parties will use to customize the system to their particular needs and integrate it into their enterprise environment. This commitment manifests itself in several ways, including the following:

- A rich set of deep customization capabilities and open interfaces to allow adaptation of the COTS system
- Excellent documentation of and training on these capabilities and interfaces
- Very few changes to the capabilities and interfaces over time that disturb customizations of third-party extensions to the system
- A continuing and demonstrable commitment to keep the open interfaces open, even when closing an interface may give a competitive advantage to the vendor trying to sell additional capabilities in a particular circumstance

The sales engagement model is different for an adaptive COTS vendor. Rather than focusing just on the feature/functionality and the compelling features of the user interface, sales and pre-sales folks must help the prospective customer see beyond the now, to the possibilities that the system opens up to them and their business—possibilities coming from the multiple uses that the adaptive COTS system can perform, the multiple constituencies that it can address, and the future flexibility that it can bring. Thus, the sale of the adaptive COTS product is not a point event, but a journey that the new customer and the vendor take together—more of a partnership than a customer/vendor model. But this also opens up another sales channel for the adaptive COTS vendor, through systems integrators, who can derive significant revenue from the continuing adaptation of the system in partnership with the customer—their normal mode of operation.

Implementation requires a different way of thinking about the problem of adaptation to a prospect's environment. Rather than meeting with the prospect and reviewing the minimal set of changes that the product needs to have made to it (probably by the vendor) to fit the prospect's environment, modeling must be done upfront to configure the product for initial implementation, and then planned extensions and customizations by the vendor or the prospect scheduled and programmed.

The support structure of an adaptive COTS vendor is much richer than normal. The customers, or third parties, will adapt the system and build extensions to the product. But sometimes these will not work as planned, and the customer will request help from the vendor, which will require the vendor to have access to highly skilled development and deep domain knowledge personnel to work on-site with the customer to discover their mistakes and put them right. This is both a problem and an opportunity for the vendor.

The successful adaptive COTS vendor must also plan changes to their product in a new way. Changes, first and foremost, must not disturb the extensions and customizations that have already been created by the customers or others. Secondly, the vendor must look at the requested features at a much higher level of abstraction than is usually the case—balancing the feature/functionality's benefit to a particular customer with the overall benefit to the multiple customers who use the system in very different ways. But this also opens up a new possibility—vendors can plan extensions to their systems through watching what the customers and the third parties are adding to their systems, thereby adding new subsystems and "technology packs," pre-built customized configurations that encapsulate the best common practices of an industry sub-segment.

So the "business surround" for an adaptive COTS vendor is more difficult, but it is also richer than that of a normal COTS vendor or software development shop.

The Future of Adaptive COTS

Adaptive COTS is showing itself to be a very effective model of software delivery, both for the vendors that have been able to master the technical and business complexity and the CSPs that have been able to see beyond the next system procurement cycle and the current problems. But it has required significant education of both the vendor and CSP communities, and will require more. As the software technology continues to evolve and as the adaptive COTS vendors prove themselves to be more commercially successful than their competitors over the long run, adaptive COTS will help continue the trend toward less custom development in the CSPs and more commercial off-the-shelf (but adaptable, this time) software procured.

Acknowledgments

The authors thank the designers and developers of the various adaptive COTS systems for demonstrating their technical expertise in the competitive marketplace and their marketing groups who have released information on their products into the marketplace to promote better understanding of their products and approaches. Also, we would like to give special thanks to John Langley of Telcordia Technologies for his insight and help in preparing this paper.

Making Innovation Happen

Eckart Pech
President and Chief Executive Officer
Detecon, Inc.

Telecom companies continue to be under tremendous pressure. Wireline operators are losing margins and face increasingly fierce competition from new entrants on their turf. With a continued loss of voice revenues, they are moving toward broadband infrastructures at full steam. Wireless operators appear to be more and more threatened by extinction as well. In spite of growing markets, average revenue per user (ARPU) for voice is declining, and the only way to fill the gap is to introduce operationally costly data services. All in all, it has become imperative for telecom companies to reinvent themselves and adopt the nimbleness that makes their new competitors so successful. A very tangible path in this direction is a radical farewell to rigid product development processes, replacing them with much more agile, exploratory, and entrepreneurial approaches to materialize innovation.

Innovation is no easy task. It requires embracing a new way of doing business, where spotting and cultivating new technology is fundamental. This includes the systematic identification of new technologies and business models, detailed evaluation, aggressive prototyping, and, last but not least, implementation and operations. These are key differentiators where tomorrow's successful telecom leaders need to succeed.

The Pressure Points

Both wireline and wireless telecom companies are being increasingly pressured by market saturation, new competitors, eroding margins, and technological disruptions. The pace of change in each of these areas is significantly more intense than in other industries. Keeping up with this acceleration is a challenge. The pressure on wireline telcos intensifies with continued price erosion, mainly driven by new entrants that attack the incumbents on their home turf— voice communications. As voice becomes just another application that can be delivered over an Internet protocol (IP) environment, the potential for new competition is enormous. These trends, including the following, are happening globally:

- Market liberalization, new technologies such as voice over IP (VoIP), and new competitors are all shaping a potentially lethal environment for traditional landline operators.

- Access line erosion is continuing at a depressing rate. For example, the total number of residential lines in

the United States today has already decreased to 140 million and is expected to fall to 120 million by 2010. The bet for the telcos is to fuel growth by getting a larger share of their customers' telecom/communication/entertainment wallet to compensate for a loss in market share and revenues.

- Enormous capital expenses (CAPEX) have been committed by the regional Bell operating companies (RBOCs) to escape this pressure and make triple play a reality. We estimate that more than $15 billion of additional CAPEX will be allocated by the U.S. TriBOC within the next three years. In order to materialize a return on these investments, it is imperative that the RBOCs launch an innovative and sophisticated portfolio of value-added services that is appealing to their clientele.

The pressure on wireless operators is apparent and results mostly from increasing market saturation. The following are trends related to decreasing overall ARPU, mainly caused by erosion in voice, which cannot be substituted for by data services:

- In the most attractive markets that are characterized by high ARPU, growth is starting to slow down and competition is getting increasingly fierce.

- New entrants, especially mobile virtual network operators (MVNOs), threaten the existence of wireless operators beyond being a pipe.

- Voice ARPUs are stagnating based on increasing price pressure.

- The development and diffusion of data services were thought to be able to compensate for these reductions but have proven to be more costly than expected.

- The targeted data ratios of the total ARPUs, initially planned at around 40 percent in 2005, are still far behind initial expectations and linger around 20 percent in western Europe and 8 percent at maximum in the United States.

- Even in the most advanced wireless markets such as South Korea, mobile data services have not been able to create incremental ARPU; in fact, they are barely making up for the voice and interconnection erosion that is happening at the same time.

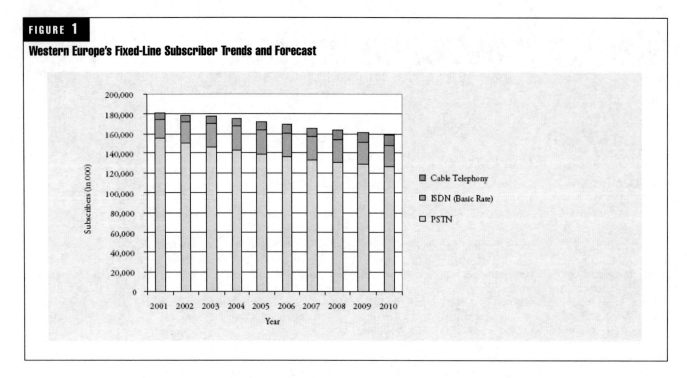

FIGURE 1

Western Europe's Fixed-Line Subscriber Trends and Forecast

- In addition, investments in third generation (3G) infrastructures are increasingly threatened by new entrants—large and small municipalities, unconstrained by profitability requirements, are offering Wi-Fi at a fraction of the cost of a 3G data subscription, mostly for about $10 a month and in some cases even for free.

New competitors such as the MVNOs or municipalities in the wireless case, or new entrants such as Skype in the wireline case, jeopardize the formerly dominant positions of market incumbents and are living proof that it is now imperative for telecom companies to differentiate themselves in order to avoid being perceived as a mere pipe. They have to be quicker and more efficient in order to stop losing ground. This means a shift in the legacy innovation paradigm, away from that characterized by an overall conservative approach, plus rigid and highly formalized processes, and toward agile and more entrepreneurial approaches where failure is an integrated part of the learning process that will ultimately foster innovation and differentiation.

Convergence Will Happen—With or Without You

As technologies, devices, markets, and competitors all converge, service providers of every type are facing a very new

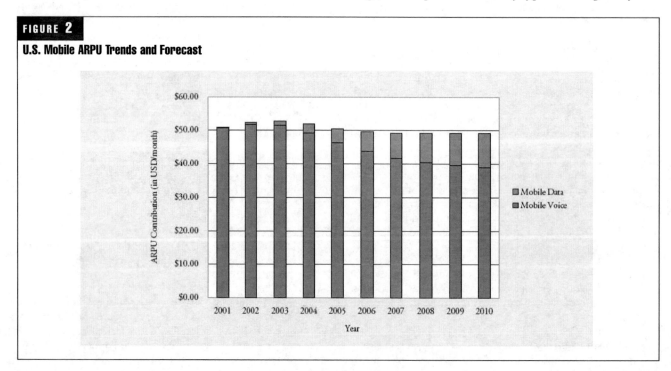

FIGURE 2

U.S. Mobile ARPU Trends and Forecast

future. Cable and Internet players such as Yahoo, Google, and Microsoft are offering voice services, which in turn pressures wireline operators to offer video and Internet service just to stay at par. Convergence touches every aspect of how multiple industries do business, including managing the investment community, deploying new infrastructure, training the sales force, and integrating customer care.

The last area, customer care, is a significant area worth focusing on. Telecom companies will need to develop methods and procedures that encompass a wider spectrum of functionality. The knowledge base used by customer care reps will need to store info on a wider variety of services. Billing systems will need to be fine-tuned in order to integrate services and pricing plans, rather than merely printing multiple charges on a single page. The reps themselves will need to be cross-trained on multiple services. From billing inquiries to tech support for feature-rich devices, the customer care challenge is staggering.

As triple and quadruple play become real, the players will need to develop not just new processes for the technology they are deploying, but also new skills and strategies for the business they are entering. For example, cable operators entering the voice arena will need to learn new things such as how to develop calling plans or deal with customers who deny they made those 57 calls overseas. Likewise, telcos will need to learn about TV programming and selling commercials quickly.

Companies that can adapt and grow will win, offering a seamless, integrated, and innovative new portfolio of services that make their customers' lives easier and more cost-effective. Those that do not adapt will suffer dearly at the hands of competitors that were more successful at reinventing themselves.

Reinvention and Nimbleness: Learn from the Leaders

Successful firms have embraced new paradigms to speed up the materialization of innovation, thereby building and maintaining a significant competitive edge. This has the potential to fundamentally shift the entire way in which innovation is materialized. The following are some case studies of rather big and successful companies that have embraced this new paradigm:

- Microsoft has developed a dedicated blog to better embrace customer feedback and opinions related to their products. Blogs are gathering momentum as a communication tool on the Internet—the numbers have increased from 5 million users at the end of 2004 to more than 50 million active bloggers during the summer of 2005. Microsoft deploys its blog in a systematic fashion to foster the interaction with its user base. The blog is being used more and more as a substitute for the annual user conferences, as it provides a much more timely and up-to-date platform for user interaction. For any new product initiative at Microsoft, it is now imperative to share new functionalities or significant changes with the MS blogging community. Microsoft's blog has more than 50,000 active users.

- The operating system Linux is another very tangible example of traditional approaches for innovation being made obsolete. Since the original placement of the raw software kernel by Linus Torvalds in 1991, the program has continued to evolve and, in 2003, achieved a 25 percent market share as an operating system (OS) for servers. Institutions and companies such as NASA, eBay, Mitsubishi, Sony, Google, and Motorola have deployed Linux on their servers and helped to make it a much tougher competitor for Microsoft Windows. A continually increasing user community contributes to the optimization and improvement of the program. Based on these continual improvements, there is an updated Linux version available on the net for download every four to six weeks. Similar on-line networks are being used by companies such as Procter & Gamble or Audi in order to speed up time to market and improve access to prospective customers.

- Cellon is a wireless phone design and systems integration company targeting the wireless handset industry. Cellon has helped phone manufacturers trim the time from design to market down to five months. The basis for this speed is modular designs, including chipsets, circuitry, seamless interaction, and cooperation with manufacturers. In this specific case, such an approach is a key differentiator, as the life cycle for a wireless phone today is only nine months. Some 25 million phones have been shipped to date based on Cellon's design and approach.

The impact of this is apparent—anticipating and embracing these new approaches will allow telecom companies to take the lead rather than fall behind. It enables them to face the challenge before them in the shape of new competitors that are eroding their margins. Ultimately, a more agile approach will allow them to regain the lead based on their superior and rich infrastructure inventory that promotes service innovations.

The Model: Tangible Innovation At Hand

In the course of our project work in the past few years, Detecon has designed and implemented a very nimble approach that has been proven to deliver the desired results in various client engagements. This approach consists of the following five steps:

- *Innovating*: Based on a long track record of respective engagements, we have developed a dedicated approach to not only identify and assess technology and business model innovations for telecom companies, but to also "cut through the mist" and provide a thorough assessment of those technologies that are potentially disruptive. This scouting and identification work is based on a methodology that we have developed over the past eight years and a comprehensive database that features in excess of 1,000 innovative companies, technologies, and business models.

- *Detailing and impact analyzing*: In many cases, a more detailed and thorough analysis of a new technology or business model is required before the actual development dollars are committed. This optional step needs to happen in an efficient fashion in order to not delay the innovation momentum, providing the required data points based on the 80-20 rule.

- *Prototyping*: Today's Web technologies and increasingly commoditized electronic components have created a whole new range of opportunities to make innovation tangible. Rapid prototyping of both physical and virtual products will allow an audience to embrace new and revolutionary ideas much more quickly than a slide deck. This is especially true for a board-level audience, which will appreciate an illustration that not only shows certain features, but also inspires creativity related to additional features and functionalities.

- *Implementing*: This step requires more hands-on expertise and pragmatism than any other, which implies that it is also the inflection point where the "operations guys" take over to make sure that the concept and prototype go beyond imagineering. Detecon has been in the industry for many years, and it is a fundamental element of our approach that disruptive ideas will only flourish and live up to their potential if the individuals handling the implementation phase can rely on hands-on expertise.

- *Operating*: In many cases, new technologies fail due to a lack of capabilities and tools to operate them. For example, many mobile network operators have seen exuberant growth in their early years, and then neglected the importance of the evolution and sophistication of their platform and back office operations. The impact is often costly, and in order to keep subscribers happy, the operator employs an increasing number of customer care reps and incentives in order to retain customers. This is, of course, not a long-term strategy.

Conclusion

It is vital to remember that the innovation process is cyclical. Once a new service has been brought to market, the quest for even further innovation and improvement needs to begin. There is no status quo, given the increasingly fast pace of change in technology, competition, and customer needs.

Life Organization, or How Communication Providers Can Become Co-Managers with the Customer

Thomas Jürgen Quiehl

Executive Vice President, Innovation/Technology Strategy
T-Com

Summary

As life becomes more complex, users demand a new quality in the services provided. These have to not only contain individual innovations, but also embody an overall concept to make everyday life easier and save time. Communications services are a key component of this. The addition of "human services" is the next step.

Basic Motivation

Let us begin with a common enough real-life example—putting together a new piece of furniture. How often does it happen, with all the steps that are needed, that even the most hardened customer loses track of things, groans over confusing assembly instructions, and longs for help?

Isn't this rather like the perception our customers have of individual life in general and their specific communication needs in particular?

Functional products and services have turned into lifestyle choices, interaction in everyday life (private and professional) has increased enormously, and the rate of change in technology demands constant attention. The old-style telephone service with its gray-dial telephone has grown into today's fixed network with its multitude of handsets and tariffs. Mobile telephony has opened up another world with an even faster rate of change, and the Internet touches all areas of life via the personal computer (PC) and TV.

All these examples enrich and challenge users. New activities are possible, and there is greater freedom in terms of space, time, and, above all, scope for fun and entertainment. At the same time, this innovation confronts people with demands, sometimes excessively. A constant influx of new technical devices (with and without interoperability), complex installation tasks, a drastic increase in demands in the professional and private sphere, and a flood of information are signs of this.

So we cannot keep putting more new products on the market and creating more and more complexity for the user. We have to put together a package for customers that makes their life easier, helps them organize it, and gives them more disposable time. This will give private customers the same benefits offered to business customers by business process outsourcing.

Customer/Market View

This approach to life organization, or acting as "co-manager to the customer," is not new, but we are a long way from it in the real world. Who has not sworn at technical hurdles such as programming a new satellite receiver, installing a wireless local-area network (WLAN) access point, or configuring a BlackBerry client? Who has not wished for a call center with fast, expert resolution of problems, a helpful elf to run private errands during working hours, or structured training in new areas (Internet, home networking, etc.)?

Customers are less and less interested in combining services by themselves and are looking more and more for "one-stop shopping." The key elements of this are networking, interaction, and integration—elements that can be best achieved by a communications provider.

What can help cover the customer's everyday needs and be sensibly incorporated into a service offering?

First, the basic product and service requirements have to be met. This begins with product design and pricing, takes in

FIGURE 1

Life Organization Logic

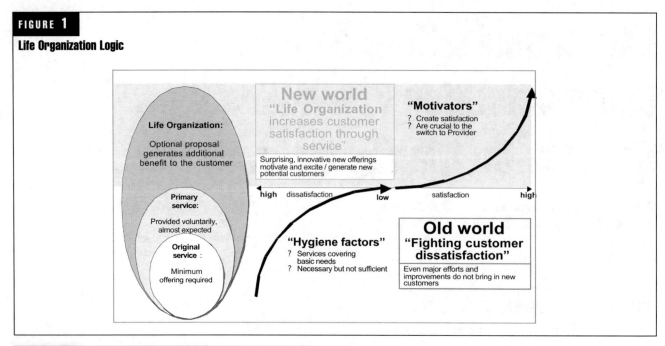

FIGURE 2

Producer/Service Provider

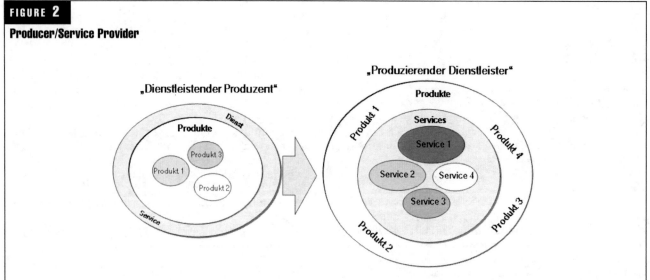

technical functionality and usability, and covers sales and customer support. It must be possible to commission a telephone at home without any effort, networking PCs and printers should be a matter of "plug and play," and the call center should resolve any queries in a competent manner. These dissatisfiers mainly address customer (dis)satisfaction. They are taken for granted, though not always provided. Stable customer satisfaction only comes when customers are offered unexpected or particularly useful additional services.

So what are these unexpected services? From the preceding comments, it is clear that restricting ourselves to purely technical services or support services alone is not sufficient. We need to take a broad view. Life organization services are designed to make everyday life easier and turn the provider into a problem solver for the customer's everyday organization. They reach right into people's daily lives, so they have to reflect social developments and trends.

To make this clear, it is helpful to consider both "automated" services, i.e., more functional issues such as remote-controlled management of roller blinds or voice-controlled information to callers, and "human" services, i.e., services provided by people, such as car maintenance or taxis.

Service Portfolio

Today, the first approach to integrated life organization services can be seen in the marketplace, although the German market is still a service vacuum, especially in comparison to the United States.

Some examples of possible innovations in this area are shown in *Figure 4*.

Complexity can be reduced and security increased with network-centric services, which not only allow data to be stored but also keep it automatically updated to the latest

FIGURE 3

Customer Needs

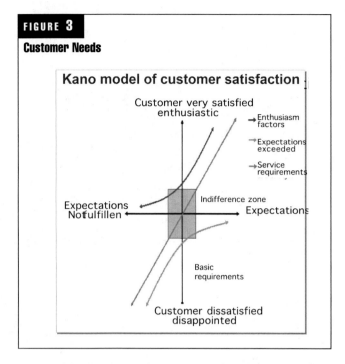

The area of location-based services includes animated maps that would enhance locations or route plans with built-in images, videos, and additional information. Similar enhancements can be made in the way of a personal diary to include the user's own photos, voice recordings, and notes, all wrapped up in a seamless service.

Specific help, training, and support products are also useful to assist users pre-emptively and directly with specific problems. Expert call centers or extended services in the form of blogs (text, images, and video) on the network are one aspect of this.

Approaches

Implementing this new, comprehensive life organization approach needs new business models, because service components from different "producers" need to be combined and offered to the customer as an integrated product.

A health service for people with allergies is one example: Mobile sensors placed by the customer in his environment send pollen counts to the access point via an ad hoc network. Within the communication network, the figures are compared with the user's individual profile, and recommended actions are derived. These may be implemented as a daily message on the bathroom mirror to read while shaving, or as a direct connection to the medical call center.

This involves terminal manufacturers, network and information technology (IT) providers, health organizations, and editorial functions. The challenge is to couple the technical/functional service elements with the direct "human service." A variety of options can be supported, including a voice announcement or Web link via an expert call center to dedicated support staff providing the appropriate services to the customer.

This creates a comprehensive value-added network, generating an overall life organization service for the customer. Dual business models, as well as franchising and partnering

state of the software. In this way, wedding photos or dissertations can be available even 20 years later at the touch of a button.

This type of facility can be extended with a reminder service that holds payment dates for insurance policies, elections of new club members, or the expiration of the fixed-interest phase of a loan and reminds users in every conceivable way.

In everyday life (private and professional), a network-based shopping list, accessible from any device, would make shopping easier. Anyone in the family could access it at any time and from anywhere via a mobile handset or information point, pick up information, and enter confirmation details for every other user.

FIGURE 4

Services and Facets of Life

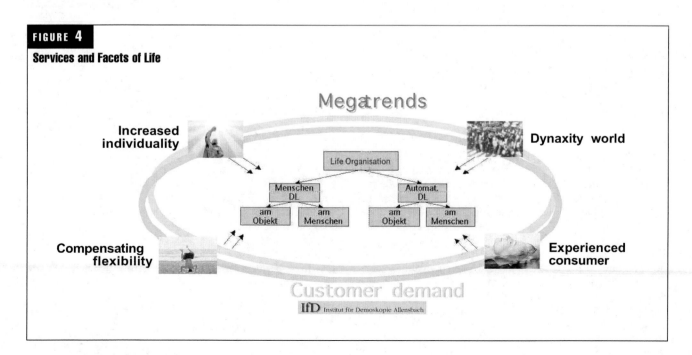

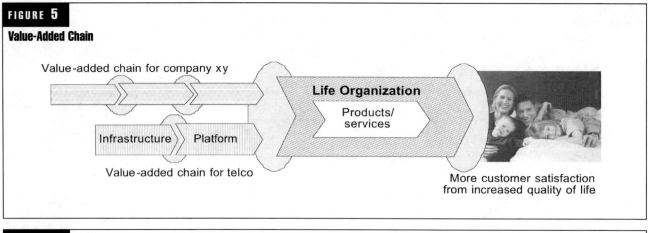

FIGURE 5

Value-Added Chain

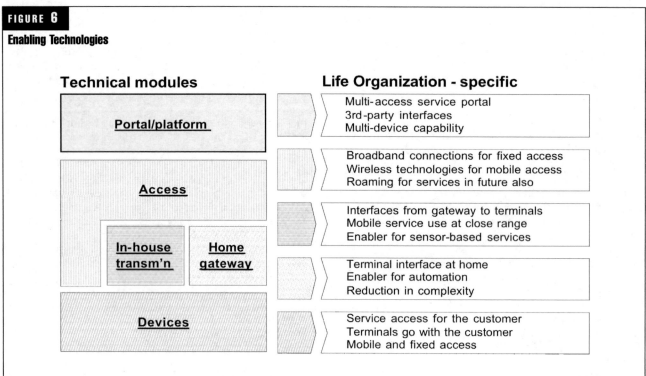

FIGURE 6

Enabling Technologies

arrangements, are needed to provide an end-to-end service. So far, only the beginnings of this can be seen in the marketplace. A strategic, crafted design process has yet to emerge.

Technology

From the point of view of a communications provider, the key elements of a life organization concept are available or will be market-ready in the near future. Apart from an application-oriented range of terminals, the aspect of in-house transmission and general access is important. Application platforms and portals complete the set of building blocks.

The design challenge remains to integrate the various components and to network all those involved in the service process. The technical convergence of communication networks should make it possible to achieve usability and seamless access in the foreseeable future. For the customer, complexity barriers can be overcome as they arise with a step-by-step provision of services.

Conclusions

The new life organization approach allows increasingly demanding customers to be served with demand-driven integrated products, providing direct benefits by optimizing life and gaining time.

Along the way, new "service stories" should be created from existing individual products, enabling the user to be provided with a genuine full service. New business models create the necessary preconditions to bring together the various players in the value-added chain.

Technically, the relevant components are already available or will be perfected in the near future, so "all" we have to do is integrate them.

Original Design and Manufacturing Services: Selecting the Right Partner

T. Sridhar

*Vice President of Technology and Software
Architecture, Corporate Technology Group*
Flextronics

Deepak Satya

*Director of Product Management, Corporate
Technology Group*
Flextronics

Introduction

The original design and manufacturing (ODM) services industry has witnessed explosive growth in the past decade. The market research firm iSuppli estimates that the total worldwide revenues for ODM products will exceed $60 billion in 2005, with a compound annual growth rate (CAGR) of 19 percent for the next few years. ODMs have begun to offer more complex products that target enterprise and service provider markets; until recently, several of these products had been designed internally by the original equipment manufacturers (OEMs). The explosion in consumer electronics and Internet-based technologies and applications has led to quick product turnovers, rapid price erosion, and accelerated time-to-market requirements. This has forced OEMs to consider ODM programs as part of its long-term business strategy. This article addresses some of the factors OEMs should consider when selecting an ODM for its future product strategy. It also describes how ODMs can address the requirements of OEMs.

Evolution of Electronic Manufacturing Services

In the past decade, the manufacturing services industry has evolved from traditional electronic manufacturing services (EMS) for OEM–developed products to design and manufacturing services. Such product design services are offered on a one-time contract basis, as in the case of contract design and manufacturing (CDM) services, or as ODM services, where the manufacturer is responsible for the majority, if not all, of the efforts in the product development. Such a business model can offer significant time-to-market and cost benefits to the OEMs.

Advantages of Using ODMs

There are several reasons for ODMs to be attractive under the current market conditions. OEMs face increasing competition to offer cost-effective products in rapidly growing markets. Any related ground-up product development effort is expensive and time-consuming. In several instances, the OEMs may be interested in enhancing their product portfolio or introducing products that do not necessarily take advantage of their core competency. In other cases, the OEMs may need quick launching of new products to address adjacent or newer market segments. If such products are successful, the OEMs may have to ramp up production even more to meet the market demands. ODMs are ideally suited to address these factors by offering design and manufacturing services that are flexible to suit the various needs of the OEMs.

ODMs typically offer already designed and production-ready products that can be modified to meet the requirements of OEMs. These modifications could be simple, such as aesthetic changes to the product, or much more complex, such as modification or replacement of one or more core hardware and software components. Such modified products can be introduced into the market within a very short time—sometimes six months or less—while a ground-up development of such a product might take two to three times longer with a much higher development cost. In addition, the burden of maintaining and supporting the new products in-house may further drain precious resources from OEMs.

Products That Are Well-Suited for ODM Programs

While the bulk of ODMs' business focuses on high-volume and consumer electronics products such as personal computers (PCs), cell phones, and audio-visual equipment, any product portfolio that meets the following criteria can be an ideal candidate of an ODM model:

- The product is in a mature market, has already been tested and accepted in the marketplace, and is comparable to the offerings of the competitors.

- The product requires quick entry into the market and demands rapid adaptation of standards and industry-accepted functions and features.
- The product can be designed with readily available hardware and software components and has efforts in design that are more about function and feature integration that require little of core technology differentiators.

Products addressing the enterprise and service provide markets, including Ethernet switches and routers and broadband and wireless customer-premises equipment (CPE), are complex but often meet the requirements listed above and therefore are well suited for ODM business models.

Market Trends That Are Helping ODM Business Models

Several market and technology trends are fueling the growth of the ODM business model. The following are some examples of industry-wide trends and initiatives that are leading to rapid adaptation of ODM business models:

- Rapid growth in voice over Internet protocol (VoIP), broadband, wireless, and consumer electronics markets
- Standardization of several core technologies and interfaces such as advanced telecom computing architecture, and software specifications from the service availability forum
- Readily available protocol stacks and other software components
- Rapid adaptation of operating systems such as Linux
- Availability of open source software
- Mature silicon offerings such as highly integrated microprocessors and application-specific standard parts (ASSPs) from leading silicon vendors

These trends are requiring OEMs to bring in high-volume products based on standardized technologies and components to the markets quickly. ODMs that have identified these trends and invested in product developments and are capable of quickly tailoring these products to meet the OEMs' requirements are poised for success.

ODMs as Strategic Business Partners

Instead of choosing an ODM that has a product best matching the immediate needs, OEMs should consider an ODM that can bring a strategic business advantage to their business. With an objective to develop a long-term relationship, OEMs should evaluate an ODM partner based on product design, manufacturing, and market and technical knowledge.

Product Design Infrastructure

Product design infrastructure requires well-designed product development processes and methodologies, matching product road maps and proper product planning during the entire life cycle of the product.

Most successful products, during their life span, require additional functionality and feature enhancements along with good product support. The market typically demands longer life span from complex and expensive products. It is also more expensive and time-consuming to sustain the product life cycle as the complexity of the products increases. When OEMs use an ODM product to enter the market, long-term product planning becomes critical to the success of the program. To sustain the time-to-market advantage offered by the ODM products, OEMs also need to consider whether the ODM can offer solid sustainable road maps to the ODM products. The OEMs further need to evaluate whether the ODM is capable of continuing to offer the cost and time-to-market advantage during the entire life span of the product.

Platforms or Finished Products?

There are two common models used by ODMs for serving their customers. The first is when ODMs offer the complete hardware, software, and system as a complete package. With minor aesthetic changes, OEMs can introduce this product to the market. In this model, when an ODM product is used by multiple OEMs, it is difficult to differentiate products from one OEM to another.

In the second model, ODMs offer a product platform to which OEMs can provide their own custom features, including their own "secret sauce" to address the specific market. For example, an OEM requiring a product for the wireless switching market can obtain a base Layer-2 or Layer-3 Ethernet switching platform from an ODM and add its own features such as wireless LAN (WLAN) switching and radio resource management as a means of differentiation. Another OEM might take the same switching platform and customize it to turn into a switching blade on a blade server. This model is more advantageous to both OEMs and ODMs because of its flexibility and larger market reach.

Another aspect of offering platforms is the ability to mix and match—for example, one OEM may prefer to have just the base hardware, while another OEM may want the hardware and software. Even in the software, OEMs may need different operating systems and software baselines. A platform-based approach will permit these various configurations, especially if the platform was designed with these considerations in mind.

Product Road Maps

An ODM's value is in providing a range of platforms and services related to a product for an extended period—specifically in the ability to help customers with their own road maps and products. In that context, ODMs should be able to provide a solid road map of their products or platforms.

OEMs will typically need to enhance their products because of the arrival of new chipsets with their integration benefits. Also, they might need to address emerging requirements of the marketplace by including support for new interfaces (e.g., 10 Gbps Ethernet) and new protocols (e.g., VoIP signaling via SIP). It is important that ODMs be able to address the requirements of the OEMs via a road map that includes upgrades of their platforms. For example, ODMs offering a WLAN access point platform will need to incorporate functionality for the protocol to emerge from the control and provisioning of wireless access points (CAPWAP) working group in the Internet Engineering Task Force (IETF). Such enhancements may be implemented using software upgrade capabilities to the access point.

To address the needs of their customers, ODMs may also include innovative approaches in their design and implementation of their platforms as part of their road map. Via their partnerships with silicon vendors, ODMs can offer platforms that address an entire product line and multiple market segments with appropriate hardware and software configurations.

Product Design Capabilities

Several ODMs use reference platforms from silicon vendors as a base for their products and platforms. While this allows them to demonstrate a platform to OEMs with new silicon, there is a certain degree of optimization required for this platform to be considered a viable base for OEMs. As an example, a reference platform may include different types of interfaces, since it is intended to be a development platform. These add-ons will contribute to the cost of the system without providing any benefit to OEMs in a shipping product. ODMs might try to re-engineer the reference platform to remove these features, but that is usually less than optimal.

Two aspects contribute to the differentiation of the platform: differentiation of the individual components and the design of the platform with these components. These may include silicon, backplanes, power systems, thermal design, enclosures, and industrial design.

Use of optimized processes and components to achieve greater flexibility and cost advantage is important to OEMs, but is difficult for them to differentiate from their competitors in all these areas. If an ODM drives the innovation—either internally or through strong relationships with silicon and software providers—the OEM can reap the benefit.

The innovations will be tied to the road maps of the constituents. For example, the power supply road map might involve cost reduction and higher efficiency. The software road map might involve the use of a more efficient process scheduler (e.g., the use of the Linux 2.6 kernel with its real-time scheduler). It is important that the platform and the ODM be able to use the advantages provided by the road maps.

Meeting Cost Targets with Tailored Products

ODMs need to keep their architecture flexible so that it is relatively easy to add or take out some of the constituents. As indicated earlier, OEMs will have multiple requirements that necessitate the ODMs to implement custom modifications. Unfortunately, these modifications can add significantly to the cost. ODMs, which keep a focus on cost in the architectural design and implementation of the product, will find it less difficult to implement these modifications without compromising cost.

Typically, OEMs focus on time to market and quick introduction of a product, so the first platform is not usually optimized for cost. After the product hits the market, the first revision is usually a "cost reduction" design. This allows OEMs to squeeze out more margins from the product or allows them to drop prices and keep the same margins. ODMs, in contrast, will need to optimize its design for cost from the beginning without compromising time to market. It is clear that the reference design that is available from silicon vendors will not meet these cost requirements.

How are ODMs able to provide a cost-optimized design without compromising time to market? Via a platform approach. ODMs should be able to provide a platform optimized for cost on which OEMs can add their "secret sauce" and still obtain their time-to-market advantage. Often, the differentiation is via software functions, though it need not be restricted to that.

It is important that ODMs have the right teams to implement these optimizations and extensions for customers. In fact, multiple OEMs may engage with an ODM for their own extensions, so the scalability of the ODM's engineering teams is an important factor. Too often, ODMs struggle with having the right (number of) people to work on their customer engagements.

Integrating OEMs' Unique Features

We talked earlier about OEMs wanting to include their "secret sauce" on top of ODM products. The platform approach enables it, especially if it uses standard programming interfaces such as well-known APIs in software. Consider an OEM that has some proprietary applications that run on their current products under Linux. These applications can be easily ported to an ODM platform if it supports Linux—a simple case of customizing for unique features on a standard platform.

In some cases, an ODM platform may need to be enhanced for an OEM with proprietary hardware features—e.g., a WLAN access point (AP) that uses an innovative chipset for high-speed wireless access, along with enhanced (and proprietary) radio resource management features. The wireless AP platform from an ODM can be implemented with this chipset or be easily modifiable to incorporate this chipset. The key is not limiting OEMs while providing a cost-effective platform.

Market Knowledge and Technical Capabilities

In-depth market knowledge is critical for predicting and introducing products to match market needs. ODMs targeting OEMs should not limit themselves to just understanding the needs of their customers and competitors. To offer highly differentiable products and platforms, ODMs may need to consider several factors related to multiple technology markets, consumer behavior, service provider behavior, application-specific integrated circuits (ASICs), and other component markets that are striving to bring in new features and higher integration and technology initiatives and activities of standardization groups.

ODMs should not be limited by their model of engagement with their customers. While they are often lumped with the "work for hire" OEMs, they have to go further. They have to be able to monitor market and customer trends so that they can come up with cost-effective solutions for customers in a proactive manner. ODMs have the advantage of dealing with multiple customers, which enables them to distill the knowledge and experience gained by their various engagements into competitive platforms. They should also look forward in terms developing multiple platforms, thus allowing them to address a larger customer base.

There is another aspect in the value ODMs can bring to customers. Through monitoring and implementation of some of the newer standards and optimizations, ODMs can ensure that the customers get a competitive advantage. For example, they can implement process optimizations for printed circuit board (PCB) design that will lower the cost of manufacturing the product. As another example, they can implement a newer version of a protocol used to communicate with a peer networking entity. ODMs should do all of these without any customer prodding. In addition, ODMs should continue to interface with the customer for any other optimizations that they desire in the platforms provided to them.

Manufacturing as a Core Strength of ODMs

OEMs need to evaluate ODMs based on whether their core strength in manufacturing capability is well aligned to the product under consideration. Component procurement, production lines, testing and production yields, reverse logistics, return merchandise authorization (RMA), and support must be considered during such evaluation.

Component Procurement

Because of high volumes achievable by use of components in several products and for multiple OEMs, ODMs' purchasing power lets them achieve cost targets far better than OEMs could individually achieve. Some of the complex components such as ASSPs may have longer lead times. Several ASSP and silicon vendors are developing partnerships with ODMs to enter the market quickly and achieve faster ramp-up of the products. ODMs need to develop strong relationships with component vendors and maintain preferred customer status to narrow the lead-time in procuring such components. Proper planning and efficient procurement processes, combined with proactive vendor relationship development, are essential to achieving cost and time-to-market advantages.

Efficient Production Lines

An ODM product may be used by several OEMs with minor modifications in hardware and/or software components. Setting up a separate production line for each OEM may increase the cost of manufacturing and thus reduce the cost advantage offered by the ODM. ODMs need to be capable of establishing common production lines that are flexible and efficient to serve multiple OEMs. Automation of the production lines is essential to ensure that each OEM's unique product can be delivered efficiently. OEMs should understand how these processes are implemented at the ODM's manufacturing facility.

Testing and Production Yields

As the complexity of the products increase, the manufacturing tests and validation procedures become extensive. Each product requires manufacturing tests and validation at all levels, including component, functional block, stand-alone, board, and system. While design verification and testing and system quality assurance may have been implemented extensively during the product development, a separate set of manufacturing tests are required in the production lines to achieve higher volumes and yields. Many of these tests require automation and expensive test equipment. OEMs and ODMs need to work together to define and implement manufacturing tests to help achieve higher yields and enable shorter fault detection cycles. Defective products from the line with inefficient fault detection processes will eventually lead to increased product cost.

Reverse Logistics, RMA, and Support

The after-sale support and maintenance of the ODM products is equally critical. OEMs need to evaluate whether ODMs have the appropriate reverse logistics, RMA, and proper product support processes built into the program. OEMs need to ask whether ODMs have automated capabilities to download the software bug repairs to the products on deployed environment and whether they can identify the hardware faults of the returned products and refurbish them quickly. As the complexity of the products increase, usability, reliability, availability, and serviceability become critical factors to the sustained success of the products. While the products may be generic and offer minor OEMs specific modifications, these processes need to be tailored to the quality requirements of the OEMs. Lack of proper infrastructure and after-sales support may otherwise jeopardize the brand equity that OEMs have established over time.

Protecting OEMs' Intellectual Property

ODMs' products are typically generic and quickly adaptable to specific requirements of OEMs. OEMs may decide to bring in differentiable product features to the ODM product that are unique to its business and the addressed market segments. Integration of such functionality is challenging. In some cases, OEMs may work with specific component vendors to integrate their intellectual property (IP) into the products. ODMs need to ensure that such integration is not made available to other OEMs using the same ODM product.

If OEMs intend to offer these products to mission-critical and highly secure environments, the IP protection will be a far more critical issue when choosing ODMs. In such cases, the location of the business operations and the manufacturing facilities of ODMs could be a critical issue in ensuring IP protection.

If OEMs have to integrate this IP–specific functionality in-house, they may lose the time-to-market and cost advantages offered by the product. If ODMs can demonstrate that the IP belonging to the OEMs can be protected, OEMs might choose to let ODMs integrate this functionality into the product. In such cases, OEMs need to ensure that ODMs have the necessary infrastructure, including isolation of the OEM–specific development environment and secured access to the IP–related software and hardware components by the ODMs' staff. In some cases, the ODMs' design teams may operate within the OEMs' premises to ensure security. Appropriate liability and indemnification agreements between OEMs and ODMs are generally necessary to achieve a high level of confidence in securing IP. ODMs need to demonstrate to OEMs that they can be trusted and develop a long-term business relationship for success in the programs.

Conclusion

To ensure the success of ODM programs, OEMs need to establish a long-term relationship with ODMs. This article has outlined the key factors OEMs should consider when choosing ODM partners. ODM partners need to demonstrate their knowledge of the technologies, the market, and their commitment to road maps for the product development. Moreover, ODM partners should have a strong record of execution on all fronts, including design and development, customization, manufacturing, and reverse logistics. Choosing the right ODM partner is a big step in ensuring OEMs' success in addressing customer needs.

Assurance: A Past, Present, and Future Perspective

Karl Whitelock

*Lead Strategist OSS, Assurance Solutions
Division*
Agilent Technologies

Randy Custeau

*Director of Marketing, Strategy, and New
Business*
Agilent Technologies

Executive Summary

Several years ago, communications services were all about connectivity—enabling one person to talk with another. Network integrity and network functionality received all of the attention, as carriers were determined to keep their networks operating effectively and efficiently. The customer was often viewed as a burden to successful network management.

The emergence of competition in the mid-1980s started an evolution toward customer assurance as technologies drove the introduction of new services and consumers began to have choices about who or what was providing their communications needs.

Today's world of improving technologies, evolving business strategies, changing regulations, competition, and complex service offerings are finally delivering on the longtime promise for customer-centric services that support a variety of customer needs. New providers, in addition to current operators, are enabling consumers to have even more choices, especially the choices now coming from the customer-savvy virtual network operator (VNO) market.

Our industry is quickly evolving into a world about profitability, not just market share. Retaining customers in today's competitive environment requires operators to continuously deliver an acceptable service experience. Obtaining measurements about this experience on a real-time basis is the most essential part of customer assurance. Without it, operators can only guess about how well their customers are getting along. The age-old adage "If we don't do it, someone else will" is a reminder that consumers have

a choice and will exercise their option to choose the best service and/or content from the most customer-centric organization available.

Introduction

Over the past 30 years, service providers have been concerned with monitoring and managing the complex networks that underpin their businesses. Every network element within their circuit-switched voice networks, transport networks, Internet protocol (IP) data services networks, and wireless access networks were managed regarding configuration of customer services, performance of those services, and detection of fault conditions. In every case, the highest-priority focus was on effective operation of the network, with the customer experience at best a second-place concern. In addition, the static nature of network configuration and the dominance of point-to-point services made this approach satisfactory.

Evolving regulatory change, competitive forces, technology advancements, the appetite for content, and customer-use model changes have caused today's market to be more dynamic and customer-centric than in times past. Service providers globally are realizing that focusing on the customer is the highest priority and that doing business as usual with network tools is no longer enough.

> In times past, maintaining the network and monitoring its performance to assure proper network operation, was the most important carrier concern. Customers, for the most part, were an afterthought and even a burden in some cases to effective network management.

Only a true measure of the customer experience can adequately provide an understanding of how consumers use services.

This white paper looks at the progression of telecommunications technologies and the evolving need for monitoring

> A measurement point co-resident to the customer will provide operators with the most accurate information on a customer's quality of experience. It is here where the communications sector will look more like what banking and retail does today.

Today, the focal point for business operations is on the customer. This is driven from a need to support multiple supplier/partner service offerings for a continuously evolving, technology-savvy customer base. In this new environment, operators must have a way to gauge the customer experience to continue delivering the levels of service quality that they and their customers have grown to expect.

the customer experience to capture, retain, and grow an organization's customer base and revenues. It describes how the telecom market has changed from its network-focused approach of ages past to today's competitive, complex landscape. It also looks at why customer assurance is so critical for meeting the needs of today's service offerings and changing business models.

A Captive Audience (1940s through 1984)

Telephone networks initially evolved, along with the companies that owned and operated them, into communications monopolies emphasizing end-to-end voice connectivity for a growing customer base. In North America, for example, the monopolistic nature of the former Bell System, headed by AT&T, delivered a nearly ubiquitous voice service capability throughout the highest-population centers of the United States. This success eventually led the U.S. Justice Department in 1984 to break up the former AT&T Bell System. The breakup produced AT&T—an interexchange carrier (IXC)—and seven regional bell operating companies (RBOCs), which also proved to be the seven largest U.S.–based incumbent local-exchange carriers (ILECs).

Throughout the AT&T breakup process, attention was completely focused on network integrity and network functionality, as shown in *Figure 1*. The customer was, for all intents and purposes, "captive" within the geographic boundaries of one of the seven RBOC operating territories. Similar results occurred in other countries as communications services started to deregulate worldwide.

The RBOCs and many other Post Telephone and Telegraph Administration (PTT) carriers found themselves in an environment that provided the following:

- *Little regulatory change*: Prior to the 1984 AT&T breakup, there was little if any regulatory change. After 1984, however, similar deregulation of the United Kingdom telecom environment occurred and was soon followed by other European countries. Today deregulation continues worldwide, which has paved the way for more customer choices and competitive service offerings.

- *No direct competition*: Customers had only one choice for their communication services that consisted of analog, direct-dial, wireline-based voice. Even classic features such as three-way calling, call forwarding, and busy do not answer were not available.

- *Slow technology evolution*: Improvements in the public switched telephone network (PSTN) were meticulously slow in order for the ILECs to reap the highest return on their investment capital. In the case of the RBOCs, they were now free to invest without parent company approval. They accelerated replacement of electro-mechanical and stored program control analog switches with digital technology. These improvements started the evolutionary changes that have defined the

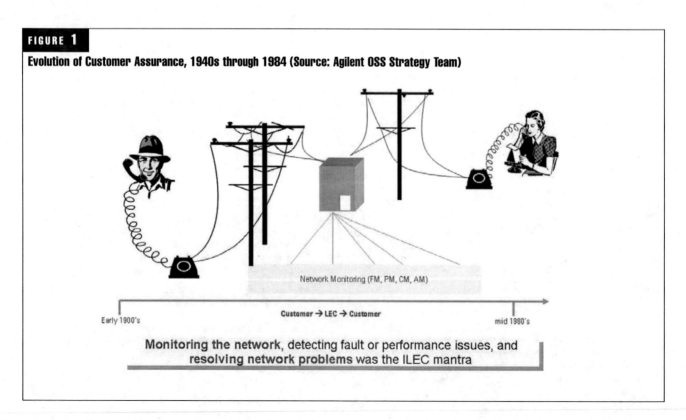

FIGURE 1

Evolution of Customer Assurance, 1940s through 1984 (Source: Agilent OSS Strategy Team)

Network Monitoring (FM, PM, CM, AM)

Early 1900's   Customer → LEC → Customer   mid 1980's

Monitoring the network, detecting fault or performance issues, and resolving network problems was the ILEC mantra

communications industry over the past 30 years. Even transmission between switching centers was mostly analog at this time. Similar replacement strategies were started at other global carriers.

- *Network-focused network monitoring and management*: Service providers constantly monitored their respective networks for fault conditions or when traffic flow crossed above performance levels that exceeded predefined engineering specifications.

Maintaining the network and monitoring its performance to assure proper network operation was the most important carrier concern. Customers, for the most part, were an afterthought and, in some cases, even a burden to effective network management.

The Emergence of Competition (Mid-1980s through the Late 1990s)

During this 15-year period, not only did new market forces gain strength, but the technologies enabling many of the competitive strategies for a variety of service alternatives also became commonplace, as shown in *Figure 2*.

This era saw a number of changing market conditions, including the following:

- *Broadened regulatory change*: In the United States, the Telecom Act of 1996—changing rules concerning last-mile access, local number portability (LNP), and the continual lobbying for classifying competitive service

offerings as "information services"—were the main drivers of regulatory activity.

- *Wireless services more easily available*: The rollout of analog wireless and later updates to second generation (2G) digital enabled expansions of wireless mass-market services across Europe, Asia Pacific, and North America. Customers accepted wireless and wireline services from different organizations.

- *Access and long-distance alternatives*: Long-distance alternatives to AT&T, including MCI, Sprint, and multiple resellers, dominated North America. In the late 1990s, competitive local-exchange carriers (CLECs) debuted and grew as LNP rules helped customers to more easily change service providers.

- *Technology convergence*: In the very late stages of this era, some organizations began "bundling" services tied around a common customer "bill" as a means for attracting end users from incumbent carriers and other competitors. It marked the early beginnings of attending to the customer rather than the network and set the stage for more significant customer-focused business requirements that face service providers today.

- *Network build-out land grab*: Larger companies with an almost unlimited pocketbook for gaining market share acquired competitors and smaller innovative companies. Establishing a market name, rather than focusing on good customer service, was the dominant theme.

FIGURE 2

Evolution of Customer Assurance, the Mid-1980s through the Late 1990s (Source: Agilent OSS Strategy Team)

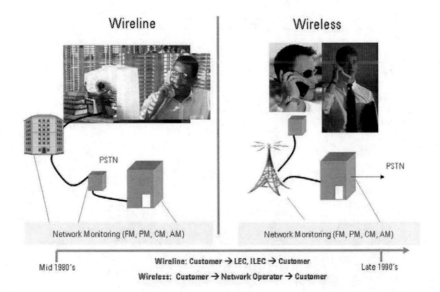

- *Mass-market broadband begins*: In the late 1990s, the roll-out of high-speed cable modem services and the mass-market offering of DSL services ushered in the age of broadband. Accessing the Internet and using it as a significant business/educational tool established several new technology and business requirements for today's services environment. Service offerings started to become something more than the bearer services of a single carrier.

Though evolution of the wireline, wireless, and some alternative services carriers was important for the long-term benefit of the industry, each of these groups placed their emphasis and business focus on capturing, constructing, consolidating, and operating network technologies.

Service-level agreements (SLAs) were often talked about but realized only at a rudimentary level because operators lacked a dependable means for measuring the customer experience. With network management data as the only source for assuring quality of service (QoS), SLA commitments unique to each customer were not feasible. Focusing on the customer experience and guaranteeing customer-level service is an objective that remains unfilled today.

The Customer Finally Receives Attention

We live in a world where evolving network technologies, business strategies, regulations, competition, and complex service offerings all are converging as shown in *Figure 3*.

More than ever, ILECs are feeling competitive pressure from alternative voice over IP (VoIP) carriers and data access carriers (cable, satellite). Wireless operators continue to feel the competitive effects of fighting for the same customers.

Market conditions in today's environment include the following:

- *Significant regulatory change*: In the United States, the Telecom Act of 1996 is being reviewed, cable companies continue to experience "information services" status, and long-held broadcast TV franchise rules are relaxing. In other areas, strong calls for carriers to maintain call/session usage data in the name of "national security" is taking center stage. The trend for less regulation as a means for continued industry stimulation is emerging.

- *The long-awaited reality of convergence*: Delivery on the longtime promise for seamless voice and data access is commonplace.

- *Significant competition as new business models emerge*: Not only must the ILECs fight competitive pressures from well-established wireless providers, but now satellite and cable operators abound with competing triple-play and quadruple-play offerings. Of most significance, however, is the emergence of VNOs that aggressively market their content to dedicated consumers using both fixed and wireless network capac-

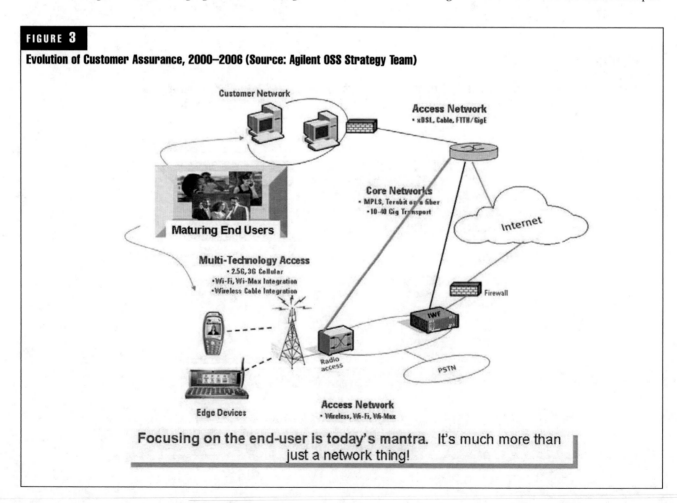

FIGURE 3

Evolution of Customer Assurance, 2000–2006 (Source: Agilent OSS Strategy Team)

Focusing on the end-user is today's mantra. It's much more than just a network thing!

ity from wholesale carriers. The emerging VNO sector is taking customer-centric to significantly advanced new levels as this concept continues to gain momentum over the coming months and years.

- *The business priority of profitability as carrier consolidation continues*: The land grab of the 1990s is over, with some of the most aggressive players from that era in North America (MCI, AT&T) now consumed by lesser players from the same period, e.g., SBC acquiring AT&T and Verizon acquiring MCI. Profitability and accountability are finally becoming important as operators realize that competitive forces with specialized attention to the end user are taking market share.

- *The proliferation of multi-supplier/partner strategies*: Service offerings today are complex and involve multiple suppliers working together. Gone are the days when a single carrier can deliver end-to-end services for its customers. Handing off information and assuring the quality of that information take on greater significance. Complexity from behind the scenes to the end user continues to increase and begs for a means to monitor and assure the end-to-end customer experience.

- *Customer services on any media*: Long a promise of the late 1990s, the delivery of content to any media is a critical business driver for new networking technologies such as the IP multimedia subsystem (IMS). It is the basis for enabling the customer terminal device to seamlessly roam across multiple wireline and wireless networks to provide the best customer experience.

The complex service offerings that meet customer needs today involve an intricate behind-the-scenes set of partner/supplier relationships. Managing the customer rather than just the network finally takes on significance with all major carriers, as new competitive threats from advanced enabling technologies provides end users with more service capabilities.

The focal point for business operations is rapidly moving toward a spotlight on the end user rather than on the net-

work. This is driven from a need to support multiple supplier/partner service offerings for a continually evolving, technology-savvy consumer base. In this new environment, operators must have a way to gauge the customer experience to continue delivering the levels of service quality that they and their customers have grown to expect.

Assuring the Customer Experience

Only a true measure of the customer experience can adequately provide an understanding of how services are used by consumers, as shown in *Figure 4* below. Assuring a high-quality customer experience can be accomplished in two ways: directly measuring how end users experience their service or inferring by using complex relationships that combine network configuration, network inventory, performance measurement, and fault management data. The measured customer experience data is further used to prioritize resources in the "detect, isolate, and repair" process that forms the basis of assurance.

The inferred approach, implemented by traditional network systems vendors, assimilates fault and performance details from individual resources into a complex services model that projects a view of the health and status of services and customers. Correlating this information with service subscription data from the customer care system, service providers can infer that some customers are experiencing less than optimal conditions whenever network issues arise. This method can be useful in determining customer impact during major network outages but is insufficient for more mature, redundant networks and for monitoring non-network-related problems (e.g., when configuration issues with the service or customer terminal device are undetected). As service and network complexity increases, so does the correlation complexity and effort. In many cases, it cannot be performed in a timely fashion, if at all.

In addition, non-network-related problems account for the bulk of technical complaints that enter call centers today. This is due to increased sophistication of the end terminal (i.e., mobile phone, set-top box, etc.) and increased comfort of the consumer to customize their services.

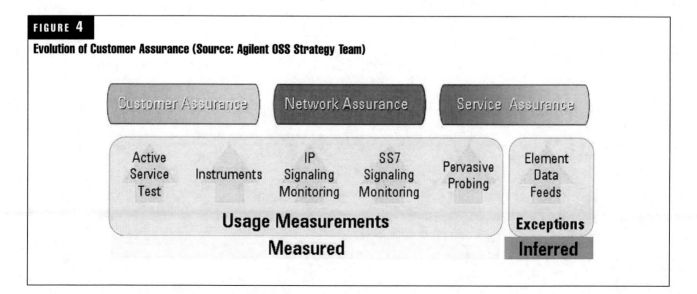

FIGURE 4

Evolution of Customer Assurance (Source: Agilent OSS Strategy Team)

The measured approach involves network-signaling and customer session data to provide a solid readout of a customer's usage experience, which is used to prioritize service provider resources for assuring services. This direct measurement of what the customer actually experienced on a voice call, short message service (SMS) or multimedia message service (MMS) message send, video session, e-mail, or even Web access attempt will detect configuration problems, which have been noted by all service providers as the most significant source of technically related customer complaints to date.

A Look to the Future

What does the future hold? A glimpse through the looking glass shows that the maturation of technology and end users will encourage the emergence of new business models that are highly specialized and content-driven. For example, the following may occur:

- You will buy communications services at the same place you buy coffee, toilet paper, toys, groceries, sporting goods, etc.

- Bundled services will not mean local and long distance—instead, it will refer to purchasing several small, cheap things rather than a few big things.

- You will buy services independently from network access, and your network will figure out how to deliver them to you, wherever you are or whatever the services may be.

- Your kids will think the phone company is Disney, ESPN, Harley-Davidson, NASCAR, 7-Eleven, the cable company, AOL, eBay, or one of many other companies.

The supplier world will also change. For example, the following may occur:

- Some people (other than the customer) will actually care about QoS, but it may not be because they want to make it better. Most will take voice quality for granted, as long as it works.

- Fewer people will care about traditional resource management, since alarm and event monitoring cannot directly determine the customer experience and provisioning resources does not mean the customer has service.

- Customer experience and care will rule operations, thereby providing significant differentiators for wholesalers, retailers, and VNOs.

- Operations spending will shift from the back office to customer-facing systems.

More customer intimacy with the supplier (and supplier does not imply network operator) will drive the need for a more direct measurement focus through more pervasive and distributed measurements. In this approach, a measurement point co-resident to the customer will provide the supplier with the most accurate information on the quality of experience. It is here where the telecommunications sector will look more like what banking and retail does today.

With the advent of converged wireless/wireline access and transport capabilities, combined with ongoing sophistication of the customer terminal device, there is no turning back to the times when a single operator or supplier addressed all of the needs of its customer base. The genie has been let out of the bottle, and customer-centric assurance is likely one of the three wishes!

Network
Management

Beyond Access: Raising the Volume of Information in a Cluttered Environment

Saul Berman

BCS Global and Americas Business Strategy Leader
IBM Corporation

Adam Steinberg

Partner, IBM Business Consulting Services
IBM Corporation

Louisa Shipnuck

Global Media and Entertainment Industry Leader
IBM Corporation

Executive Summary

From carrier pigeons to the advent of mass communications, the business information industry has successfully adapted to many changes in its long and storied history. Today, however, the US$70 billion industry [1] may face its toughest transition yet: competing in a digital world with a digitally savvy user.

New online information providers and free services—including online newspapers, expert blogs, targeted search engines, low-cost research sites, and niche providers—have begun to pull audiences from incumbents. With knowledge seekers atomizing and loyalties shifting, new competitive and substitution risks may be driving the industry toward an inflection point. Industry players need to adapt swiftly or risk losing ground in key areas.

Near term, the industry's two historical barriers of entry—proprietary content and proprietary analysis—are expected to hold up. However, in the long term, the "opening" of source information will most likely drive heightened competition and the erosion of barriers. If these trends continue, future competitive advantage will go to firms that provide value-added services and expert insights, not access.

A Pedigreed Legacy

The information provider industry has weathered many changes in its long and storied history. In the 19th century, Reuters used carrier pigeons to disseminate stock quotes, and the Associated Press created a collective of rowboat reporters to "scoop" news from incoming ships. Later came decades of industrialization and the advent of the telephone, radio, and television. Business information providers have managed to adapt successfully through each successive wave of fundamental change.

A Digital World Drives Fundamental Change

Within this US$70 billion dollar industry [2], the explosion of the Internet has given rise to multiple new sources of information, many of which are technologically innovative and often available without fees for the user. The Internet has also made it possible for primary information suppliers—from news outlets to law courts to manufacturers—to go directly to consumers, thus creating an additional new layer of de facto competition. In the past, these suppliers would provide original source information to journals or information services. Now that one can offer supplier information directly online, some information providers face the very real prospect of disinter-mediation.

The wave of change engulfing information providers includes another distinctive element: digitally savvy end users. Broadband adoption continues its rapid growth in major markets around the world (see *Figure 1*), driving users' appetites for new, online content, and services. Incumbents are often behind the curve in terms of adopting new technologies for compelling and robust information services. In most cases, non-traditional players have simply been quicker to innovate and are more agile in leveraging new technology for users' benefits.

News and Newswire Services Face New Competition as Suppliers Go Direct

Today, some information industry segments face the threat of disinter-mediation more acutely than others. In the news and newswire segments, for example, traditional information services are feeling the bite of commoditization as users gain direct access to myriad digital alternatives. Internet distribution is making it possible for brands to reach their constituencies easily, cost-effectively, and directly. In order to gain advertising revenue—and, to a lesser degree, online subscription and archive access fees—more than 5,000 newspapers and 2,000 trades magazines are now available on the Internet. [3]

This explosion of content availability is also giving rise to a new breed of aggregators intent on competing in a space once reserved for traditional and fee-based information providers. A prominent example of this new form of aggregator is Google News, which offers access to 4,500 daily global news sources—as a free user service updated every 15 minutes and customizable by the user. With competitors able to reach audiences as never before, fee-based business models that rely primarily on aggregation services are most susceptible to competitive pressure.

These dual trends—open supplier information and the emergence of new aggregation points—are conspiring to suppress business performance in the newswire category. For example, Dow Jones Newswires revenues were down US$22 million, or 9.3 percent, over the last two years, while PR Newswire saw its revenues decline 7.1 percent between 2002 and 2003. [4] Even as the economy recovery helped to push subscription news services back into positive territory in 2004, growth remained slow to moderate. Subscription

growth for the top ten services was estimated at 3.9 percent between 2003 and 2004. [5]

However, even the new class of aggregators is not free of the competitive threats posed by technological change. New tools like really simple syndication (RSS) are allowing users to create their own services—in effect disinter-mediating the aggregators. RSS feeds allow a user to go directly to the sources filtered by an aggregator—be it Google News or a traditional provider—in order to create their personalized news services from the Internet.

As knowledge seekers embrace diverse information channels, user demand is trending toward disparate and niche information sources. This flowering of choice has put customer loyalty in a state of flux. To some degree, users are conferring to free, online sources levels of credibility and confidence they once reserved only for fee-based services. In a 2005 survey of corporate workers (from 20 industries and 10 functional areas of business), 73 percent responded that they rely on the public (free and open) Internet as a primary research tool, with 69 percent relying on the open Web for daily decision making in their job. Furthermore, 77 percent utilize search engines to get information for the job; in fact, the percentage was even higher for workers belonging to such fields as science, engineering, or information technologies (IT). Specifically, 72 percent responded that the public Internet contains information from credible sources. [6] The numbers (see *Figure 2*) demonstrate the resilience of free Internet sources (versus traditional fee-based services), a trend which continues to change the economics of the industry.

For incumbents in the information provider industry, these results speak to a risk of substitution—more acute in some

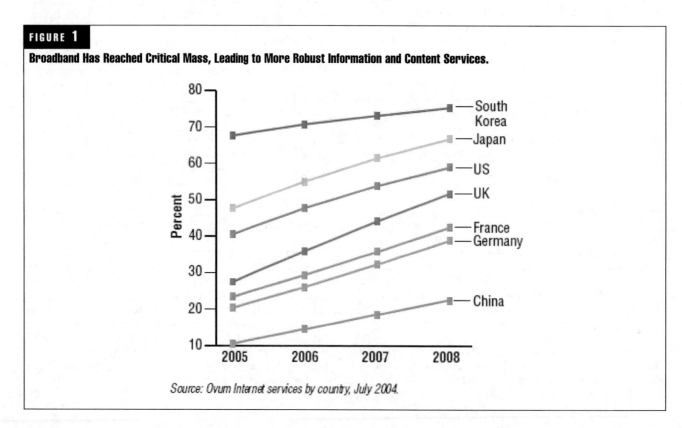

FIGURE 1

Broadband Has Reached Critical Mass, Leading to More Robust Information and Content Services.

Source: Ovum Internet services by country, July 2004.

content categories than others—as well as an end-user perception gap. The results also suggest that users are taking their consumer habits into the workplace, all while blurring the distinction between consumer and business users. Information product bundles are unlikely to satisfy these empowered end users, especially as they gain more say over purchasing decisions. As this trend plays out, information providers must learn to appeal to a highly diverse set of needs. The user-centric trend also has pricing implications: if existing product bundles are no longer adequate, providers will be challenged to segment and monetize their products in new ways.

Efforts to adapt to these unprecedented changes are being hampered by legacy issues, including a persistent product and copyright focus, manufacturing-age processes, and legacy and piecemeal technologies and cultures that are oriented around the physical, rather than the digital world.

Current trends suggest that the market may be headed for an inflection point (see *Figure 3*). The industry is far from monolithic, and different content segments will be impacted in different ways. Overall, however, the combined risks of substitution, evolving user needs, technology, and new competition threaten to create a growing chasm between leaders and laggards.

An Uncertain Outlook

Through 2008, the information provider industry is forecasted to grow a moderate 5.2 percent [7], a rate just above the projected consumer price index. Large players will most

---

FIGURE 2

High Dependence on the Free, Public Internet by Corporate Users Is Challenging Traditional Fee-Based Information Services.

	Total corporate respondents	Finance, HR and Legal	Information Technology	Sales and Marketing	Science and Engineering
Search engines allow me to get the information I need for my job	77%	71%	79%	76%	82%
I use the Internet primarily as a research tool	73%	70%	71%	76%	71%
The Internet contains information from credible and known sources	72%	72%	71%	73%	67%
The Internet provides information that I use to make daily decisions	69%	75%	70%	66%	51%

Source: "Hot Topics: 2001 vs. 2005: Research Study Reveals Dramatic Changes Among Information Consumers." Outsell, Inc. May 2005.

FIGURE 3

The Information Provider Industry May Be Approaching an Inflection Point. Different Segments Will Be Impacted at Different Rates.

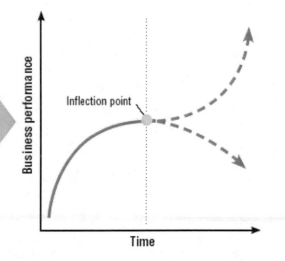

- **Industry issues**
  - Low growth industry history
  - Legacy culture and business models
  - Fragmented audience
  - Shifting user loyalties

- **Competition and substitution challenges**
  - Suppliers releasing information
  - Free ad-supported services
  - Move to open access sources

- **Technology challenges**
  - Legacy technology
  - Distribution challenges

Source: IBM Institute for Business Value analysis.

likely continue to fight for share within the mature market while grappling with new competitive dynamics. In the near term, category and geographic expansion are expected to drive revenues. Existing products will likely move into fast-growing markets like China and India, as well as Central and Eastern Europe and Latin America. Both emerging market and financial data are expected to provide a new source of revenue for meeting regulatory requirements and identity data for security analysis. An ongoing recovery in advertising also promises to boost revenue from its present anemic state. Longer term, however, the industry may face more fundamental challenges.

In order to evaluate competitive advantage—now and in the future—IBM has created a model defined by two historical barriers of entry (see *Figure 4*). The first of these barriers is created by *proprietary source content*, which can include court rulings, breaking news, equity quotes, company financial data, technical specifications, and scientific research. Such information creates a barrier of entry when the provider exclusively owns the source data. Aggregated information and archives are of great value to professional users, but providing these services hinges on the availability of foundational data and information.

The second barrier to entry is *proprietary analysis*, the expert analysis, commentary and interpretation that providers wrap around the source content. This barrier is maintained through branded, copyrighted analysis—ranging from strategic investment analysis to legal opinions—that competes for audience mindshare. As a component of competitive advantage, proprietary analysis is particularly pronounced in content areas where customers rely more heavily on expert opinion in the course of doing business.

A more detailed look reveals how each sector falls at a different point along the two proprietary axes:

- *Marketing* (top right quadrant): Marketing research and user behavior pattern information is expensive to collect and interpret. Thus, these assets are of high value along both axes. Companies like Taylor Nelson Sofres (TNS), Nielsen Media Research, and Arbitron hold unique source information as well as the rating or cross-tabular analysis that gives rise to insight.

- *Science, medical, and technology* (top left quadrant): Science derives great strength from its peer-to-peer quality control processes and copyrighted journal articles. Thus, it rates highly along the content axis, but less so on the analysis axis. While industry leaders like Elsevier's ScienceDirect and Thomson's ISI Web of Knowledge have rolled out abstracting and cross-referencing capabilities, ancillary analysis and editorials (over and above the research itself) are not traditionally a part of the offering.

- *Legal and regulatory* (bottom right quadrant): In this vertical content, most source documents—court rulings, for example—are available in the public domain. Thus, value resides in analysis, editorials, and opinion-making, as well as in the case summaries, archives, and cross-referencing of large databases.

- *General news and newswires/Business and trade* (bottom left quadrant): According to the present analysis, the news and newswire categories are exposed to the greatest threat from commoditization. As suppliers continue to offer content directly to customers, these

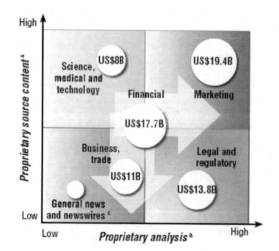

FIGURE 4

Each Content Category Currently Derives Competitive Advantage from Two Barriers to Entry. A Company's Strength in the Future is Dependent on Category and Portfolio Mix.

[Chart: Y-axis "Proprietary source content A" (Low to High), X-axis "Proprietary analysis B" (Low to High)]

- Science, medical and technology — US$8B
- Marketing — US$19.4B
- Financial — US$17.7B
- Business, trade — US$11B
- Legal and regulatory — US$13.8B
- General news and newswires C

Legend:
- ☐ Constrained market position
- ☐ Area of concern for information commoditization
- ☐ Strongest market position
- ☐ Relatively strong position, reliance on analysis

A *Proprietary source content refers to the foundational information used for professional service information; this includes court rulings, breaking news, equity quotes, com, financial returns, technical specifications and scientific research results.*
B *Proprietary analysis refers to the level of with analysis/editorial/interpretation wrapped around the source content for the average customer.*
C *General news and newswires estimated at US$2.5M-3.5M including only the top subscriber services. All market size data is as of year end 2003.*
Source: IBM Institute for Business Value analysis.

industry verticals must evolve quickly into new business models and partnerships. Business and trade information providers face similar threats because the data creators now compete directly with the traditional aggregator for the consumer eyeball. The providers most at risk are the data aggregators rather than the analysts.

- *Financial information* (middle): This category is bifurcated, with some content—such as company records, earnings trends, and stock quotes—easily found via free services. However, investment reports, real-time data, and uniquely collected industry and company information continue to garner premium prices.

While the two historical barriers of entry—proprietary source content and proprietary analysis—seem safe for the short term, only the latter appears likely to survive long term (a trend illustrated in *Figure 4* with directional arrows). IBM believes that source content will continue to "open" due to regulatory change and technological advancement. As sources open, their value as a protected stream of revenue declines.

Against this backdrop, traditional players are likely to seek growth by invading each others' markets. For example, Bloomberg, as a finance provider, has moved into law, while Thomson has moved into finance broker stations. Such maneuvers will likely shake-up share between the major incumbents within defined markets. In addition, new entrants will continue to exert competitive pressure, as relative newcomers like Google, Yahoo! and CNN.com are joined by vertical information providers vying for the specialized user. These niche players include such diverse companies as compliance information provider Complinet, as well as Google Scholar and the U.S. National Institutes of Health (NIH) PubMed Central for scholarly publishing and scientific research. In addition, free and low-fee information providers—ranging from CyberLaw expert Lawrence

Lessig's blog to subscription archive HighBeam Research to "do it yourself" RSS technology—will provide options for more price-sensitive customer groups (see *Figure 5*).

Moving Beyond Access

Faced with these dynamic changes, business-to-business information providers must seek to offer services and solutions that move far beyond access. With myriad touchpoints for information, access will remain plentiful, and it might even proliferate in ways we cannot imagine today. Similarly, breadth and depth of content via archives and aggregation will continue to be a vital part of the business. Yet, as the industry evolves—and mere access to content erodes as the basis for competitive differentiation—a firm's ability to attract and retain customers will shift toward the delivery of value-added expertise and insight (whether provided outright or in the form of services that extend the expertise and insight of the customer).

To succeed in this environment, information providers will have to learn from traditional and nontraditional competitors and adopt new business models and service dimensions. In many cases, partnerships will be required to deliver broader solutions, such as workflow integration and value-added services, including consulting and client-customized services. One will need to consider the growing competition among emerging platforms, such as portals or integrated search tools on the professional's desktop, as information providers roll out new enterprise solutions and services.

To help information providers adapt to these new conditions, the IBM Institute for Business Value recommends six specific strategies (see *Figure 6*):

1. *Protect the core business by providing solutions.* With access to information commoditizing on many fronts, providers must move beyond their traditional models toward solution

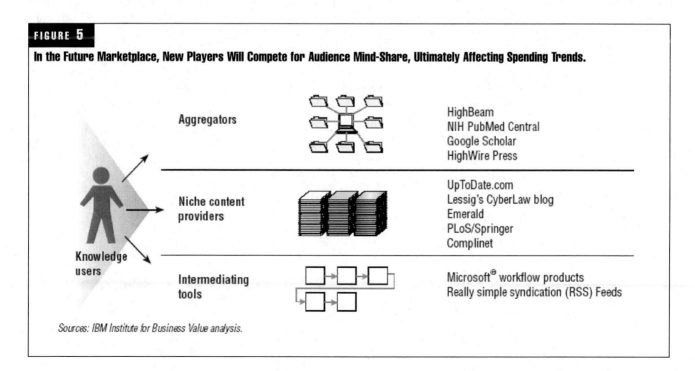

FIGURE 5

In the Future Marketplace, New Players Will Compete for Audience Mind-Share, Ultimately Affecting Spending Trends.

Aggregators
HighBeam
NIH PubMed Central
Google Scholar
HighWire Press

Niche content providers
UpToDate.com
Lessig's CyberLaw blog
Emerald
PLoS/Springer
Complinet

Knowledge users

Intermediating tools
Microsoft® workflow products
Really simple syndication (RSS) Feeds

Sources: IBM Institute for Business Value analysis.

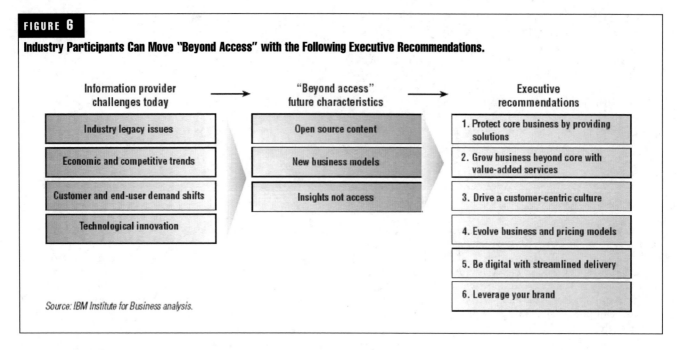

FIGURE 6

Industry Participants Can Move "Beyond Access" with the Following Executive Recommendations.

Source: IBM Institute for Business analysis.

delivery. The key is to deliver greater relevancy and expertise in a format that is compelling, seamless, and easy to use. Solution capability, along with analysis, will become the second barrier of entry to the future.

For this to happen, third-party information providers must recognize that they reside in larger ecosystems of information that encompass several dimensions: the enterprise's own work-product and information repositories, other third-party information providers (including public sources and the public Internet), and end-user's desktop information. Playing in the information ecosystem will require providers to overcome their siloed structures and propensity to "go it alone." As *Figure 7* illustrates, information providers need to partner to deliver full, next generation solutions to demanding customers.

Customizing the user experience through workflow integration is an important strategy in moving towards a solutions-based business model. Value-added services should seamlessly blend with user applications and common touchpoints and should be customized by a functional user group. Such workflow integration creates high customer switching costs by linking into the user process flow and by making third-party content integral to doing business.

Another fruitful area is the development of ontologies, which have the potential to make the Web a far richer, semantically-based system by providing a sophisticated sense of users' needs. As the user interfaces with the system over time, the ontology develops a "mind map" of that user, which, in turn, it uses to deliver focused information. While search is an important component of doing business today, developing future solutions will require a long-term commitment to both ontologies and the "mind" of the user.

2. *Grow beyond the core business with value-added services.* Value-added services expand analytical capabilities into new service realms. Possibilities include providing innovative visions, quality expertise and insights; delivering expert product extensions; offering time-saving services and solu-

tions; and driving insights from multiple and unique data sources.

3. *Drive a customer-centric culture.* In the past, information providers often went in search of customers. However, given the growing sophistication of information buyers and heightened demands on service providers, it is imperative that industry participants move to a customer-driven product development process. Knowing the customer—through ongoing feedback loops, analysis, segmentation, and customer relationship management—is key. Indeed, customers should take a central role in the service development process.

Customer-centricity should also focus on end users. As users become ever more sophisticated, their needs will ultimately be represented in customer contracts. Therefore, aligning products to user needs—both practical and emotional—is an important element in creating loyalty and stickiness. Becoming customer-centric also requires investment in data collection and analysis. Sophisticated insights based on good data will underpin many of the segmentation and service customization capabilities of the future.

4. *Evolve business and pricing models.* Different strategic differentiators imply different business and pricing options. As a company seeks to evolve its strategy, it should consider its current mix of proprietary information and proprietary analysis, as well as the industry's expected shift toward solutions as a strategic dimension. For example, a company that enjoys defensible positions along both axes today (see *Figure 4*) may be able to leverage pricing and service stratification—growing beyond the traditional niche—to forestall new competitors that enter at lower price points. Companies in more vulnerable positions (the bottom right quadrant) will have fewer options and may need to focus on volume versus price plays. The bottom line is that learning across categories is the new imperative. Each service within a company's portfolio needs to audit its unique competitive advantages and learn from other areas within the parent company.

FIGURE 7

In the Future Information Ecosystem, Information Providers Are One of Many Key Participants.

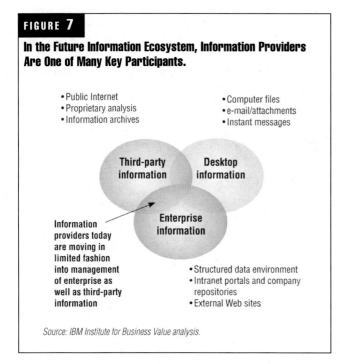

- Public Internet
- Proprietary analysis
- Information archives

- Computer files
- e-mail/attachments
- Instant messages

Third-party information

Desktop information

Information providers today are moving in limited fashion into management of enterprise as well as third-party information

Enterprise information

- Structured data environment
- Intranet portals and company repositories
- External Web sites

Source: IBM Institute for Business Value analysis.

To drive cost-effectiveness, information providers can assess their overall company structure to identify opportunities for leveraging third-party specialists. In certain cases, services can be handled more cost-effectively through an outsourcing partnership, particularly in arenas facing commoditization and compressing margins.

5. *Be digital with streamlined delivery.* Digital delivery continues to transform the workplace, with doctors using portable devices for mobile diagnostic analysis, financial professionals demanding constant alerts on the data they are tracking and personal digital assistant (PDA)-wielding litigators garnering briefing materials on the go. For information providers, delivering the right information to the right user on the right device—at the right price—requires adopting industry standards, optimizing digital supply chains, and building in cross-platform portability where appropriate. Mobility—of both people and content—is unlikely to fade as a feature of the workplace, and information providers need to find ways to help professionals obtain content and analysis through targeted channels and devices. This process needs to be customer-driven and tailored to meet and exceed expectations.

6. *Leverage your brand.* As users face a deluge of information, branded content grows in importance. Information overload will only worsen as the volume of online documents continues to double every six months or more. In this age of "overchoice," a brand that needs to be nurtured and fully exploited for advantage is a valuable asset.

Many information providers have established brand equity among their core constituency; in fact, some companies even enjoy brand penetration outside their core. However, brand equity can erode as new providers and aggregators "pull their eyeballs" in new directions. For example, database services that merely aggregate documents make easy targets for ad-driven free aggregators (such as Google

Scholar, My Yahoo! and RSS feeds) that provide the same content.

As information providers seek to focus on delivering answers, for instance, through value-added services and workflow integration, it is crucial that the brand stay aligned with the differentiating business features. The good news is that today's information providers have the ability to grow and leverage the brand to create an anchor position in a flash flood of untested content.

How Will Your Firm Move Beyond Access?

The six strategies outlined in this paper can provide your firm with insights for competing in a changing information environment. The following questions are designed to stimulate your thinking about issues that will likely arise as your organization evolves beyond access.

- How solutions-focused is your enterprise? What collaborative opportunities can your firm pursue with other ecosystem players, including other information providers? How effective is your ability to integrate with customer workflows?

- What gaps exist in serving customer demand today? How robust is your company's ability to scale up new services areas? What new kinds of skills and capabilities will you need to build new services?

- How effectively does your firm include the customer in product development? Does your development process include systematic feedback loops?

- How strongly is your firm positioned in terms of the industry's traditional barriers to entry (proprietary source content and proprietary analysis)? What are the immediate open source threats? Where do opportunities exist for deeper customer segmentation and pricing stratification? Where might advertising be integrated into current offerings? How mature is your firm's ability to innovate by collaborating across business units and content categories?

- How will emerging delivery channels affect your firm's ability to reach end users? How might they expose your firm to new competitive pressures? Which users require multiplatform access and what are the pricing bundle possibilities?

- How valuable is your brand as a beacon in a crowded market characterized by "overchoice"? What level of investment is appropriate to make your brand stand out? How often do you refresh and update the user brand experience?

As they face a new wave of fundamental change, today's information providers are well positioned to leverage their clout with customers and their vast market and user expertise. But in a future marked by competition to wrap value around information, success will require change. Access to information alone will not be enough to secure competitive advantage. Leading information providers will deliver and extend expertise and insights.

Acknowledgements

The author team wishes to thank additional interviewees, including Chris Charron, Forrester Research; Professor Mark McCabe, Georgia Institute of Technology; Michael Nathanson, Sanford Bernstein; Anthea Stratigos, Outsell; and David Worlock and Nick Dempsey, Electronic Publishing Services.

References

[1] "Global Entertainment and Media Outlook: 2004-2008." PricewaterhouseCoopers. June 2004.

[2] Ibid.

[3] "Facts About Newspapers 2004: A Statistical Summary of the Newspaper Industry." Newspaper Association of America. May 26, 2004. http://www.naa.org/info/facts04/interactive.html; "Harness the Power of B-to-B Media." American Business Media. May 2005. http://www.americanbusinessmedia.com/abm/Default.asp

[4] Company annual reports, 2002 and 2003; IBM Institute for Business Value analysis.

[5] "EIR's Current Awareness News & Research Online Subscriber Survey, 2004 vs. 2003." Electronic Information Report. April 2005.

[6] "Hot Topics: 2001 vs. 2005: Research Study Reveals Dramatic Changes Among Information Consumers." Outsell, Inc. May 2005.

[7] "Global Entertainment and Media Outlook: 2004-2008." PricewaterhouseCoopers. June 2004; IBM Institute for Business Value analysis.

Performance Evaluation of Analog Fiber-Optic Links

Pushkar Chennu

Research Assistant, Electrical and Computer Engineering
University of Texas at San Antonio

Medí Shadaram

Briscoe Distinguished Professor and Chairman, Electrical and Computer Engineering
University of Texas at San Antonio

Abstract

Fiber optics constitutes an attractive alternative to conventional wiring for numerous analog applications. Indeed, in addition to its small size and low mass, the fiber does not interfere with electronic devices and provides excellent isolation of the transmitted signal. One of the targeted applications of our work is the reference frequency signal distribution to various subsystems of a telecommunications and satellite system. Another application could be a remote antenna synchronization and control. However, to meet our system requirements, the fiber-optic link should not degrade the signal quality, in particularly its phase noise. The digital band pass modulation techniques such as M-ary quadrature amplitude modulation (M–QAM) are being used for the fiber-optic transmission of video signals and the transmissions of digitally modulated radio frequency carriers. These signals have a higher spectral efficiency and are more robust than the conventional amplitude modulated signals with respect to noise and non-linearities. In links with high bit rates, direct optical amplification is preferred because of the bottleneck introduced by the electro-optic conversion. The amplitude spontaneous emission (ASE) noise generated from the amplifier—along with thermal noise, shot noise, relative intensity noise (RIN), and other noises—affect the carrier-to-noise ratio (CNR) of the link and ultimately degrades the link performance. The QAM technique in particular changes both the phase and amplitude of the radio frequency (RF) signal. Thus, as the number of levels of QAM increases, the signal becomes more susceptible to phase noise. In this article, we study the effect of optical amplification on a simple fiber-optic link and examine the effect of phase noise on an M–QAM fiber-optic link. Bit-error rate (BER) will be considered as the performance measure. Also considered are the limitations caused by the amplifier gain and noise figure saturation characteristics.

Introduction

Phase noise is a random modulation of the phase of a source's output due to a variety of internal processes within the source circuit. This random modulation manifests itself in the time domain as jitter, the random displacement of clock edges from a nominal position. In the frequency domain (*Figure 1*), the phase noise is seen as a broadening of a source signal's frequency spectrum due to modulation sidebands. It can be the result of thermal noise, shot noise, and/or flicker noise in active or passive devices. It is a significant performance factor in many application fields such as space telemetry systems, Doppler radar, radio, and communication systems. Analog fiber-optic links are used for transmitting reference signals [2, 3, 8, 11] in various applications such as standard frequency distribution systems, Doppler radar, and phased array radar. These systems typically use fiber-optic links for transmitting reference signals. Therefore, the phase stability of reference signals is an important issue in such applications.

One way to improve the phase stability of the system is to use very precise signal sources. However, using very precise signal sources are costly; therefore, reference signals are transmitted to remote locations. In many defense applications, an optoelectronic system distributes a reference signal of low noise with a highly stabilized phase and frequency from an atomic frequency standard to a remote facility at a distance up to tens of kilometers. The reference signal is transmitted to the remote station as the amplitude modulation of an optical carrier signal, which propagates in an optical fiber. The stabilization scheme implemented in this system is particularly intended to suppress phase and frequency fluctuations [8, 4] caused by vibrations and by expansion and contraction of the optical fiber and other components in diurnal and seasonal heating and cooling cycles. Fiber-optic microwave links have the potential to be used in a large number of defense applications (radars, radio communication, etc.), ranging from simple point-to-point microwave links for reference signal distributions to advanced signal processing in phased array antennas (i.e., optical beam forming, adaptive jammer nulling, filtering, delay lines). Radars operate in different frequency bands, typically ranging from 500 MHz to 35 GHz.

Direct current modulation is one of the major assets of semiconductor lasers. It has enabled rapid developments in fiber

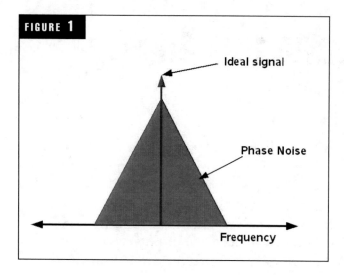

FIGURE 1

Ideal signal

Phase Noise

Frequency

optics for telecom and datacom with bit rates in the multi–Gbps range. Laser diodes are also used to transmit analog signals over optical fibers with applications in radar systems, cable television systems, cellular systems for mobile communications, as well as others.

Importance of Phase Noise

Phase noise affects the performance of analog systems. Phase noise on such reference signals considerably affects applications that require frequency and phase conversion. Some applications that are affected by phase noise are discussed in the following section.

Doppler radar detects the speed of a moving target by measuring the frequency shift, which is proportional to the speed of the target from the returned signal. At the receiver's end, intermediate frequency conversion is carried out for further signal processing. A critical parameter in the performance of airborne radars is the phase noise of radar's carrier frequency. Low phase noise is important for accurate long-range detection of a target. The return clutter signal from stationary earth is generally large enough such that when decorrelated by the delay time difference, the phase noise [25] from the local oscillator can partially or even fully mask the target signal. Thus, the phase noise of the local oscillator sets a limit on the detectability of the target.

Fiber-to-the-Home

BellSouth [17] had operated the first asynchronous transfer mode (ATM) passive optical network fiber-to-the-home (FTTH) system. Initially, FTTH was deployed as a more cost-effective architecture for providing familiar voice, video, and data services. As FTTH becomes more widespread, it will be exciting to watch many new applications emerge to make use of the increased available bandwidth. Digital enhancements, including high-definition TV (HDTV) and the ability to associate Internet content, are expected to be prominent in forthcoming projects. Voice over IP (VoIP) or voice over ATM (VoATM) are two other applications that are being deployed. Most of these applications, which require the conversion of frequency to phase (such as QAM and other data transmission applications), are very sensitive to variations in phase. Phase noise can prove very fatal to the performance of these types of systems.

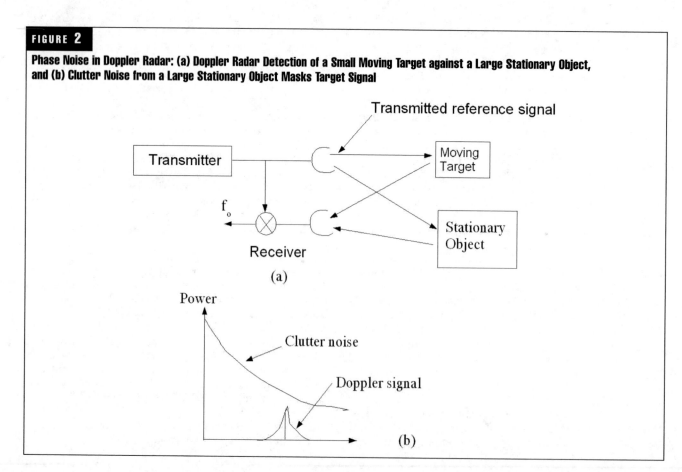

FIGURE 2

Phase Noise in Doppler Radar: (a) Doppler Radar Detection of a Small Moving Target against a Large Stationary Object, and (b) Clutter Noise from a Large Stationary Object Masks Target Signal

Optical Communications for Deep-Space Missions

The Jet Propulsion Laboratory (JPL) [16] has had a longtime interest in optical communications for use in deep-space exploration missions. The National Aeronautics and Space Administration (NASA) also uses optical communications for deep-space missions. To provide TV, high-definition TV (HDTV) and image maximum (IMAX) images from planets and telecommunications systems must operate at unprecedented data rates. Instruments required for these systems must have the ability to sustain high data rates as well as have a minimal impact in terms of mass, size, and power of spacecraft. Meeting these seemingly conflicting requirements has driven mission designers to higher telecommunications carrier frequencies. To respond, JPL deep-space missions will have to evolve from the current X-band to Ka-band and eventually to optical frequencies. Free-space laser communication (lasercomm) technology [15] must follow the requirements to meet the demand for high-bandwidth, low power–consumption, low-mass telecommunications subsystems. Free-space lasercomm tech has been used in NASA missions and has been under development at JPL for the past 15 years. In these cases, phase noise is a very sensitive parameter, because of the long transmission distance associated with deep-space missions. If a slight change in phase were to occur, it would result in a large error, so to keep the error minimum, accurate measurement of phase noise is of utmost importance in such applications.

Spacecraft Navigation

Antennas used in deep-space networks (DSNs) for spacecraft navigation are separated by continental distances. One of the navigation techniques used in the DSN is the very long baseline interferometry (VLBI). In this technique, a signal from the spacecraft is received by the two widely separated antennas. Global positioning system (GPS) satellites are used to accurately synchronize time standards in these antennas. The signal at each antenna is time-tagged and auto-correlated with the time-tagged signal received by the other station. The difference in arrival time of the two signals between the two stations is used to find the exact location of the spacecraft. Phase noise in leads to miscalculations of the position of the spacecraft. Fiber-optic systems are used to keep the local timing at each station very accurately. This distributed reference frequency is used by the station to maintain its internal time and synchronize its internal frequency references to the centralized frequency standard.

Crosstalk in DWDM Systems

Dense wavelength division multiplexing (DWDM) is being rapidly deployed in high-speed digital networks. In addition to providing higher capacity over single link, WDM provides for networking flexibility such as wavelength add-drop. Optical transmission systems using WDM techniques and operating in the 1550 nm fiber transmission window are being prototyped for use in the telecom fiber backbone networks. One state-of-the-art research system has a channel bit rate of 20 Gbps and eight signal channels for transmission over hundreds of kilometers of conventional fiber [30]. Another example system has 30 signal channels (channel bit rate 40 Gbps and accumulated bit rate of 1.2 Tbps) and transmits over 85 km dispersion flattened fiber [31]. In the practical design of these WDM systems, it is important to have a rigorous system model that accounts for the use of erbium-doped fiber amplifiers (EDFAs) as power amplifier,

in-line, and preamplifiers. An important point is to account for the effect of optical filtering in the transmission path. In the design of the WDM systems, the practical transmission window is frequently in the 1,540–1,560 nm range and optical noise filtering may be applied to diminish the influence of amplified spontaneous emission (ASE) noise from the part of the 1,510–1,590 nm EDFA gain band, which is outside the transmission window. The accumulated ASE noise may limit the effective signal amplification of the last amplifiers in a cascaded link due to saturation effects and may also influence the BER performance of the receiver. In addition, optical demultiplexing (DEMUX) filter also called arrayed waveguide grating (AWG) type [32, 33], which has a periodic optical frequency transfer function with free spectral range (FSR) that allows detection at the receiver part of the ASE noise from the whole 1,510–1,590 nm band. The DEMUX characteristics thus cause enhanced influence of the ASE noise and lead to signal crosstalk.

Evaluation of Phase Noise Using Experiments

Experimental Setup for Measuring Phase Noise in the Fiber-Optic Link under Investigation

Figure 3 shows the block diagram of the experimental setup used in a laboratory. It uses an RF oscillator to generate a reference signal very close to 100 MHz mainly because the photodetector we are using in this experiment has a bandwidth limitation of 125 MHz. The MZ modulator changes the phase of this reference signal, which modulates the intensity of the laser transmitter. The MZ modulator output travels through a 16 km spool of single-mode fiber and is amplified by an EDFA. The amplifier output is received by an optical receiver, which then converts the optical signal into an electrical signal, which is of interest to us. An optical attenuator is used to verify that the photodetector is not overpowered and that the minimum power required by it is always present. Ideally, we can look at the power spectral density of the receiver signal to measure phase noise. This requires a spectral analyzer with a high bandwidth and very high resolution. In the absence of such an analyzer, a different measurement technique is devised. A phase detector is used to mix the reference signal with the far-end receiver signal. The low-frequency AC output of this phase detector is proportional to the phase at the receiver. The output of the phase detector is received by an oscilloscope. The signal from the oscilloscope is copied into an Excel file and the power spectral density of the signal is analyzed and graphed using algorithms developed in Matlab. Using Virtual Photonics Incorporated (VPI, or an optical communication simulation package) software, simulations were also performed with similar specifications used in the laboratory-based experiments. The phase noise in the above described fiber-optic link is analyzed in three EDFA configurations: power amplifier configuration, preamplifier configuration, and in-line amplifier configuration. In each of these configurations, the EDFA gain is varied and the variation in RF phase noise at the receiver is analyzed.

Power Amplifier Configuration

In the power amplifier configuration, the optical amplifier is placed right after the transmitter to boost or increase the signal power. It is used when the optical light source has limited output power. The amplifier may saturate and cause distortion if the optical light source has higher power. In this

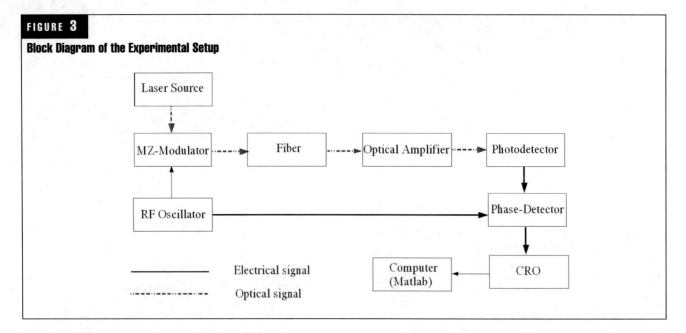

FIGURE 3

Block Diagram of the Experimental Setup

experimental setup, an optical attenuator is placed right before the EDFA and the input power to EDFA is changed, thereby changing the EDFA gain. The change in phase noise by varying the EDFA gain is analyzed using algorithms developed in Matlab. This experiment was repeated for different wavelengths between 1,540 nm and 1,560 nm. In all the experiments, phase noise frequency is band-limited to around 50 Hz. Phase noise estimations for different wavelengths and for different amplifier gains, along with the phase noise of the link without an amplifier, are shown in *Figure 4* through *Figure 8*. A comparison of the phase noise energies (1,000 Hz bandwidth) at different wavelengths is shown in *Figure 9*.

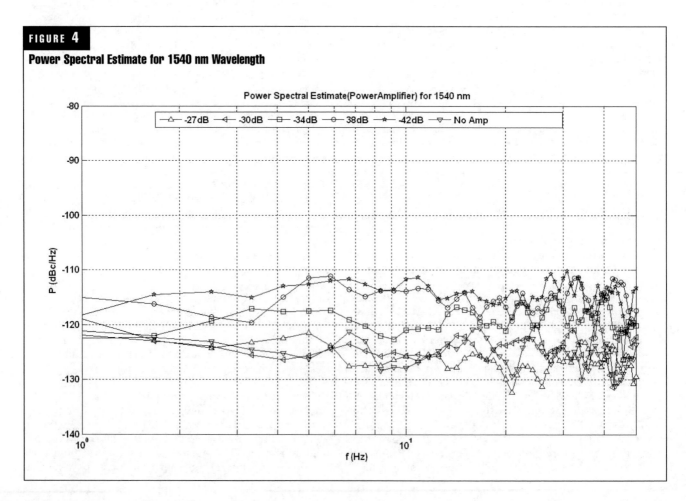

FIGURE 4

Power Spectral Estimate for 1540 nm Wavelength

FIGURE 5

Power Spectral Estimate for 1545 nm Wavelength

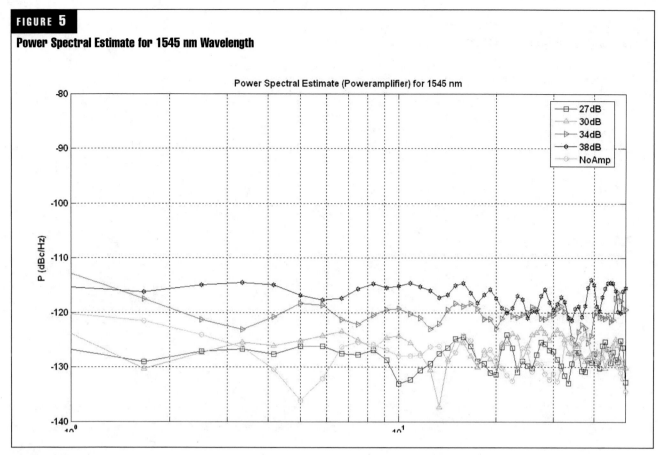

FIGURE 6

Power Spectral Estimate for 1550 nm Wavelength

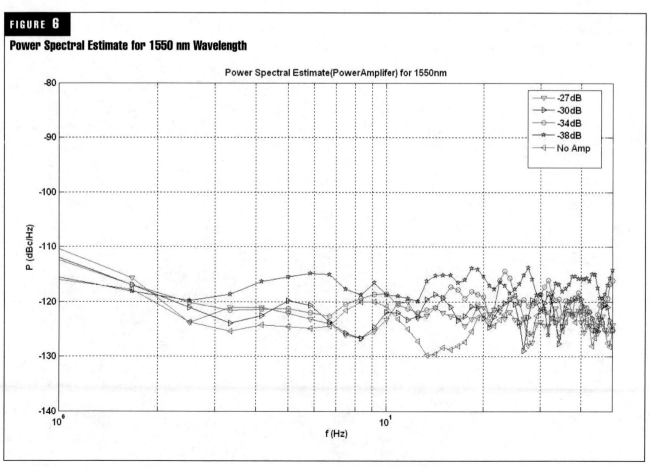

FIGURE 7

Power Spectral Estimate for 1555 nm Wavelength

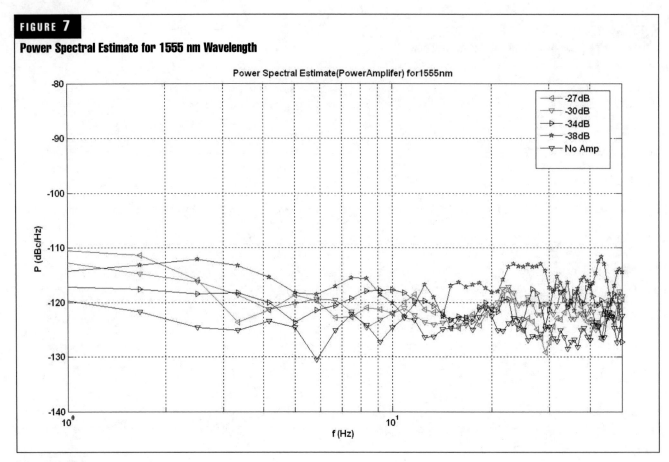

FIGURE 8

Power Spectral Estimate for 1560 nm Wavelength

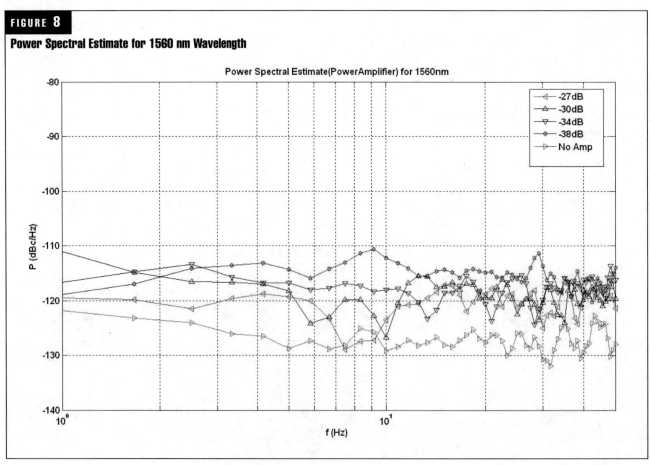

FIGURE 9

FIGURE 9

Comparison of Phase Noise Energies (Power Amplifier) for Different Wavelengths and Different Amplifier Gains in 1000 Hz Frequency Band

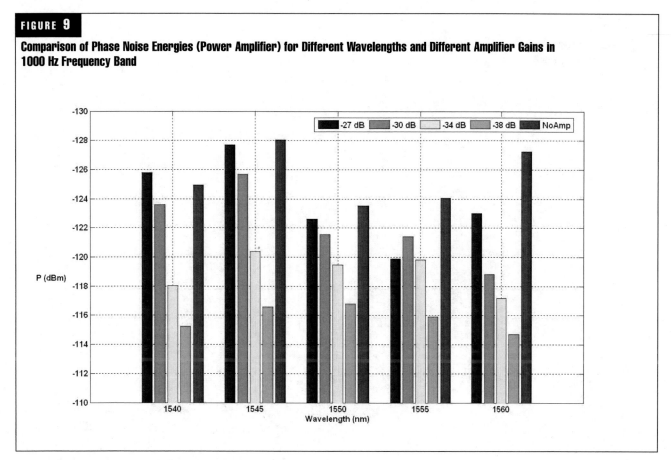

Preamplifier Configuration

In the preamplifier configuration, the optical amplifier is placed just before the photodetector to amplify a weak signal. There is no problem of saturation introduced with this type of optical amplifier because the signal power is very weak at this point in the link. A minimum signal power, however, is required at the input of this amplifier to maintain a good signal-to-noise ratio (SNR). Otherwise, amplified spontaneous noise will dominate the output signal. In this experimental setup, an optical attenuator is placed right before the EDFA and the input power to EDFA is changed, thereby changing the EDFA gain. The change in phase noise by varying the EDFA gain is analyzed using algorithms developed in Matlab. This experiment is repeated for different wavelengths between 1,540 nm and 1,560 nm. A comparison of phase noise energies (1,000 Hz bandwidth) at different wavelengths and different amplifier gains is shown in *Figure 10*.

In-Line Amplifier Configuration

In-line amplifiers are used when the transmission distance of a system is very long. These amplifiers can be inserted periodically along the link to boost signal power when needed. The optical amplifier is placed right between two 8 km spools of fiber and the output of the second 8 km fiber is detected by the photodetector. In this experimental setup, an optical attenuator is placed right before the EDFA and the input power to EDFA is changed, thereby changing the EDFA gain. The change in phase noise by varying the EDFA gain is analyzed using algorithms developed in Matlab, and is repeated for different wavelengths between 1,540 nm and 1,560 nm. A comparison of phase noise energies at different

wavelengths and different amplifier gains is shown in *Figure 11*.

In all the above three EDFA configurations, the laser source output is always kept constant to its maximum output power. *Figure 12* through *Figure 16* show the comparisons of the phase noise energies between the three amplifier configurations at different wavelengths and different EDFA gains.

Effect of Phase Noise in QAM Fiber-Optic Links Using Computer Simulations

M–QAM signals are widely being used for fiber-optic transmission of video signals and the transmission of digitally modulated RF carriers. An M–QAM system has been considered a promising method for fiber-optic video transmission because digital signals are less sensitive to noise and non-linear impairment, than their analog counterparts. Fiber amplifiers are used to amplify the weak signals in long-distance transmission links. Direct optical amplification is preferred in high bit-rate links due to bottleneck, introduced by the electro-optic conversion. Although the signal gets amplified, the optical amplification introduces spontaneous emission noise, which will be amplified along with the signal. The amplifier generated noise affects the CNR of the link. As the information in a QAM signal lies in amplitude and phase, any kind of noise introduced will change both the phase and amplitude of the signal, thereby affecting the BER. As a result, as the number of levels of QAM increases, the signal becomes more susceptible to the phase noise. ASE noise generated from the EDFA can worsen the situation. This part of the article discusses the

FIGURE 10

Comparison of Phase Noise Energies (Preamplifier) for Different Wavelengths and Different Amplifier Gains in 1,000 Hz Frequency Band

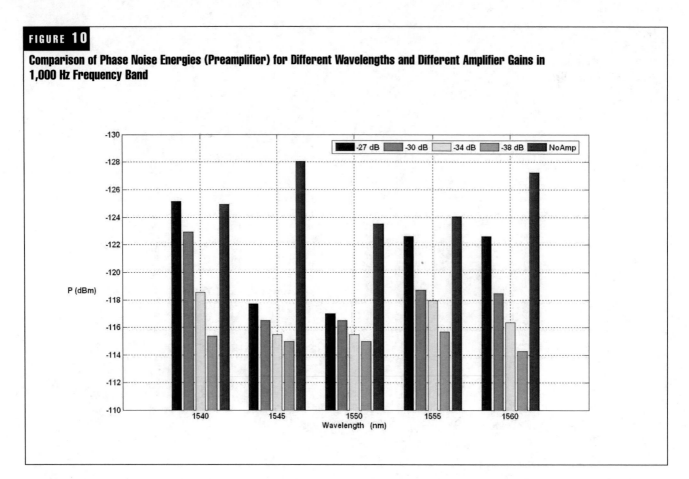

FIGURE 11

Comparison of Phase Noise Energies (In-Line Amplifier) for Different Wavelengths and Different Amplifier Gain in 1,000 Hz Frequency Band

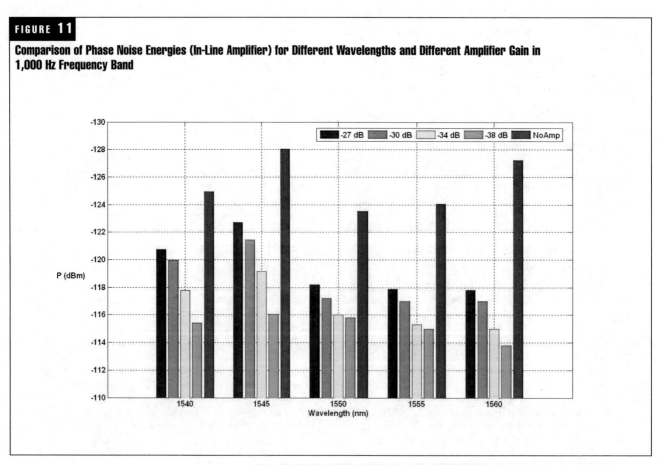

FIGURE 12

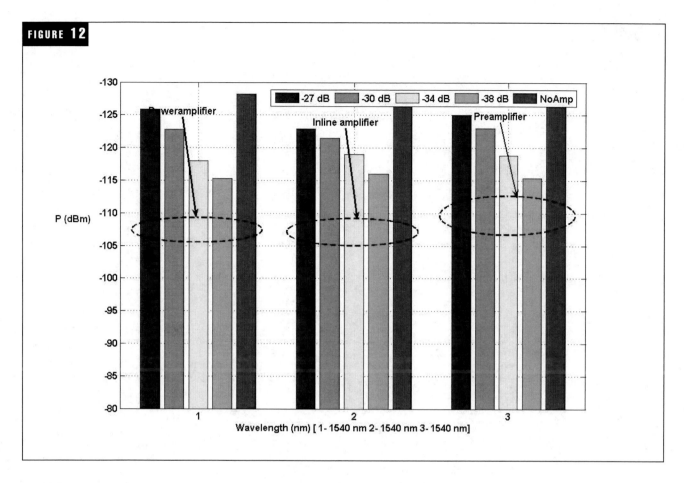

FIGURE 13

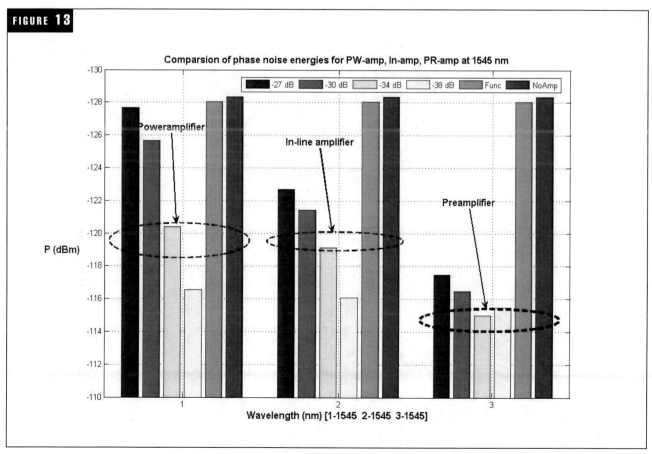

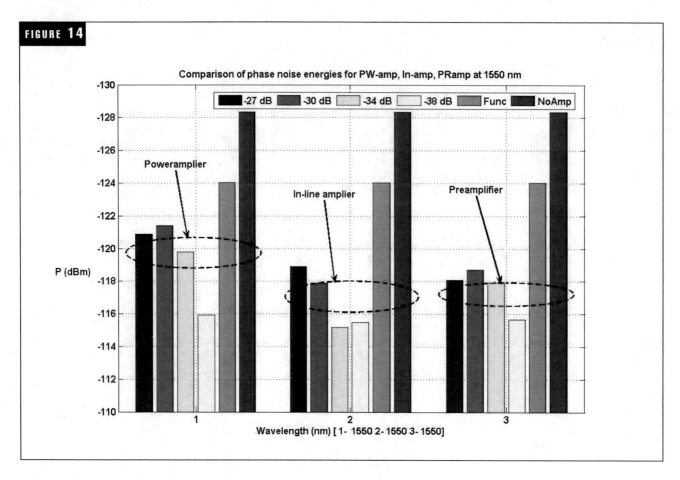

FIGURE 14

Comparison of phase noise energies for PW-amp, In-amp, PRamp at 1550 nm

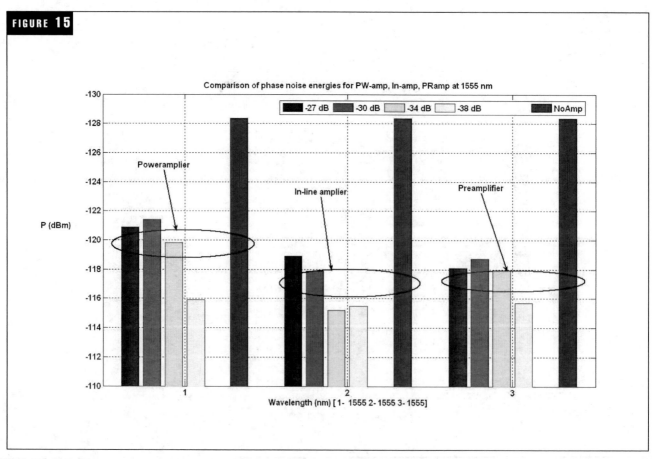

FIGURE 15

Comparison of phase noise energies for PW-amp, In-amp, PRamp at 1555 nm

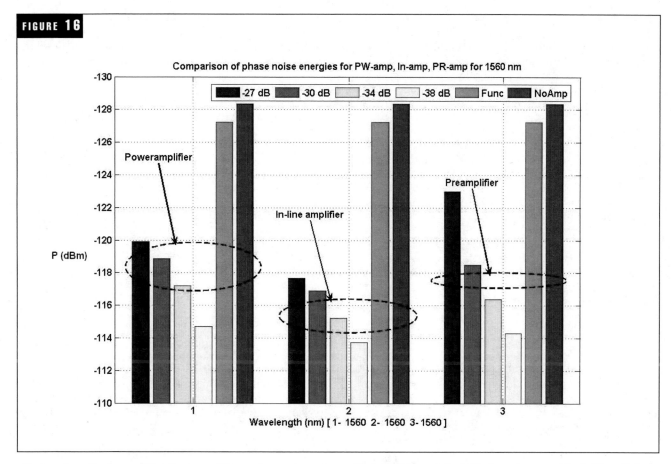

FIGURE 16

Comparison of phase noise energies for PW-amp, In-amp, PR-amp for 1560 nm

effects of optical amplification on the performance of M–QAM fiber-optic links. Different signal constellations and levels of QAM exhibit different performances. The simulation was performed using VPI optical simulator.

Simulation Setup of M–QAM Fiber-Optic Link

The simulation setup for the M–QAM fiber-optic link shown in *Figure 17* has been considered for performance evaluation. The link includes a 1,550 nm laser source, guarded by an optical isolator to prevent back reflection, single-mode fiber, EDFA, and the PIN photodetector. The MZ modulator is used as an external modulator to generate the QAM signal. The following link parameters have initially been chosen and maintained throughout the calculations: absolute temperature (T = 300 K), optical wavelength of 1550 nm, fiber cable attenuation of 0.2 dB/km, peak modulation index (m = 0.25), EDFA noise figure (NF = 5) and photodetector responsivity of 0.8 W/A is used. A bit rate of 40 Mbps and a carrier frequency of 751 MHz are used in the simulation. The link includes a 1,550 nm laser source that

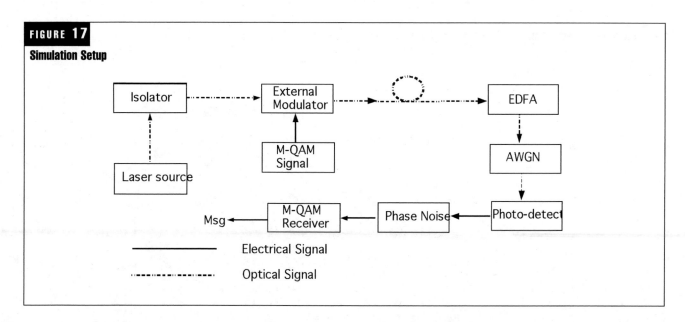

FIGURE 17

Simulation Setup

emits light into an MZ modulator via an optical isolator. An RF QAM signal then modulates the intensity of the light through the modulator and is then transmitted toward a PIN photodetector through a single-mode optical fiber. The photodetected signal is then down converted and applied to the QAM receiver. A 256–QAM signal is assumed unless otherwise specified throughout the article.

Modeling Phase Noise for M–QAM Simulations

A simple model for generating phase noise is presented in this section. First, generate a complex addictive white Gaussian noise (AWGN), which is spread over the required frequency band. Then, design a filter that has $1/f$ type of transfer function. Pass the complex AWGN noise through this $1/f$ filter. In this way, we can model phase noise for any power and any frequency offset. The main reason for using complex AWGN noise is that the QAM is also a complex signal. That is, the generated phase noise can be easily added to the phase of the QAM signal. Another way of generating phase noise is to generate AWGN noise in the required bandwidth with $1/f$ type of characteristic and then phase modulate this signal with the QAM signal. The second approach of adding phase noises to the QAM signal is used in this article.

Performance Evaluation of M–QAM Fiber-Optic Links Using BER Curves and Constellation Diagrams in the Presence of Phase Noise

Phase noise or phase jitter is a statistical quantity that affects the I and Q channels equally. In the constellation diagram, phase jitter shows up by the constellation points being shifted around their ideal positions. *Figure 18* shows the constellation diagram with phase noise in the FO link. We can see from the figure that points in the outer periphery rotate more than the inner points.

Simulation setup for evaluating the performance of M–QAM fiber-optic link in the presence of phase noise is shown in *Figure 17*. Phase noise with different phase noise levels were introduced in the link for different EDFA input powers and the performance of QAM signal was analyzed

using BER curves (*Figure 19*). The lower and upper limits of increasing the noise power density were also simulated. Performance of the FO link with same phase noise level but different QAM levels were also studied.

Observations

Since the results have already been plotted for experiments on three amplifier configurations and simulations for QAM FO link, we compare them to make some useful observations.

Comparison of Three Amplifier Configurations

Figure 4 through *Figure 11* show that phase noise increases with the presence of optical amplifier in the link and phase noise increases with the increase of amplifier gains for all the three amplifier configurations. From *Figure 12* through *Figure 16*, we can observe that at 1,540 nm wavelength, power amplifier and preamplifier configurations have less phase noise than the in-line amplifier configuration, for lower gains such as 27 dB and 30 dB, and for higher gains (34 dB and 38 dB) as well. For 1,545 nm wavelength, power amplifier and in-line amplifier have less phase noise than the preamplifier configuration. For 1,550 nm wavelength, the power amplifier configuration has less phase noise than the other two configurations. For 1,555 nm wavelength, power amplifier configuration has lesser phase noise than the other two configurations. For 1,560 nm wavelength, the power amplifier configuration and preamplifier configuration have less phase noise than the in-line amplifier configuration.

BER Curves with Induced Phase Noise

Figure 19 shows the BER curves for a QAM fiber-optic link in the presence of phase noise. We can see from the figure that BER increases as the phase noise increases with decreasing input powers to EDFA. At phase noise level of -100 dBc/Hz, 256–QAM FO link almost failed compared to -200 dBc/Hz phase noise level. The figure also shows 64–QAM FO link performs better than 256–QAM in terms of bit error rate.

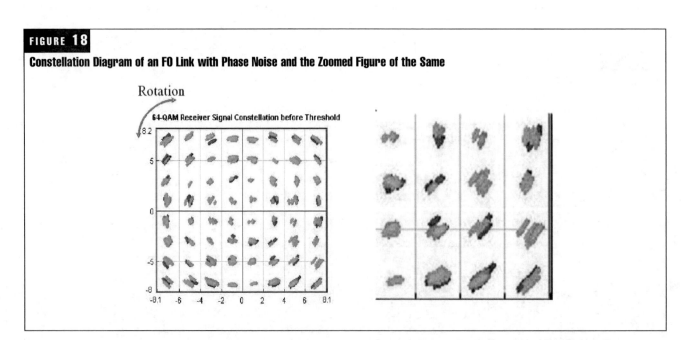

FIGURE 18

Constellation Diagram of an FO Link with Phase Noise and the Zoomed Figure of the Same

FIGURE 19

BER Curve with Induced Phase Noise

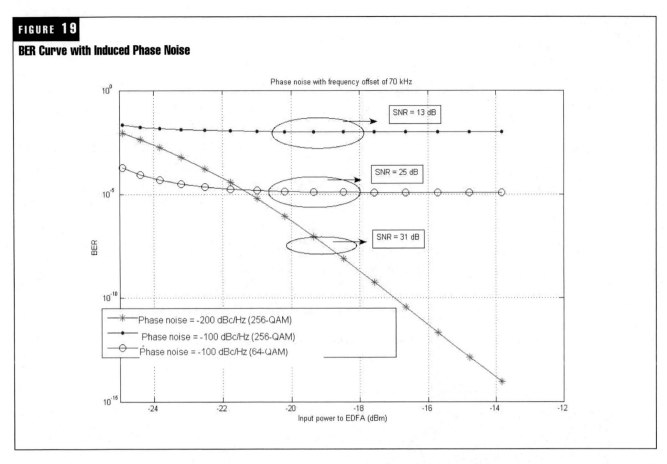

FIGURE 19

BER Curve with Induced Phase Noise

Conclusions

The results can be summarized using the above observations as follows:

Conclusion on Experimental Results

- The presence of EDFA in the FO link introduces phase noise on the reference signal and the phase noise increases as the gain of the EDFA increases. This can be explained as follows: The dominant source of noise in the amplifier is the amplified spontaneous emission (ASE). Some of the excited erbium decays to the ground state with spontaneous emission before it has time to meet with an incoming signal photon. As a result, a photon is emitted with random phase and direction. A very small proportion of emitted photons will occur in the same direction of the fiber and be captured.

- We also observed that all wavelengths at higher gains have almost the same level of phase noise. This can be explained as follows: one of the effects of ASE is to reduce the available gain of the signal field. In addition, ASE effectively lowers the saturation gain level since excited erbium ions are being taken away. EDFA has constant noise figure at higher gains. Therefore, at higher gains all the wavelengths exhibit the same amount of phase noise.

- We observed in the previous section that different wavelengths exhibit different amounts of phase noise.

This can be explained as follows: MZ modulator modulates each wavelength differently, that is different wavelengths have different gains. As a result, the phase noise is different for different wavelengths.

- We have compared the performance of three amplifier configurations in the previous section and confirmed that power amplifier configuration is best configuration for 1,540 nm, 1,545 nm, 1,550 nm, 1,555 nm, and 1,560 nm wavelengths. Preamplifier configuration is better for 1,540 nm and 1,560 nm wavelengths. In-line amplifier configuration is better for 1,545 nm. This gives the designer an idea about the phase noise at different amplifier configurations and wavelengths.

Conclusion on Simulation Results

- *Figure 19* shows the BER curves for phase noise levels of -200 dBc/Hz and -100 dBc/Hz. From the figure, we can conclude that BER increases as the phase noise level increases with decreasing input powers to EDFA. At phase noise level of -100 dBc/Hz, 256–QAM FO link almost failed compared to -200 dBc/Hz. Generally, any digital video transmission requires a minimum BER of 10-9. As we can see from the figure, the 256–QAM link, when affected with phase noise of -100 dBc/Hz, is not able to provide the minimum BER required for video transmission. *Figure 15* also shows the SNR requirement for each of the BER curves. So, 256–QAM, with phase noise of -100 dBc/Hz, has to try almost 18 dB harder to achieve the target SNR of 31 dB for digital video transmission. *Figure 19* also shows the BER curves for 64–QAM and 256–QAM for the same phase

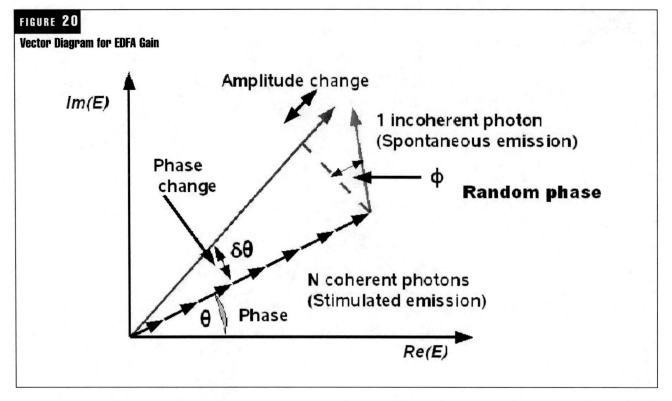

FIGURE 20

Vector Diagram for EDFA Gain

noise of -100 dBc/Hz. We can see from the figure that 64–QAM FO link performs better than 256–QAM FO link. The reason for this trend can be explained as follows: For 64–QAM, the constellation points are farther apart, compared to the 256–QAM, which makes 256–QAM more susceptible to phase noise. So, a 256–QAM link gives the worst performance compared to 64–QAM in the presence of phase noise.

• *Figure 20* shows the effect of phase noise on the constellation diagram. We can see from the figure that points in the outer periphery rotate more than the inner points. Amplification in the optical amplifier takes place through two processes: stimulated emission and spontaneous emission. During the stimulated emission, the output signal of the amplifier is amplified compared to the input signal and has same phase as that of the input signal, but during spontaneous emission, the output signal has random amplitude and phase compared to the input signal. So the points on the constellation diagram rotate.

References

[1] M. Shadaram, C. Thomas, J. Summerfield, P. Chennu, "RF phase noise in WDM fiber optic links," *7th International Conference on Transparent Optical Networks* (Barcelona, Catalonia, Spain, 3–7 July 2005)

[2] G. F. Lutes," Reference frequency distribution over optical fibers, a progress report," *Proceedings of the 41st annual frequency control symposium,* (May 1987): 161–166.

[3] M. Shadaram, V. Gonzalez, J. Ceniceros, N. Shah, J. Myres, S. A. Pappert, D. Law, "Phase stabilization of reference signal in analog fiber-optic links," *Proceedings of International Topical Meeting on Microwave Photonics* (Kyoto, Japan, 1987).

[4] M. Shadaram, J. Medrano, S. A. Pappert, M. H. Berry and D. M. Gookin," Technique for stabilizing the phase of the reference signals in analog fiber-optic links," *Applied Optics,* 34 (1987): 8,283–8,288.

[5] M. Shadaram, S. A. Pappert, and D. Lafaw, "Performance evaluation of a 64–QAM microwave fiber-optic link with a remote external modulator," *Proceedings of the Seventh Annual DARPA Symposium on Photonic Systems for Antenna Applications* (Monterey, CA, 13–16 January 1997): 105–110.

[6] D. Y. Chiang and D. I. Way, "Ultimate capacity of a laser diode in transporting multichannel M-QAM signals," *Journal of Lightwave Technology,* 15(10) (October 1997).

[7] A. N. Mody, M. Shadaram, and S. Pappert, "Effects of RF phase noise on the performance of the QAM fiber-optic links using Taylor series approximation," *Electronics Letters,* 34 (November 1998): 2,261–2,262.

[8] S. Nimesh and M. Shadaram, "Phase stabilization of reference signals in analogue fibre optic links," *Electronics Letters,* 33 (June 1997): 1,164–1,165.

[9] M. Shadaram and M. Kaiser, "Performance evaluation of M-QAM fiber optic link with EDFA," *Proceedings of CSNDSP 2002 Symposium* (Staffordshire University, Stafford, UK, July 2002): 140–143.

[10] M. Shadaram and M. Kaiser, "Effect of amplified spontaneous emission noise on the phase of reference signals transmitted through fiber amplifiers," *Proceedings of ICAPT 2002* (Quebec City, Canada, June 2002).

[11] L. E. Primas, R. T. Logan and G. F. Lutes, "Applications of ultrastable fiber optic distribution systems," *Proceedings of IEEE 43rd annual symposium on frequency control,* (1989): 202–210.

[12] C. K. Sun, G. W. Anderson, R. J. Orazi, M. H. Berry, S. A. Pappert and M. Shadaram., "Phase and amplitude stability of broadband analog fiber optic links," *SPIE,* 2560 (1995): 50–56.

[13] M. C. Wu, J. K. Wong, K. T. Tsai, Y. L. Chen and W. I. Way, "740-km transmission of 78-channel 64-QAM signals (2.34 Gb/s) without dispersion compensation using a recirculating loop," *IEEE photonics technology letters,* 12 (9) (September 2000): 1,255–1,257.

[14] N. Kanno and I. Katsuyoshi, "Fiber-optic subcarrier multiplexing video transport employing multilevel QAM," *IEEE journal on selected areas in communications,* 8 (1990): 1,313–1,319.

[15] Robert A. Ewart, M. Enoch, "Free space laser communications," *IEEE communications magazine,* 38 (August 2000): 124–125.

[16] Keith Wilson, "Optical communications for deep space missions," *IEEE communications magazine,* 38 (August 2000): 134–139.

[17] Keith Wilson, "Driving fiber to the home," *IEEE communications magazine,* 38 (November. 2000): 124–125.

[18] J. Proakis, *Digital Communications* (New York: McGraw-Hill, 2004).

[19] Gerd Keiser, *Optical Fiber Communications* (New York: McGraw-Hill, 2000).

[20] J. Senior, *Optical Fiber Communications, principles and practice* (Englewood Cliffs: Prentice Hall, 2003).

[21] G. Agrawal, *Fiber optic communication subsystems* (New York: McGraw-Hill, 2000).

[22] A. Mody, *Performance evaluation of M-QAM fiber optic links*, MS thesis submitted to the department of ECE, UTEP, July 1998.

[23] Djafar K. Mynbaev, Lowell L. Scheiner, *Fiber-Optics Communications Technology*, (Englewood Cliffs: Prentice Hall, 2000).

[24] Max Ming-Kang Liu, *Principles and Applications of Optical Communications* (Chicago: Irwin, 1996).

[25] S. J. Goldman, *Phase noise analysis in radar systems using personal computers* (Canada: John Wiley and Sons Inc., 1989).

[26] Dr. Kenneth S. Schneider, *Fiber Optic Data Communications for the Premises Environment*, www.telebyteusa.com/foprimer/foch1.htm, accessed June 21st, 2005.

[27] X. Zhang and A. Mitchell., "A simple black box model for erbium-doped fiber amplifiers," *IEEE photonics technology letters*, 12(1) (January. 2000): 28-30.

[28] Gerard Terreault, *QAM signal impairments and their effects on MER and BER*, www.sunrisetelecom.com/broadband/QAMImpairmentEffectson MERBER104.pdf, accessed May 1st 2005.

[29] Mini-Circuits, *VCO Phase noise*, www.minicircuits.com/appnote/vco15-6.pdf, accessed September 1st 2005.

[30] L. D. Garrett, A. H. Gnauck, F. Forghierri, V. Gusmeroli, and D. Scarano, "8 X 20 Gbps 480 km WDM transmission over conventional fiber using multiple broad fiber gratings," *Proc. OFC'98* paper PD18.

[31] C. D. Chen, I. Kim, O. Mizuhara, T. V. Nguyen, K. Ogawa, R. E. Tench, L. D. Tzeng, and P. D. Yates, "1.2-Tb/s WDM transmission experiment over 85 km fiber using 40 Gb/s line rate transmitter and 3R receiver," *Proc. OFC'98* paper PD21.

[32] M. K. Smit, "New focussing and dispersive planar component based on an optical phased array," *Electron. lett*, 24 (1998): 385–386.

[33] M. K. Smit, "PHASAR-based WDM devices: Principles, design and applications," *IEEE J. Select. Topics Quantum Electron*, 2 (1996): 236–250.

Equivalence of Fast Circuit Switching and Connection-Oriented Packet Switching

Kevin DeMartino

Independent Consultant

Abstract

Fast circuit switching and connection-oriented packet switching are two sides of the same coin. Fast circuit switching can closely emulate the characteristics of connection-oriented packet switching, and vice versa. Most communication functions can be efficiently supported by either technique. From a theoretical point of view, network convergence can be based on either fast circuit switching or connection-oriented packet switching. From an implementation perspective, the best approach for convergence involves a tradeoff between efficiency and performance and compatibility with legacy networks.

The Development of Packet Switching

For more than 100 years, circuit switching was the primary mode of operation for communication networks. With circuit switching, connections are established through the network and network resources are dedicated to a particular traffic flow for the duration of the connection. With packet switching, on the other hand, network resources are dynamically allocated to various traffic flows based on demand. Consequently, the network capacity is utilized more efficiently with packet switching than with circuit switching, especially if the traffic is bursty. Packet switching has been viewed as fundamentally different from circuit switching, and the shift by the telecommunications industry from circuit-based transport to packet-based transport has been seen as a paradigm shift.

TCP/IP and the OSI Reference Model

Packet switching came upon the scene when the ARPANET was launched in 1968 [1]. The original packet-switching protocols were superseded by the transmission control protocol (TCP) [2] and Internet protocol (IP) [3]. IP is a connectionless protocol that enables packets in a data stream to be treated independently. TCP is an end-to-end connection-oriented protocol designed to compensate for the limitations of IP.

The open systems interconnection (OSI) reference model [4], which has been widely accepted as the ideal model for data communications, focuses mostly on packet switching. The short shrift of circuit switching by the OSI model signaled a break with the prevalent mode of communications at the time the model was introduced (the early 1980s). The writing on the wall was that, for many in the industry, packet switching was the preferred mode of data transport.

Packet switching based on the OSI model was supposed to supersede packet switching based on TCP/IP, but TCP/IP won out instead. The connectionless property of IP facilitated the interconnection of packet-switching networks and was largely responsible for the success of the Internet. Packet switching based on TCP/IP has proved to be very effective for Internet applications involving file transfers such as e-mail and surfing the Web. However, the TCP/IP approach is not well suited for applications involving continuous data flows such as voice and video, and the connectionless property of IP makes the TCP/IP approach problematic for applications requiring a guaranteed quality of service (QoS).

Connection-Oriented Packet Switching

With connection-oriented packet switching, virtual connections are established for particular traffic flows prior to transferring data. Establishing a virtual connection usually entails setting up the route along which the traffic will flow and establishing parameters, including virtual connection identifiers, along this route. With TCP, end-to-end virtual connections are established. However, TCP operates on top of the lower-layer protocols (the lower three layers of the OSI model) seen by the network, and the network is unaware of and operates independent of the TCP virtual connections. This approach simplifies network operation and makes it easier to interconnect packet-switching networks. However, it precludes some advantages offered by connection-oriented packet switching such as the potential for providing a higher QoS than connectionless operation.

Asynchronous transfer mode (ATM) [5] is a connection-oriented packet switching protocol. With ATM, virtual connections are established and data blocks are broken up into small (53-byte) packets called ATM cells. The virtual connections and small cell size enable delays through the network to be minimized, which is particularly important for voice applications. Although ATM–based networks can provide better QoS than IP–based networks, it is difficult to achieve interoperability among ATM networks of different carriers and among ATM networks and other types of packet-switching networks. With the global deployment of the Internet, which consists of a large number of packet-switching networks interconnected via IP routers, ATM was relegated to a limited role.

Label switching (or tag switching) [6] is a technique that can be used to incorporate connection-oriented features into IP networks. With label switching, virtual connections are established through the network and labels (connection identifiers) are assigned. Packets are switched (forwarded) through the network based on the labels rather than on the complete destination address. Labels are similar to the connection identifiers in ATM and frame relay (FR) headers, and thus, ATM and FR can be viewed as forms of label switching.

Label switching based on multiprotocol label switching (MPLS) standards [7] has been widely accepted throughout the industry. Normally with MPLS, a label is attached to an IP packet. However, as the name implies, MPLS labels can be used in conjunction with other protocols, and MPLS labels can instead be attached to Layer 2 (L2) frames or ATM cells. MPLS provides a connection-oriented mechanism for IP networks, which can enhance the capabilities of IP networks.

Convergence

In the past, networks were designed to support a particular type of communication application. The telephone networks were designed to support voice, the cable network to support broadcast video, and the Internet to support computer communications—particularly file transfers. Although these networks can support other applications, they are optimized to support their primary applications. Thus, the telephone networks are primarily circuit-switching networks, and the Internet is a collection of packet-switching networks.

An integrated network that can handle a wide range of communication applications is clearly desirable. One network that can accommodate multiple applications would be more efficient and cost-effective than separate networks designed around particular applications. With an integrated network, the full range of services can be offered over the entire coverage area, rather than being limited to the areas where the separate networks overlap. Also, achieving interoperability among the networks of different carriers is facilitated if all applications are handled the same.

A truly integrated communication network should be able to support all the important applications, including voice, video, and computer communications, which collectively are often referred to as the "triple play." Video includes broadcast video, video on demand (VoD), and the audio normally associated with video signals. Computer communications, which primarily involves file transfers, is often referred to as "data communications." However, this is mis-

leading since, as discussed below, all communications can be treated as data communications.

In an integrated network, functions associated with transport and applications may be integrated. Integration of transport functions (or integration of bearer services) involves handling different types of communication data (voice, video, and computer data) in the same way. This is referred to as "transport convergence" and is one of the key concepts addressed by this paper. Integrated applications (or integrated value-added services) and other functions and services not associated with transport are not addressed in this paper.

Roots of Convergence

For a long time it has been recognized that integrated communication networks are possible.

The Nyquist Sampling Theorem [8], which was formulated in 1928, provides the theoretical basis for an integrated communication network. Nyquist's theorem states that band-limited signals can be accurately represented by discrete samples as long as the sampling rate is greater than twice the signal bandwidth. If samples are digitized, any practical communication signal can be accurately represented by a finite number of bits per second. In this case, transport refers to functions associated with moving bits through the network.

Sampling at the Nyquist rate (or higher) can result in very high data rates so that in many cases direct transmission of digitized samples is impractical. Shannon's Source Coding Theorem [9] implies that data rates can be reduced below the rates associated with Nyquist sampling and direct digital encoding. This theorem provides the basis for data compression. Shannon's theorem and related information theory concepts imply that all types of communication are fundamentally the same and provide the theoretical basis for integrated communication networks. When Shannon's seminal paper [10] was published in 1948, transmission and switching technologies could not support an integrated network. However, over the following decades, transmission and switching technologies improved dramatically so that current technologies are more than adequate to support transport convergence for a wide range of communication applications and services. The industry has been slow to exploit these technological advances and develop truly integrated networks.

Attempts at Convergence

In the 1980s, the first systematic efforts were made to develop truly integrated networks and bring about transport convergence. Integrated services digital networks (ISDNs) [11] are techniques and associated standards that were developed to support these efforts. Narrowband ISDN (N–ISDN) supported integrated voice and data over circuit-switched channels. Broadband ISDN (B–ISDN) attempted to extend integration to include video and other services, which would require considerably higher data rates. B–ISDN moved away from circuit switching and adopted a transport approach based on ATM. [12] For a while, it appeared B–ISDN and ATM might be the keys to convergence (more about this later), but they failed to live up to their advance billing.

In the 1990s, when there was a strong push for ATM, the Internet really took off. Regardless of its advantages and disadvantages, ATM was crowded out by TCP/IP. Although ATM has had considerable success, its deployment has been limited, and it is no longer considered to be the key to convergence. Instead, IP has replaced ATM as the basis for transport convergence.

Almost all computer data is transported via IP packets. The industry is moving away from traditional circuit-switched voice and toward voice over IP (VoIP). With the move toward digital video and video compression, video over IP is gaining acceptance. Some believe that, in the future, all data will be IP data. Since IP by itself is not well suited for applications involving continuous data streams and/or high QoS, convergence based on IP alone is problematic. However, MPLS can compensate for the limitations of IP, and IP combined with MPLS can provide a sound basis for convergence.

It is generally accepted that convergence should be based on packet switching rather than circuit switching. The industry endorsed ATM for B–ISDN, and now a consensus appears to be forming around IP/MPLS. This paper takes a different, but not contradictory, view.

Convergence based on packet switching is addressed in detail in the next section. It is then demonstrated that fast circuit switching is equivalent to connection-oriented packet switching and can provide an efficient basis for transport convergence. A proposed converged network is presented, and the process of evolving from the current network to the proposed network is addressed.

Convergence Based on Connection-Oriented Packet Switching

As stated in the previous section, it is generally accepted that transport convergence should be based on packet switching—specifically IP/MPLS, which is form of connection-oriented packet switching. This section addresses how connection-oriented packet switching can be made equivalent to circuit switching for continuous data flows and how it can achieve a comparable QoS.

With IP, routing (forwarding) of packets is based on the destination address in the IP header. Since IP is a connectionless protocol, packets in a data stream can be routed independently of each other. Connection-oriented features are provided by TCP, which runs on top of IP, and which is implemented in the end-user equipment. This approach is suitable for file transfers but not very suitable for continuous data streams such as voice and video data streams.

With the IP/MPLS approach, virtual connections called label-switched paths (LSPs) are established, and labels are attached to IP packets to identify particular LSPs [13]. Forwarding of a packet is based on the 20-bit label in the MPLS header rather than on the 32-bit IP address (for IP version 4 [IPv4]) in the IP header, which makes forwarding easier to implement. With MPLS, network capacity can be reserved along the LSP so that the QoS for a traffic flow can be guaranteed. Reserving network capacity usually means

ensuring that a particular LSP will be guaranteed a certain number of time slots over some interval of time.

The label distribution protocol (LDP) [14] and/or the resource reservation protocol (RSVP) [15, 16] are employed to establish LSPs and reserve network capacity along these LSPs. The Internet Engineering Task Force (IETF) in developing a new extensible IP signaling protocol suite (NSIS) [17], which will provide additional capabilities and more flexibility in establishing and controlling virtual connections. This paper is not concerned about the details of these protocols, but rather about how signaling protocols can be used and extended to achieve the desired result. More specifically, the issue is how signaling can be used in conjunction with MPLS to enable a QoS comparable to that provided by circuit switching.

QoS is generally not an issue for file transfers, which are the mainstay of computer communications. TCP/IP supports reliable file transfer. The delays through the network may be significant, but applications involving file transfers, including Web browsing and e-mail, can tolerate significant delays. On the other hand, QoS is a significant issue for continuous data streams, particularly for voice. QoS primarily involves delay, jitter (delay variations), and data loss. For transmission of data streams over landlines, which generally support very low bit-error rates, the primary QoS parameters are delay and jitter.

With IP/MPLS protocols and the signaling protocols for reservation of network capacity, continuous data streams and other delay-sensitive applications can be accommodated. Sufficient capacity needs to be reserved along an LSP so the transport delays and jitter are tolerable. Ideally, the delays through the packet-switching networks will not be significantly greater than the delays associated with transport via circuit switching. This is feasible with IP/MPLS networks with resource reservation.

Circuit Emulation

With circuit switching, network capacity is dedicated to a particular traffic flow. A particular circuit-switched connection can support a continuous data stream with a rate up to (and including) the assigned capacity (C) of the connection. As long as the instantaneous rate of the data stream does not exceed the capacity of the connection, the data stream can be transported through the network without distortion. If the instantaneous data rate exceeds C, then the data must be buffered at the input of the network, which causes the waveform of the data stream at the output of the network to be different than waveform at the input. If the input data stream consists of regularly spaced data blocks (packets), each containing the D bits of data, then the blocks will be stretched out as they are transported through the network, as shown in *Figure 1*. However, if the data rate does not exceed C, the interval between blocks will remain the same. At the output of the network, the waveform will look similar to the waveform of the stretched-out blocks at the input, except for a fixed delay. The delay, τ_{CS}, in transporting each block through the network is given by the following equation:

$$\tau_{CS} = D/C + \tau_F$$

where τ_F is a fixed delay (independent of D), which includes propagation delays (proportional to the path length) along the circuit path and fixed delays through the switching elements.

Circuit-switching operations can be emulated by a packet-switching network. Circuit emulation could be implemented in a label-switching network as follows. The signaling protocols would be used to reserve capacity along the entire LSP. Specific channels (time slots) would be reserved for a particular traffic flow for an extended period of time. With this approach, packet switching (or, more precisely, label switching) becomes equivalent to circuit switching in that it can match the performance of circuit switching. Implementing circuit emulation via this approach would require an extension of the existing (and proposed) signaling protocols.

With circuit emulation, some of the efficiency and flexibility of packet switching would be sacrificed. If the data rate of the source falls below C, some network capacity will go unused. If the rate of the source exceeds C, then some data may be lost. To prevent the loss of data, the source would need to be rate-limited. Additional capacity would need to be reserved to accommodate the packet overhead, which makes circuit emulation somewhat less efficient than circuit switching.

Thus, circuit emulation is not the complete solution, but rather the solution for applications that required firmly guaranteed QoS. It is envisioned that circuit emulation would be used for a relatively small subset of the overall traffic.

Performance of Packet-Switching Networks
It would be desirable to be able to assure adequate performance with packet switching without resorting to circuit emulation. As described below, packet switching can come close to matching the performance of circuit switching under certain conditions.

Figure 2 shows a simplified packet-switching network with two packet-switching nodes and two end-user nodes. Although a traffic flow often traverses multiple packet-switching elements, the packet-switching concepts can be more easily illustrated by the network of *Figure 2* rather than by a network with multiple packet switches. Typically, there are dedicated (or circuit-switched) access lines between the end users and the packet-switching network. Each packet switch would support connections to many end users. The packet switches in *Figure 2* are connected to each other by a dedicated connection through the circuiting switching network. There would also be connections to other packet switches, which are not shown in *Figure 2*. A virtual connection would be established between the end-user nodes, capacity would be reserved for this connection. The dedicated connection between the packet switches could support many virtual connections, as well as connectionless traffic.

For the connection between an end user and a packet switch, the situation is similar to the circuit-switching situation described above. For a packet size of D bits and a link capacity of C bits per second, the delay in transmitting the packet will be D/C. In propagating from the end user to the packet switch, the packet will incur a delay that depends on the length of the access path. Unlike the case with circuit switching, the packet switch will not begin forwarding the packet until the entire packet is received. On the line connecting the two packet switches, capacity is reserved for many virtual circuits. Thus, there can be packets from multiple virtual connections waiting in the output queue to be transmitted on this line, and there will be a queuing delay. This delay, τ_Q, depends on packet size, and percent utilization of the capacity, C_T, of the line connecting the two packet switches. This assumes that there is sufficient processing capacity in the switch so that the switch itself does not introduce significant delays. *Figure 3* shows the waveform at the output of the packet switch for a particular virtual circuit. Even if packets enter the switch at a fixed rate, the packets

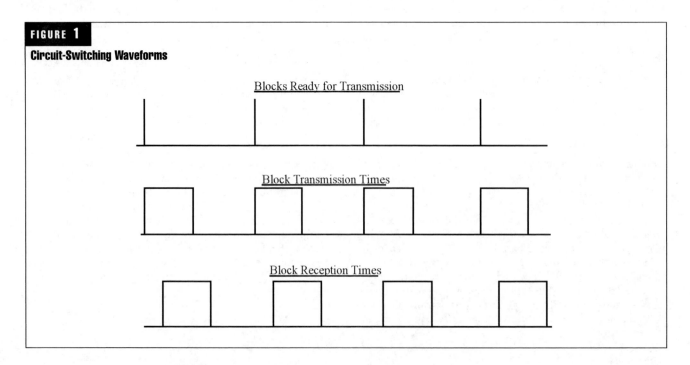

FIGURE 1

Circuit-Switching Waveforms

Blocks Ready for Transmission

Block Transmission Times

Block Reception Times

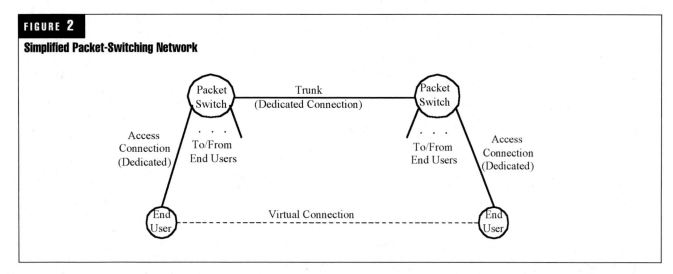

FIGURE 2

Simplified Packet-Switching Network

will encounter queuing delays, which will cause jitter at the output. At the output of the packet switch, the time it takes to transmit the packet will be D/C_T.

In traversing the network, a packet will incur additional delays. There will be propagation delays on the line connecting the packet switches and on the line connecting the second packet switch to the end user and another transmission delay (equal to D/C) at the output of the second packet switch. The fixed delays, which include the propagation delay along the entire path of the virtual connection and other fixed delays (independent of packet size and utilization), can be lumped together in a term τ_F. The total delay, τ_{PS}, in delivering a packet through the network is given by the following equation:

$$\tau_{PS} = \tau D/C + D/C_T + \tau_F + \tau_Q$$

From the first and second equations, it can be seen that the delay through the packet-switching network is greater than

the delay through a circuit-switching network with the same path length. Assuming the access-line capacities of the packet-switching network are the same as dedicated capacity in the circuit-switching network, the first term in the second equation $(2D/C)$ is twice as large as the corresponding term in the first equation. The fixed-delay term in the second equation (τ_F) is generally greater than the corresponding term in the first. Also, the second equation includes two additional terms, D/C_T and τ_Q.

Under certain conditions, the delay through a packet-switching network will not be significantly greater than the delay through a comparable circuit-switching network. The first term $(2D/C)$ on the right side of the second equation can be minimized by making the capacity of the access lines sufficiently large. Packet switches are usually interconnected by high-capacity trunks, which make the second term (D/C_T) in the second equation small. The next term (τ_F) in the second equation can be significant if the distance is large. However, in this case, the term will not be signifi-

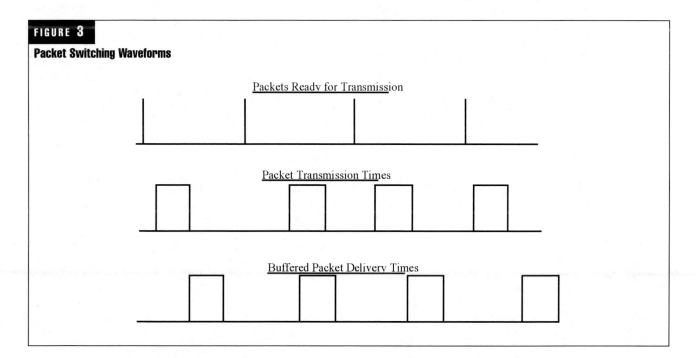

FIGURE 3

Packet Switching Waveforms

cantly different from the corresponding term in the first equation. If the utilization is close to 100 percent, then the last term (τ_Q) in the second equation can be very large. However, if the reserved capacity is a small percentage of capacity of the trunk interconnecting the packet switches, then the queuing delay can be small. To ensure that particular traffic flows incur small delays, the packet switch must give priority to these flows over connectionless traffic flows and traffic flows that exceed their reserved capacity. If the traffic flow traverses n packet switches instead of two, then the D/C_T and τ_Q terms would be multiplied by (n-1). However, the delay can still be comparable to the circuit-switching delay if the capacity on the trunks interconnecting the packet switches is large enough and the percent utilization of this capacity is low enough.

As illustrated by *Figure 3*, queuing in the packet switch produces jitter in the traffic stream. The queuing delay and the amount of jitter introduced depend strongly on the percent utilization of the trunk connecting the packet switches. The amount of jitter would increase if the traffic flow traverses more than two packet switches. For particular traffic flows, the jitter can be controlled by keeping the total reserved capacity small compared to the trunk capacity. However, even if the reserved capacity is low, there will be some residual jitter, and the waveform at the output of the packet-switching network will be different from the waveform at the input. However, this situation can be rectified if the delay is low enough. In this case, the data stream can be buffered and most of the jitter removed so that the data stream presented to the user looks the same as the input data stream. Since buffering introduces an additional delay, it should not be employed if it causes the overall delay to exceed the specified limit.

Equivalence of Packet Switching with Circuit Switching
As discussed earlier, packet switching based on label switching with circuit emulation is equivalent to circuit switching from a performance perspective. However, as described above, it is possible for a packet-switching network to come close to matching circuit-switching performance without the restrictions of circuit emulation.

Packet switching can be said to be equivalent to circuit switching for the transport of continuous data associated with a specific application if delays do not significantly impact the application and if the jitter is minimized at the output of the network. In this case, the packet-switched data stream will be practically indistinguishable from the corresponding circuit-switched data stream. To achieve this equivalence, the following conditions must be met:

- The reserved capacity must be small compared to the network capacity.
- Priority must be given to traffic flows with reserved capacity over connectionless traffic and flows that exceed their reservation limits.
- Traffic flows seeking equivalence must stay within their reservation limits.
- The data stream must be buffered at the output of the network to remove most of the jitter.
- The packet switches must have sufficient processing capacity so that the overall delay is limited by network capacity rather than by the capacity of the switches.

- Circuit emulation should be employed for traffic flows requiring a firm QoS guarantee.

Summary of Convergence Based on Packet Switching
For circuit switching, the utilization can be 100 percent. However, if the utilization is close to 100 percent in a packet-switching network, the delays can be very long. Thus, a packet-switching network cannot handle as much continuous traffic as a circuit-switching network. Current circuit-switching networks cannot handle data rates higher than the capacity dedicated to the connection, so data sources must be rate-limited. On the other hand, packet-switching networks can efficiently handle bursty traffic and accommodate rates that exceed the reserved capacity if the network is not heavily loaded. Rates beyond the reserved capacity could be handled on a "best-effort" basis, which means that the QoS would not be guaranteed.

Circuit emulation could be used for applications requiring a firm QoS guarantee. Connection-oriented packet switching could be used for applications that require a certain QoS level, but only where there is some flexibility in the performance requirements. Connectionless packet switching (conventional IP routing) could be used for applications that do not require QoS or where the QoS requirements are very loose.

In summary, the efficiency in handling bursty traffic, the flexibility in handling variable rates, and the ability to support high QoS are strong arguments that packet switching should form the basis for transport convergence. It appears that there is a consensus on convergence based on packet switching. However, the next section challenges this conventional wisdom.

Convergence Based on Fast Circuit Switching

Although it has been widely assumed that packet switching would be the foundation for convergence, convergence based on fast circuit switching is also feasible. For a long time, continuous traffic streams, including voice and certain types of data, have been accommodated by circuit-switched or permanent connections. The peak data rate must be limited for traffic flowing through circuit-switched connections. This results in an inefficient use of network capacity for burst sources, where the average data rate is much less than the peak data rate. Restrictions on the data rate can be loosened if the amount of capacity assigned to a connection can be changed rapidly. Fast circuit switching can accommodate bursty applications involving file transfers, such as e-mail and Web surfing. For e-mail applications, a circuit-switched connection could be established with an e-mail server for a brief period of time (a few seconds or less) so that e-mail messages could be downloaded and uploaded. Similarly, a brief circuit-switched connection could be established with a Web server to download files.

Fast Circuit Switching Viewed as an Extension of ISDN
Originally, ISDN efforts were an attempt to bring about convergence via circuit switching. N–ISDN supports voice and data communications. It employs common channel signaling to set up connections and control the bandwidth (capacity) of these connections. N–ISDN provides a separate signaling channel (the D channel) for the user-network

interface (UNI) and employs Signaling System Number 7 (SSN7) for signaling among ISDN switches [18]. Unlike associated channel signaling, where the signaling links to and controls a particular connection, common channel signaling decouples data transport and signaling and can provide a shared mechanism for controlling multiple connections and traffic flows.

A previous paper [19] addressed how ISDN could have been enhanced to efficiently handle file transfers and video. Faster switching and more capacity (more DS0 channels) per connection, among other things, were needed to make ISDN and circuit switching the basis for convergence. The telecommunications industry promoted B–ISDN as the way to achieve convergence. It would have made sense to extended extend N–ISDN and make fast circuit switching with common channel signaling the basis for B–ISDN. Instead, transport was based on packet switching, specifically ATM. Thus, the industry moved away from circuit switching, the transport approach that had been so effective for more than 100 years.

Generalized MPLS (GMPLS) [20] is an extension of MPLS, which includes a feature that enables label switching to be used in a circuit-switching mode. With GMPLS implicit labeling, the LSP is identified by the physical channel (such as a time slot) containing the data, rather than by an explicit label in the MPLS header. This is similar to a DS0 channel being identified by the position of the data in a DS1 frame or a synchronous optical network (SONET) channel being associated with a series of time slots within a SONET frame.

As described in another previous paper [21], GMPLS with implicit labels can enable convergence based on fast circuit switching, which was referred to in the paper as "dynamic channel switching." Signaling protocols used in conjunction with GMPLS to establish and control LSPs are fundamentally similar to common channel signaling with SS7.

Migrating from Packet Switching to Fast Circuit Switching
Although the move from circuit switching to packet switching is often viewed as a paradigm shift, packet switching and circuit switching are not fundamentally different. The previous section described how packet switching could emulate circuit switching. The reverse is also true, as illustrated by *Figure 4*.

Figure 4A illustrates a packet-switching operation. Normally, the information required to switch (or forward) a packet is in the packet header, which is attached to the data block, as shown in *Figure 4A*. This means that switching/forwarding decisions must be made on the fly, which imposes strict latency requirements on the switch. Suppose the headers are sent in advance of the data, as illustrated by *Figure 4B*. This would allow the switch to determine the switching pattern before receiving the data, which would allow the latency requirements to be relaxed. It would also allow the switch to initiate flow control before the onset of congestion.

Next suppose that instead of sending a separate header for each data block (or packet), multiple headers are combined into a single header, as illustrated by *Figure 4C*. This would

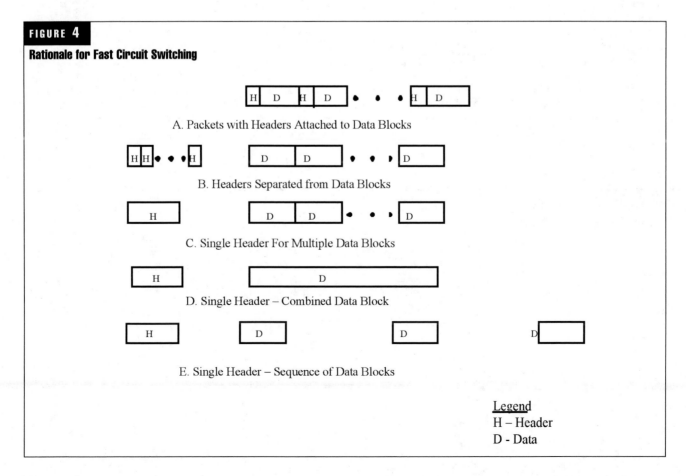

FIGURE 4

Rationale for Fast Circuit Switching

A. Packets with Headers Attached to Data Blocks

B. Headers Separated from Data Blocks

C. Single Header For Multiple Data Blocks

D. Single Header – Combined Data Block

E. Single Header – Sequence of Data Blocks

Legend
H – Header
D - Data

be particularly advantageous when multiple successive packets are part of the same data flow and are to be sent to the same destination. If multiple headers are combined, the controller could determine the switching patterns for multiple packets at the same time. This would significantly reduce the computational burden, and the throughput requirement on the controller could be significantly relaxed.

Next, suppose that multiple data blocks are combined into a single data block, as illustrated by *Figure 4D*. At this point, it would no longer be necessary to break source data into packets. Consequently, it would be possible to directly transmit data as it flowed from the data source. If the bit-error rate is low enough and sufficient network capacity is provided, then the data stream delivered by the network to the receiver would be the same as the source data stream, except for a small fixed delay.

Finally, suppose that a header can be used to control the switching of multiple data blocks that are separated in time, as shown by *Figure 4E*. This would allow a sequence of data blocks to be controlled as a group. At this point, packet switching has become very similar to circuit switching with common channel signaling, and fast circuit switching can be viewed as an extension of packet switching.

The above discussion implies that label switching with common channel signaling is a form of fast circuit switching that can match the performance of connection-oriented packet switching. With the approach described above, which is described in detail in the next section, network capacity can be reserved for brief periods to efficiently accommodate bursty data such as file transfers. Also, this approach can be used to accommodate continuous traffic streams with variable data rates, including compressed video streams. Compared to packet switching, fast circuit switching can provide higher utilization efficiency. Overhead for headers is not required, and by reserving specific channels (time slots) for particular traffic flows, the utilization can approach 100 percent in some cases. On the other hand, connection-oriented packet switching is, in a sense, more flexible and more compatible with legacy networks.

In summary, connection-oriented packet switching and fast circuit switching are basically equivalent. It is not a question of one or the other. Instead, it makes sense to exploit the capabilities of both approaches. The next section describes how the fast circuit switching and connection-oriented packet switching approaches can be implemented by common network elements and how the advantages of both approaches can be realized.

Proposed Converged Network

This section proposes an implementation for a network that provides a common approach for transporting all types of data, including voice, video, and computer data. This converged network can operate in both circuit-switching modes and packet-switching modes. To accommodate different traffic flows, the network can support simultaneous operations in multiple modes.

If packet switches are replaced by label switches, then *Figure 2* can be used as the starting point for the proposed converged network. In a simplified label-switching network similar to the network of *Figure 2*, end-to-end virtual connections (LSPs) are established through the network. A typical traffic flow would involve two LSPs—one in each direction. The signaling protocols previously discussed would be used to establish LSPs and reserve network capacity for these LSPs. In the simplified network, each end user would be connected to a label switch by a dedicated (circuit-switched) connection. Each label switch would be connected to multiple other label switches. A connection between a pair of label switches would support many virtual connections.

A Common Network Element for Circuit Switching and Packet Switching

The heart of the proposed network is the label switch, which is illustrated in *Figure 5*. The label switch is a common network element that supports circuit switching and packet switching. Data on the N input lines are switched onto the N output lines.

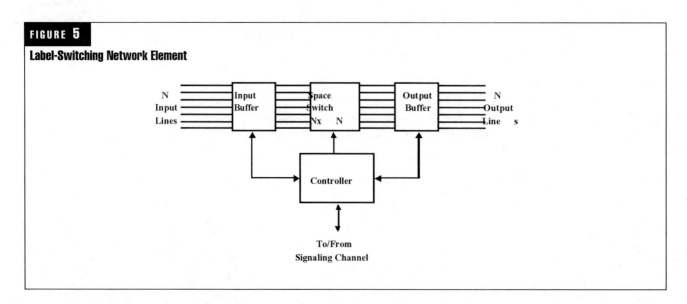

FIGURE 5

Label-Switching Network Element

The label switch contains an input buffer, an output buffer, a space switch, and a controller. The input buffer receives data on each of the N input lines (space cells) and stores this data, in some cases for only a very short time. Data on these lines are contained in time slots (time cells) associated with particular LSPs. The output buffer stores data for each of the output lines and transmits data over each of the output lines at the appropriate time. The space switch connects the input buffer to the output buffer. It switches data received from each of the input lines to storage elements in the output buffer connected to each of the output lines. In the most straightforward implementation, the space switch is an N-by-N crossbar matrix, so that each of its N input lines can be connected to each of its N output lines. Data is switched at every time slot so that for every time slot, data from a particular input line could be transmitted over a different output line. The primary function of the controller is to generate the switching matrix relating the switch input to the switch output for every time slot. As described below, the controller uses signaling information and labels to generate the switching matrix.

Figure 6 illustrates the operation of the label switch for the case of eight input lines and eight output lines. In this case, there are 23 time slots, each of which corresponds to a situation for primary-rate ISDN [22]. Each ISDN time slot contains one byte of data. The ISDN frame containing the time slots is repeated 8,000 times per second. The label switch maps time-space cells at the input onto time-space cells at the output. The switching matrix contains 184 (8 x 23) elements, with each element identifying the output line that is the destination of the data contained in a particular time slot on a particular input line. The figure illustrates how time slots on the input lines can be mapped to time slots on the output lines. Interchanging of time slots can be performed in both the input buffer and output buffer. Thus Figure 5 can be viewed as a time-space-time switch, i.e., a switch that performs time switching, space switching, and time switching.

The switch of Figure 5 can operate in both circuit-switching and packet-switching modes. In the circuit-switching mode of operation, the switching pattern is predetermined. The controller uses information from the signaling channel to determine the mapping from time-space cells at the input to time-space cells at the output before the data to be switched arrives at the switch. Since data is switched based on the time slots containing the data for each input line, time slots can be viewed as implicit labels. Implicit labels avoid the overhead associated with MPLS headers. With conventional circuit switching, the mapping of time-space cells may be fixed for the duration of a connection, which could last for a long period of time. With ISDN, the mappings can be varied during the connection lifetime as the capacity of the connection is changed, but these changes are usually infrequent. With fast circuit switching, on the other hand, the mappings can be varied quickly so that the same time slots in successive frames may contain data associated with different traffic flows.

In the packet-switching mode of operation, labels and indications of where the packet begins and ends are transferred from the input buffer to the controller. Providing indications of the beginning and end of a packet (packet framing) is a function of the L2 protocol. The controller then uses the labels and framing information to determine the switching

pattern on the fly. There will be delays at the input as the buffer waits for the entire packet to arrive before sending it to the output buffer. There will be an additional delay at the input if multiple packets destined for the same output line arrive at the same time. Similarly, there will be a delay at the output, which is usually larger than the delay at the input, as data in the output buffer, waits to be transmitted over the output line.

As presented above, the fast circuit–switching and connection-oriented packet-switching modes of operation are not fundamentally different. In both cases, the switching of data is based on labels, either implicit or attached to the packets. In both cases, network capacity is reserved, either by specific time slots or a number of time slots per second. By scheduling time slots before the data arrives, the fast circuit–switching mode avoids some delays associated with the packet-switching mode. Also, fast circuit switching can potentially use bandwidth more efficiently, since it has lower overhead and can operate with a higher utilization. However, as discussed earlier, there will not be a significant difference in performance if the utilization is low enough so that the additional delays are tolerable.

The discussion above deals with data (packets) that have labels, either implicit or explicit. What about packets that do not have labels, including packets from legacy networks and users? To accommodate unlabeled packets, the label switch would act like a conventional router. In this case, global IP addresses instead of labels would be sent from the input buffer to the controller. The controller would need to establish and maintain routing tables, which would be used to determine the switching of unlabeled packets from the input to the output of the label switch. Forwarding of unlabeled packets would be a "best-effort" basis. Labeled packets would be given priority over unlabeled packets. QoS would not be guaranteed for traffic consisting of unlabeled packets.

Elements of the Proposed Network

Figure 7 shows the network elements supporting a typical label-switched virtual connection between a client and a server in a metropolitan area. The network of Figure 5 includes equipment at the premises of the client and of the server; multiplexers at the central offices (COs), which are referred to as serving wire centers (SWCs), directly connected to clients and to servers; and label switches at hub COs. In this case, the virtual connection traverses only two label switches. Connections over a larger area may involve additional label switches.

Like in the network of Figure 5, there is dedicated physical-layer connection between label switches, in this case a SONET [23, 24] connection. Like DS1 (T1) and ISDN frames, SONET frames contain time slots with one byte of data and are repeated 8,000 times per second. SONET frames contain many more time slots than ISDN and DS1 frames, and SONET time slots can be grouped to form payloads containing many bytes. This SONET connection may involve SONET switching elements, which are not shown in Figure 7, and may traverse additional COs, which are also not shown. Since this connection supports many virtual connections, a dedicated connection is relatively efficient and cost-effective. Unlike the network of Figure 5, dedicated access to the label switches is not provided. Instead, traffic from mul-

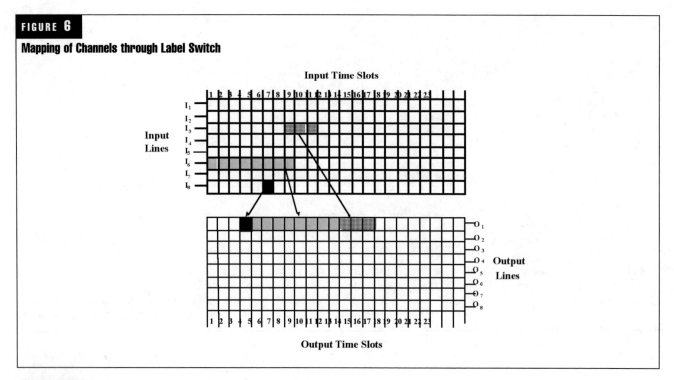

FIGURE 6

Mapping of Channels through Label Switch

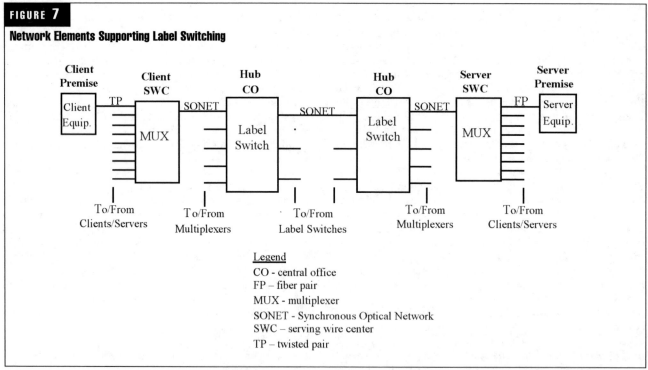

FIGURE 7

Network Elements Supporting Label Switching

Legend
CO - central office
FP – fiber pair
MUX - multiplexer
SONET - Synchronous Optical Network
SWC – serving wire center
TP – twisted pair

tiple users is multiplexed at the SWC and sent on to the label switch over a common physical connection. Similarly, traffic from the label switch is demultiplexed at the SWC and sent to users over separate access lines. Each multiplexer in *Figure 7* can be viewed as a 1 by N (instead of N by N) label switch. *Figure 7* shows a twisted-pair access line for the client and a fiber access line for the server. Generally, servers operate at relatively high data rates, which may require fiber, while clients may be satisfied with data rates that can be provided over twisted pairs using digital subscriber line (DSL) techniques.

With the approach shown in *Figure 7*, only the relatively short access lines from the customer premises to the SWCs are dedicated, while interoffice lines are shared. This makes the approach of *Figure 7* more cost-effective than the approach of *Figure 5*. Full-blown label switches rather than multiplexers could be provided at all the COs rather than only at the hub COs. This would reduce the need to back-haul all the traffic to hubs and would allow the capacity of the connection between label switches to be reduced. However, it would require more complex and more expensive network elements at the SWCs. Alternatively, label switches could be placed at a larger number of COs, which is equivalent to creating more hubs. There is a tradeoff between the transport cost, which decreases as the number of label switches (or hubs) increases, and switching costs, which increases with the number of label switches. This and other tradeoffs are addressed in another paper [25].

Communication Protocols

Figure 8 shows the communication protocols that would be used in the network described above, specifically the protocols for the connection-oriented packet-switching mode of operation. These protocols are employed in the data channel primarily to support the transfer of data between the client and the server. A different set of protocols would be used in the signaling channel (or the so-called control plane) to support the establishment and management of virtual connections and reservation of capacity for these connections. For fast circuit–switching mode, the signaling channel protocols would be similar, but the data channel protocol could be greatly simplified.

As stated above, the typical physical-layer interface at the client premises would be some form of DSL running over twisted pair (one or more). *Figure 8* shows Ethernet [26] as the L2 protocol at the client premises. Ethernet is the protocol of choice for most network customers; however, other L2 protocols could be employed. MPLS, which runs on top of the L2 protocol, supports LSPs and virtual connections. IP, the user datagram protocol (UDP) [27], and the application protocol run on top of MPLS. UDP is appropriate for applications involving continuous data flows (streaming applications). For bursty applications such as file transfers TCP would be used instead.

At the server premises, the physical-layer interface is SONET over fiber. The generic framing procedure (GFP) [28] provides L2 functions associated with packet framing, i.e., determining where packets begin and end. MPLS supports the end-to-end virtual connections with multiple clients. Again, IP, UDP, and the application protocol run over MPLS.

The label switches and multiplexers implement MPLS over GFP over SONET. The application protocol and UDP run over the heads of the label switches and multiplexers. Similarly, for label packets, IP would run over the heads of the label switches and multiplexers. LSPs (in both directions) would be established between each multiplexer and its corresponding label switch. Each of these LSPs would contain multiple end-to-end LSPs. Similarly, LSPs would be established between the label switches, and multiple end-to-end LSPs would be nested within each of these LSPs.

For transporting labeled packets through the network, IP is superfluous. Thus for labeled packets, the IP and UDP headers could be stripped off at the sending end and re-inserted at the receiving end. This would reduce the overhead and improve transport efficiency. For unlabeled packets, the IP headers would be used in the label switches, and possibly the multiplexers, to determine where to route the packets.

Services Supported by the Proposed Network

The network of *Figure 7* can support a wide range of communication services, including voice, video, and computer communications. These services can be supported using either the connection-oriented packet-switching mode or the fast circuit–switching mode. The connection-oriented packet-switching mode provides more flexibility in handling variable rates, while the fast circuit–switching mode provides a 100 percent guaranteed rate.

With the connection-oriented packet-switching mode, variable rate services can be accommodated by reserving a minimum capacity and providing best-effort service for data that exceeds the minimum rate. With this approach, a minimum level of performance can be guaranteed, and a higher level of performance can be provided if the capacity is available. For example, consider the case of compressed video, which has a variable rate depending on the scene content [29]. With the Moving Pictures Expert Group (MPEG) compression, the data stream can be divided into a base layer, which is required to maintain a continuous image, and an enhancement layer, which can be used to provide a higher-quality image [30]. Sufficient capacity must be reserved to

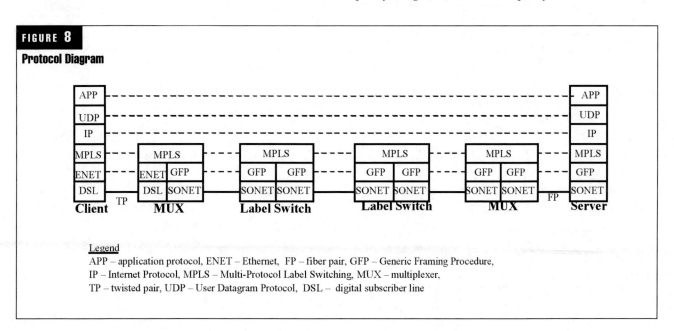

FIGURE 8

Protocol Diagram

Legend
APP – application protocol, ENET – Ethernet, FP – fiber pair, GFP – Generic Framing Procedure,
IP – Internet Protocol, MPLS – Multi-Protocol Label Switching, MUX – multiplexer,
TP – twisted pair, UDP – User Datagram Protocol, DSL – digital subscriber line

accommodate the base layer, but the enhancement layer can be transported on a best-effort basis. Occasionally, there may be some image degradation, but this degradation would be tolerable and may not be noticeable.

With the fast circuit–switching mode, variable rate services can be accommodated efficiently by requesting additional capacity for a data flow in advance of when it is needed, and releasing the capacity when it is no longer needed. Similar to the packet-switching case, there is a potential problem if capacity is not available when it is needed. To get around this problem, a minimum capacity must be assigned to a data flow and additional capacity requested and released as the data rate varies. For example, sufficient capacity must be assigned to accommodate the MPEG base layer, while the capacity for the enhancement layer would be requested only when needed. Again, this approach may occasionally result in some tolerable degradation of image quality.

Currently, transport of data, with the exception of voice, is mostly IP–based. With the move toward VoIP, almost all data will consist of IP packets, which would seem to make IP–based transport even more important. However, with the advent of MPLS, transport based on label switching is beginning to replace IP–based transport.

Security Considerations

Fast circuit switching has an advantage over connection-oriented packet switching with respect to security. IP headers are not required for circuit-switching operation. If the signaling channel is secure—which is clearly a requirement—then the communication endpoints could not be determined from the traffic in the data transport channels. This would prevent an intruder from performing a traffic analysis. Also, an intruder would not be able to direct packets to a specific endpoint, which would frustrate a denial-of-service (DoS) attack. Of course, end-to-end encryption would be required to protect the application data.

With connection-oriented packet switching, security could be enhanced by suppressing the IP headers. The MPLS headers have only local significance. This prevents an intruder from performing a traffic analysis and complicates a DoS attack. However, an intruder could more easily exploit the explicit labels associated with connection-oriented packet switching than the implicit labels associated with fast circuit switching.

Network Evolution

The industry is moving toward transport of IP data using MPLS. IP/MPLS, which is a form of connection-oriented packet switching, can accommodate current and future applications and is the recommended mechanism for transport convergence in the near term.

Using current technologies, label switches and multiplexers capable of connection-oriented packet-switching and fast circuit–switching modes can be implemented in a straightforward manner. Consequently, network elements that can handle both packet- and circuit-switching modes should be deployed. Initially, transport would be based on connection-oriented packet switching. Circuit emulation would be supported for applications that required a firm QoS guarantee.

IP–based transport could be gradually phased out as label switches are deployed. IP and UDP headers could be suppressed to reduce overhead and enhance security. Eventually, fast circuit switching could replace packet switching for most applications. This reduces overhead and enhances security.

Summary

Connection-oriented packet switching and fast circuit switching are fundamentally similar, and in a certain sense, they are equivalent. Under certain conditions, connection-oriented packet switching can provide QoS comparable to circuit-switched transport. Conversely, fast circuit switching can provide a transport efficiency that is comparable to the efficiency associated with packet switching.

Both connection-oriented packet switching and fast circuit switching can support a wide range of communication services, including voice, video, and computer communications. Either connection-oriented packet switching or fast circuit switching can form the basis for a converged transport network.

Both connection-oriented packet switching and fast circuit switching can be readily implemented in the same network using similar protocols and common equipment. GMPLS can accommodate connection-oriented packet switching via external labels (MPLS headers attached to packets) and fast circuit switching via implicit labels (time slots containing the data). The key to the proposed converged transport network is a label switch that can handle connection-oriented packet switching and fast circuit switching simultaneously. For the packet-switching mode, the routing of data by the label switch is determined on the fly based on information in the packet headers. For the fast circuit–switching mode, the switching pattern of the label switch is determined prior to the arrival of the data, based on control information in the GMPLS signaling channel. Otherwise, operation of the label switch is similar for the packet-switching and circuit-switching modes. To accommodate legacy users and enable interoperability with legacy networks, the proposed label switch could also operate as a conventional router to transport unlabeled packets based on IP addresses.

In summary, connection-oriented packet switching and fast circuit switching are fundamentally equivalent, and either or both can form the basis for a converged transport network.

References

1. M. Hauben, "History of ARPANET," www.dei.isep.ipp.pt/docs/arpa.html, downloaded February 2006.
2. "Military Standard – Transmission Control Protocol," U.S. Department of Defense, MIL-STD-1778, 12 August 1983.
3. "Military Standard – Internet Protocol," U.S. Department of Defense, MIL-STD-1777, 12 August 1983.
4. "Basic Reference Model," International Standards Organization, ISO Standard 7498, 1984.
5. S. Minter, "Broadband ISDN and Asynchronous Transfer Mode (ATM)," IEEE Communications Magazine, Vol. 27, No. 9, September 1989.
6. E. Rosen, et al., "Multiprotocol Label Switching Architecture," IETF Network Working Group, RFC 3031, January 2001.
7. R. Callon, et al., "A Framework for Multiprotocol Label Switching," IETF Network Working Group, Internet Draft, draft-ietf-mpls-framework-02.txt, January 2001.

8. "Digital Signal Processing Tutorial – DSP," Logiix4u.net, www.logix4u.net/dsp_intro1.htm, downloaded December 2005.

9. "Theory of Data Communications Compression," Groovyweb free downloads and tutorials, www.grovyweb.uklinux.net/index.php?page_name+Theory%20of%20Data%20Compression, downloaded December 2005.

10. C. Shannon, "A Mathematical Theory of Communication," *The Bell System Technical Journal*, Vol. 27, pp. 379–423, 623–656, July/October 1948.

11. T. Bartee, editor, *Digital Communications*, Indianapolis, Indiana: Howard W. Sams & Co., 1986, pp. 116–124.

12. A. Day, "International Standardization of BISDN," *IEEE LTS – The Magazine of Lightwave Telecommunication Systems*, Vol. 2, No.3, August 1991.

13. "Multiprotocol Label Switching (MPLS)," International Engineering Consortium, Web ProForum Tutorial (Trillium), blinky-lights.org/networking/mpls.pdf, downloaded December 2005.

14. B. Jamoussi, editor, "Constraint-Based LSP Setup Using LDP," IETF Network Working Group, Internet Draft, draft-ietf-mpls-cr-ldp-05.txt, February 2001.

15. R. Braden, ed., "Resource Reservation Protocol (RSVP) – Version 1 Functional Specification," IETF Network Working Group, RFC 2205, September 1997.

16. G. Swallow, et al., "RSVP-TE: Extensions to RSVP for LSP Tunnels," IETF Network Working Group draft, August 2001.

17. X. Fu et al., "NSIS: A New Extensible IP Signaling Protocol Suite," *IEEE Communications Magazine*, Vol. 43, No. 10, October 2005.

18. A. Modarressi and R. Skoug, "Signaling System No. 7: A Tutorial," *IEEE Communications Magazine*, Vol. 28, No. 7, July 1990.

19. K. DeMartino, "ISDN and the Internet," *Computer Networks – The Journal of Computer and Telecommunications Networking*, December 1999.

20. A. Banerjee, et al., "Generalized Multiprotocol Label Switching: An Overview of Routing and Management Functions," *IEEE Communications Magazine*, Vol. 39, No. 1, January 2001.

21. K. DeMartino, "Convergence of Circuit Switching and Packet Switching in Future Communications Networks," *IP Applications and Services 2003: A Comprehensive Report*, International Engineering Consortium, copyright 2002.

22. T. Bartee, editor, *Digital Communications*, op cit.

23. C. Omidyar and A. Aldridge, "Introduction to SDH/SONET," *IEEE Communications Magazine*, Vol. 31, No. 9, September 1993.

24. "Synchronous Optical Network (SONET) Transport Systems: Common Generic Criteria," Telcordia Technologies Generic Requirements, GR-253-CORE, Issue 2, December 1995.

25. K. DeMartino, "An Architecture for an Interoffice Communication Network," not yet published.

26. "Ethernet Technologies," Cisco Systems, Cisco Documentation, www.cisco.com/univercd/cc/td/doc/cisintwk/ito_doc/ethernet.htm, downloaded December 2005.

27. "UDP – User Datagram Protocol – UDP and IP Tutorial," About.com, compnetworking.about.com/od/networkprtocols/l/aa071200a.htm, downloaded December 2005.

28. E. Hernandez-Valencia, et al., "The Generic Framing Procedure (GFP): An Overview," *IEEE Communications Magazine*, Vol. 40, No. 5, May 2002.

29. P. Pancha and M. El Zarki, "MPEG Coding for Variable Bit Rate Video Transmission," *IEEE Communications Magazine*, Vol.32, No. 5, May 1994.

30. T. Sikora, "MPEG–1 and MPEG–2 Digital Video Coding Standards," wwwam.hhi.de/mpeg-video/papers/sikora/mpeg1_2/mpeg1_2.htm, downloaded October 1997.

Strategy Development for the Information Economy

A Practical Guide to Coordinated Action through Time

Jay Edwin Gillette, Ph.D.

Professor
Center for Information and Communication Sciences, Ball State University
Chairman
Advisory Council, Pacific Telecommunications Council

Abstract

This article discusses strategy development and management to help leaders in the context of the information economy. The article advances a dynamic definition of strategy as "coordinated action through time." The article outlines a six-step method that builds four foundation areas of strategic management: strategic intelligence, action options, agreement mechanisms for cooperation, and information and communication links for coordinated action.

The primary audience for the article is management with functional, research and development, or technical roles. The article's objective is to assist managers and knowledge workers to produce practical strategies that will generate leadership support, plus peer and worker cooperation, in knowledge-driven organizations.

Introduction and Statement of the Problem: Strategy Appropriate to the Information Era

Strategic management is the art of the future. It is not a task undertaken while waiting for the future to arrive; it is the way to unfold the future to reach your goals. Strategic management is a requirement for survival and prosperity for every organization doing business in the information economy. This article argues that strategic development and strategy management are more important than ever in the information economy and that managers and knowledge workers need to adapt their understanding and their tools to succeed and prosper in the information era.

Carl Shapiro and Hal R. Varian have written one of the most useful books on this era, called *Information Rules: A Strategic Guide to the Network Economy*. They advance a thesis that underscores the approach of this article: "Even though technology advances breathlessly, the economic principles you rely on are durable. The examples may change, but the ideas will not go out of date."

The world is not approaching a coming information economy. It is already here. There is interesting debate as to whether the information economy is truly new or whether the already significant amounts of information required in the prior notion of the industrial economy is simply being emphasized. For example, economist Cosmo R. Shalizi wrote, "The Great Leap Forward in information-processing took place, at least in this country, between 1880 and 1930, in which period the percentage of the workforce employed in information-handling grew from 6.5 percent to 24.5 (for scale, 35 percent of all U.S. workers were industrial in 1930).

A key indicator is that the growth rate for the information economy has slowed. In the developed countries, the high-leaping growth rates seen in the past decades have leveled off. In recent decades, the United States' information sector was "growing no faster than other parts of the economy." That is exactly the trend expected when the base number or calculation of the growth rate becomes large. In other words, the larger the information-economy base, the smaller the percentage increase in growth, precisely because the base of the information economy is so large.

Where is the economy now? What has it grown to? Researchers still do not know precisely—the information economy and ways to measure it are still too new. Yet in the biggest single economy in the world—the United States—the information sector is the largest and arguably the most dynamic sector. In the 1980s in the U.S. economy, the primary and secondary information sectors reached the level where they were reckoned at 46 percent of gross national product (essentially gross domestic product). The numbers of "knowledge workers" employed in these sectors have risen continually, to large proportions. By the mid-1980s, knowledge workers accounted for the following percentages of the workforce in the following countries:

In another measure, by 2002, 70 major industry sectors in the United States had broadly defined information and com-

TABLE 1

Country	Knowledge workers: percent of workforce
United States	41.1
Former West Germany	33.2
France	32.1
Japan	29.6

munication technologies (ICT)-skilled employment exceeding 30 percent of their workforce. The highest proportion was in "accounting, auditing, and bookkeeping services," with 82.7 percent. The lowest of the 70 sectors was "stores, furniture and home furnishings," with 30.3 percent. "Computer and data processing services" had 70.6 percent ICT–skilled employment, while "telephone communications" had 56.4 percent ICT-skilled employment.

Today, information networking—the movement and use of information, filtered and refined by knowledge workers—saturates economic activity. More than ever before, even traditional business activities such as agriculture and manufacturing are becoming information-intensive, or at least more dependent on information and much more sensitive to information flow.

Indeed, in his prophetic but unheralded book, *The Knowledge-Value Revolution: A History of the Future,* Japanese analyst Taichi Sakaiya predicts that "in the society we are moving toward … those who hope to create products that draw a good price will be mainly concerned with incorporating as much knowledge-value into a product as they can."

Even though strategic management is required of every organization, a crisis mentality has overwhelmed many leaders. The horizon has shrunk to the annual budget, decisions are measured by the quarter and the firefight of the day has overshadowed strategic development of the business in a global context.

Precisely because this is an era of structural economic change, leaders need to raise their sights to see clearly and act on a bigger picture. As a practical starting point, today's realistic strategic management horizon is three years, down from the norm of the old "five-year plan." In the rate of change the world is experiencing today, events shift too greatly to rely on a five-year time frame.

Yet too short a planning horizon is also dangerous. Some contemporary leaders actually operate on only a 12-month frame of reference. This short time horizon leads organizations to jerk their way into the future by fits and starts, because they cannot see far enough ahead to plot an informed, smooth, and effective course. For comparison, visualize the way a hopelessly near-sighted person would drive a car. If he cannot see far enough ahead, either he drives recklessly or too cautiously, and neither the driver nor the passengers will be inspired with confidence.

Strategy Is Action

What is strategy? *Strategy* comes from the Greek word *strategos,* which means "general." That historical meaning still helps us grasp what strategy means, which is "the art of generalship." It is a good perspective. And though the world of the military is a special case, many of the challenges of the military general even now are applicable to strategic management of other kinds of organizations.

Here is my definition of strategy, with four key parts, which I will analyze in turn:

- Strategy is coordinated action through time to achieve a goal.
- Strategy is action. Strategy is something you do, something you execute. It is not a fixed plan; instead, strategy is a general's outline for activity. It is appropriate for organizations to deal with a literal global information economy. You create and carry out strategy. It is a way; it is direction. Tactics are the movements you make to carry out the strategy. Strategy is the chosen path to arrive at your goal. Tactics are the footsteps that carry you along the path to get there.
- Strategy is coordinated action; for an organization, it cannot be a one-man show. Even if you choose the path, the others must come with you if you are to succeed. No one is a leader without followers.
- Strategy takes place through time; it is a continuous process. Organizations in the real world are involved in continuous strategic management. It never starts or stops at one place or one office.

The strategy, an entire coordinated action, sustained through time, aims to achieve a goal that has been set at the outset of the action. Goals are large achievements, such as achieving the lead position in a market or becoming the organization with a reputation for sustained, knowledge-value added to any product or service you touch.

Objectives are intermediate steps toward the goal, such as hiring the brightest researchers from this year's college graduates or reaching quarterly sales objectives to provide cash flow.

Goals are strategic; objectives are tactical. These two kinds of targets for action support and complement each other, but they are not the same, even though the two words are often mixed in everyday business jargon. There are generally a few important goals and many intermediate objectives continually aiming toward the goals.

It is true that strategic goals do not shift and change monthly or quarterly. However, the objectives set to reach those goals may be moved, transferred, abandoned, or linked to other objectives at will, for example, as conditions internally and externally merit change. Good strategists are firm on goals and flexible about objectives.

The Six Steps of Strategic Management

With this understanding of strategy, I offer the following six steps to guide the strategic management process, all as a way to make real strategies that achieve real goals:

Build a Strategy—an Agreed-Upon Decision to Take Action
The action to take will be one that, in the informed judg-

ment of the strategists, will achieve the goal they intend over time.

Come to Agreements with the Necessary Participants to Coordinate Their Actions with the Strategy

In principle, the people who will need to implement the strategy ought to be involved in building the strategy. In authoritarian organizations, agreement may be implied through the authority structure. In nonauthoritarian groups, work to achieve consensus and buy-in to participate in the strategy. You will need it.

In today's organizations, emphasizing total employee involvement, empowerment, quality circles and similar strategies and tactics, this agreement step is more important than ever before. In the absence of effective coercive power or two-directional corporate loyalty, strategic managers may derive authority from the informed consent of those they lead.

Communicate the Action to Those Who Need to Participate

If knowledge is power, the information needed to carry out a strategy is empowering. I will discuss communication methods further, but as the third step in the process, understanding is the springboard for effective execution of the strategy.

Since action is by definition dynamic, many tactical adjustments are required of those carrying out the strategy. If they are not informed of the objectives supporting clear strategic goals, the participants will be operating in the dark. They may make the adjustments but will be obscured by limited understanding of what is required. Communication is the way to inform those who need to know what the strategy is intended to achieve.

Begin Carrying out the Action

To launch the action the strategists intend may be an obvious step, but it requires successful execution of the first three steps in the process. Amazing though it may seem, many organizations fail at this step simply because they are not ready to carry out a coordinated action. More often, if you will examine the problem closely, it is because the people required to act truly do not understand what they need to do.

Sustain the Action through Time

This step demands perseverance and patience. Success may require a series of smaller steps or continual tactical adjustments aimed at winning a chain of objectives. Some strategies fail simply because they are not carried through. Goals take time to achieve, especially in the face of opposition. Almost all victories have required this continuous sustaining of the endeavor.

This step often is the act of faith, and the one where leaders may have to wait out the results without direct intervention, as when a battle has passed out of the hands of a general and the day's outcome depends on the people on the line. How much do you trust your organization to carry on? Is it now clear how this crucial fifth step depends heavily on communication and understanding of the strategy by others besides the leadership?

Follow through with the Strategy Until the Goal Is Achieved or Until a New Strategic Action Takes Place.

This final step is the movement that clinches victory. It is also the first step in the development of new strategies. In the third year of a strategic horizon, it is time to flex into new strategies, based on the situations that have unfolded by then. Strategy again is dynamic, not static, to match the fluid environments in which the world operates today. In such environments, you need a firm place to stand. That place is a strategic planning foundation.

Strategic Planning Is a Foundation for Strategic Management

Strategic planning is not some separate, one-time exercise in which the output is a report or plan document. A strategic plan is not a thing. It is a piece of the action. Strategic planning is itself a series of actions that provide a foundation for the initial step—building a strategy—in the strategic management process.

Strategic planning is a continual process that feeds into the decision to act. Since change in the information economy is constant, constant planning adjustments need to be introduced into the strategy to keep it viable.

In the past, strategic planning has been seen as critical because planning actions are a foundation for building strategies. But strategic planning has also been perceived as irrelevant, flawed or limited as a vehicle for change or for results, because it is. Strategic planning is never the same as executing actual strategy, any more than planning for an ordinary event such as a dinner party, a conference, or a sporting match is the event itself. Yet, as these common examples show, planning and actions taken to prepare for them affect the outcome of the events. Strategic planning is reasonable and properly an element of extended action, but it does not substitute for strategy itself.

Strategic planning is the set of inputs and management processes the strategists need to arrive at the agreed decision to act. They are divided into the following four parts:

- Strategic intelligence
- Action options
- Agreement mechanisms for cooperation
- Communication links for coordinated action

These four areas are crucial foundations for successful strategic development. Many discussions of strategic planning focus only on the first two, and some seem to stop after the first category. Each of the four areas requires activity, as follows:

Perform Strategic Intelligence Actions

These are the fundamentals of strategic planning: environmental scanning; situation analysis; and strengths, weaknesses, threats, and opportunities (SWTO) analysis.

Environment scanning is the information-gathering activity of your scouts, whoever they may be. (They ought to be most of the people in your organization; many eyes looking see more than two, no matter how dedicated.) The challenge is to select, filter, and organize information from your global operating environment in at least the following four areas, which are in order from most to least dynamic:

- Economic
- Technical
- Cultural/social
- Political/legal

Again, you can see a different operational emphasis than many organizations make. This is strategic planning, and that should reorder the leadership's priorities in that functional role.

Situation analysis is a close-up of the environmental scan, in that it is an analysis of the external situation affecting your organization—it is the context under which you operate. What is the situation you find yourselves in, as objectively as you see it, taking into account your point of view?

Analyze your organization's strengths, weaknesses, threats, and opportunities. This is the next logical view, a further close-up of the external situation analysis. But now you focus on your own organization, centered on who you are, where you are.

It is important to start with an inventory of your strengths, then weigh your weaknesses. Now, from your weaknesses, you have a clear view of threats to yourselves and your organization, and you can see where your strengths give you encouragement and protection.

Finally, after an honest inventory of the previous three elements, you can reckon your opportunities fairly and with confidence. Any strategic briefing that leads to action calls for offensive thinking. Never end with an emphasis on threats, which is defensive thinking. Start with your strengths and end on your opportunities.

Produce Action Options

Strategic planning always produces options. Planners know that one choice is no choice. Thus you always recommend several options. Three viable options is a good objective. There should never be fewer than two.

I prefer to develop options in a set that emphasizes offense-defense-overall offense. The model for this approach is any competitive sporting match. To win, you need a good offense and defense, plus a combination that takes them both into account. A one-dimensional offense will fail if its strength fails. Defense is required but never wins the battle itself. An overall offense that includes a defensive component is the key to a consolidated victory.

To develop the offense option, ask your strategists to answer the following questions for your organization:

- What does the organization do to win? (The answers are based on your strategic intelligence steps.)
- What does it take for it to do that? (These are your action options.)
- How does it prepare to take the winning actions? (This is the required set of resources, preparations, logistical demands, and personnel readiness factors—all tailored to your organization—that allows you to do the job.)

To set up your defense options, work on the following principles:

- *Design against failure*: The classic engineer's principle applied here is this: Figure out what you will do if your strategy or tactics fail. Then design options in case the failure occurs. It is the reason your car has a spare tire and fire extinguishers exist. The principle may save your strategy, your reputation, and your organization.

- *Manage risk*: Balance risks against benefits. Search out the risks that go along with your strategies. Manage those risks that will reveal themselves to your search; risks are part of any strategy. Do not assume the one-dimensional legal-office approach, which is to say, "Here is risk; avoid it or I am not responsible." There will always be risk; it is largely unavoidable. Instead, manage risk. Minimize what risks you can, and, like a wise gambler, make sure the benefits are worth the possible losses you may incur. Small benefit means undertaking small risk. Large risk means the benefit must be much larger and a potential loss must be recoverable. Desperation strategies are particularly suspicious. Henry David Thoreau wisely said, "It is a characteristic of wisdom not to do desperate things."

- *Set defense options up early*: Seek to answer, where is shelter? Routinely build it in. In the Civil War, it was said the officers realized they were finally leading veterans when the soldiers dug trenches around their camps without being told to do it. Also continually ask and answer the key question: Where is the way out? Always leave yourself an out, if you can. And ask the same question about your opposition—in conflicts, leave your opponents with a way out, so they will not oppose you ferociously, having nothing left to lose.

The following points are principles to design overall offense options to win:

- In light of current conditions, review then execute offensive tactics to achieve your objectives on the way to your goal.

- Operate on the principle that the best defense is an effective offense. Keep up an offensive momentum that contains a defensible orientation. Do not let your organization overshoot the mark and lose its potential for further offense through over-commitment, exhaustion, or letdown. Rejuvenate with rest and recreation as an offensive-defensive posture.

- Prepare for victory. Be ready to win. Have options ready when you achieve your goal. Many leaders are unprepared to cope with victory, which results in a quick loss of hard-earned credibility, especially in successful start-up companies. Plan for your victory.

- Consolidate victory. Prepare for the aftermath of victory when you have achieved your goal. Plan to consolidate your organization, mop up the details, settle into a temporary defensive position, and prepare to build new strategies based on what you have collectively learned.

Reinforce Agreement Mechanism for Coordination of the Agreed-on Options for Action

Before executing strategies, you must attend to this basic step. Management is the art of follow-up. When the leaders have agreed on the strategic options, follow up by reinforcing the means you have for coordination of actions. As you would not launch an expedition without checking and repairing your equipment, this is the time to check and repair your mechanisms for cooperative action. Focus on the following areas:

- Maintaining leadership and control systems (in business, effective decision-making and management structures)
- Ensuring that you have agreed upon decisions for action, securing commitment from those involved
- Socializing the decision with and by key leaders
- Setting up and managing by objectives that support the drive to the strategic goals

Forge Communication Links to Carry out Coordinated Actions

Communicate the required actions. Operate on the principle of need to know, but work to extend knowledge—in my view, the more broadly, the better.

A key difference in my approach to strategy from other commentators is my emphasis on strategy communication. When examining strategies that have gone awry or astray, it seems the communication process for the strategy has repeatedly gone badly, rather than that the strategic action itself was badly made.

To reinforce the point, consider this parallel concept of signaling value, by the business strategist Michael Porter. According to Porter, "Buyers will not pay for value they do not perceive, no matter how real it may be. Thus the price premium a firm commands will reflect both the value actually delivered to its buyer and the extent to which the buyer perceives this value. ... By failing to signal its value effectively, a firm may never realize the price premium its actual value delivers."

In the information economy, extra care in communicating strategy is good business. As Peter Drucker points out, today that "means of production" of knowledge workers is their knowledge itself, which makes them independent and highly mobile. Therefore, including the knowledge workers in leadership strategy gives them input upon which to apply their knowledge and helps retain their services for the organization that values them properly.

Pursue chain-of-command effectiveness. Build in feedback loops to make sure that information is flowing both ways through the chain of command. A modern leader might use e-mail as a way to generate safe feedback from all levels in the organization.

In any event, do not rely on intermediaries to manage all your internal information flows in both directions. Rely on your own impressions—not out of distrust of intermediaries, but instead to have a more direct feel for the effectiveness of the communication links that the chain of command exists to support.

Secure alignment by local leadership. Include in your communication the local leadership. Do not overlook, especially

in technical organizations, the de facto "information gatekeepers" who are centers of influence, those who locate and process information for a group. Often they are the "stars" of an organization who are seen as role models by their peers.

Summary and Conclusions

I have emphasized that strategy is coordinated action through time to achieve a goal. I see strategic management as a six-step process that begins with an agreed-upon decision to take action. Participants make agreements to coordinate their actions, and then communicate the actions required to those who will participate. Next they act in concert, sustain the action, and follow through until the goal is achieved or a new strategy is generated.

Strategic planning is the foundation for strategic management and requires strategic intelligence, action options, agreement mechanisms to ensure cooperation, and communication links to coordinate action.

The information economy is producing structural change at a rapid rate. In dynamic times, strategic management is more important than ever. The methods and information technologies exist to help you achieve your goals in this era, if you will apply them. Committed strategic management will make possible the victories you can prepare for and attain.

Notes

1. Carl Shapiro and Hal R. Varian. *Information Rules: A Strategic Guide to the Network Economy* (Boston: Harvard Business School Press, 1999), p. x.
2. See Cosmo R. Shalizi, "The Information Society and the Information Economy" (10 October 1997), http://cscs.umich.edu/~crshalizi/notebooks/information-society.html. Retrieved 04 December 2005. Shalizi's posting has a useful list of pro and con arguments and other resources on the information economy appended.
3. Michael Rogers Rubin, "US Information Economy Matures," *Transnational Data and Communications Report* (June 1986), p. 23.
4. Michael Rogers Rubin and Mary Taylor Huber, *The Knowledge Industry in the United States, 1960-1980.* (Princeton, NJ: Princeton University Press, 1986) p. 5.
5. Organisation for Economic Cooperation and Development [OECD], *Information Activities, Electronics and Telecommunications Activities, Impact on Employment, Growth, and Trade.* (Paris: OECD). Cited in Rubin and Huber (1986), p. 6. The OECD is an excellent source for such data, particularly its Information Economy unit. See http://www.oecd.org/sti/information-economy.
6. Organisation for Economic Co-operation and Development [OECD], Working Party on the Information Economy, *New Perspectives on ICT Skills and Employment.* (Paris: OECD,2005), p. 30. OECD Document STI/ICCP/IE(2004)10/FINAL.
7. Taichi Sakaiya, *The Knowledge-Value Revolution: A History of the Future.* (Tokyo, New York, London: Kodansha International, 1991), p. 62. This seminal work, first printed in Japan as *Chika kakumei* in 1985, was not published in English until 1991. A paperback version appeared in November 1992. For a comparison, see the essays of leaders in the United States information industry that discuss knowledge work and the information networking environment in Jay E. Gillette, ed., *Information Networking: Toward a Field Definition.* Morristown, NJ: Bellcore Information Networking Institute, 1991. Bellcore document ST-TEC-000132.
8. See for example Henry Mintzberg, *The Rise and Fall of Strategic Planning: Reconceiving Roles for Planning, Plans, Planners.* New York: Free Press, 1994.
9. Note concerning secrecy, the element of surprise and open strategies: In strategy, a little secrecy goes a long way. If the strategy's effectiveness depends wholly on surprise, it will be compromised by any significant

breach in secrecy. Good strategists use the *element* of surprise as a tactic, but do not gamble the success of the entire operation on it. They plan against the possible failure of surprise, and have backup tactics to shift to if surprise is not attained. Against relying on surprise, Edward Luttwak and others make arguments that the most convincing and long-lasting victories are achieved when the opposition is openly and decisively beaten "on a fair field," when they cannot blame circumstances or the fact that they were surprised, for example. And successful surprise attack, as in the tactically successful Japanese operation against Pearl Harbor in 1941, may ultimately serve to galvanize the opposition into ferocious reaction when the surprise wears off.

10. Michael E. Porter, *Competitive Advantage: Creating and Sustaining Superior Performance.* (New York: Free Press, 1985), pp. 139-140.

11. Peter F. Drucker, "The New Society of Organizations," *Harvard Business Review* (September-October 1992), p. 101.

12. There is extensive research on this significant communication role. For example, see Robert L. Taylor, "The Impact of Organizational Change on the Technological Gatekeeper Role," *IEEE Transactions on Engineering management*, EM-33 (February 1986), pp. 12-17.

Evolution Toward Converged Services and Networks

Hans Höglund

Strategic Marketing Manager, Business Unit Systems
Ericsson

Mikael Timsäter

Business Development Manager, Business Unit Systems
Ericsson

Convergence, long talked about in the telecommunications industry, is re-emerging with renewed vigor. This white paper examines which drivers and technologies enable true convergence and what operators need to consider to fully capitalize on this opportunity.

Executive Summary

Research has shown that end-user behavior is rapidly changing because of globalization, the increased value placed on individuality, emerging tribalism, new patterns of social networking and an increasingly nomadic lifestyle. From a service perspective, consumer and enterprise business users expect convenience, ease of use, reliability, security, and support to be always best connected.

The communication market is evolving rapidly. New players are emerging, and competition is increasing. As a result, new partnerships are being formed and old boundaries for conducting business are fading away. In this changing business environment, operators explore different ways to find new revenue streams, reduce operating costs, and provide solutions that create stickiness and reduce churn. The successful operator will provide a multitude of new services. Many of the services will be available by both mobile and fixed access. Others will represent a combination of TV, Internet, and telephony—all of them being converged services.

Technologies that enable converged services exist. Internet protocol (IP) and the Internet paradigm are being introduced in all areas of communication. Rapid development of radio technology, leading to increased bit rates and support for mobility, enables true converged services—the same end-user service can be reached by both mobile and fixed access via the same user interface.

Operators that adapt their strategic business plan, considering the changing environment, with an early introduction of converged services will gain a competitive edge. Furthermore, the introduction of layered architecture will improve efficiency, flexibility and enable a smooth introduc-tion of IP multimedia subsystems (IMS), a cornerstone for efficient converged service offerings.

Definition of Convergence

Traditionally, the term fixed-mobile convergence (FMC) has been used by the telecom industry when discussing the integration of wireline and wireless technologies. But it is not just about this particular kind of convergence. It is also about convergence between media, datacom, and telecommunication industries, as shown in *Figure 1*.

In this paper, convergence is considered from the following three viewpoints:

- *User service convergence*: common user service delivery capabilities with access and device awareness. This means that a multitude of services (person to person, person to content, and content to person) can be provided to the same user over different access networks and to different devices.

- *Device convergence*: common devices supporting several access types, such as code division multiple access (CDMA) 2000, wideband CDMA, Global System for Mobile Communications (GSM), fixed broadband and wireless local-area network (WLAN). Device convergence allows multiple applications to be run, reusing the same functions for identification and authentication. Furthermore, the mobile device supports more and more functions in addition to telephony, e.g. camera, TV/video and e-mail.

- *Network convergence*: consolidation of the network to provide different user services, with telecom-grade quality of service, to several access types with an emphasis on operator cost efficiency.

Convergence Is Back

A decade ago, the telecommunications industry started to discuss FMC. Even though the theory was right, the rela-

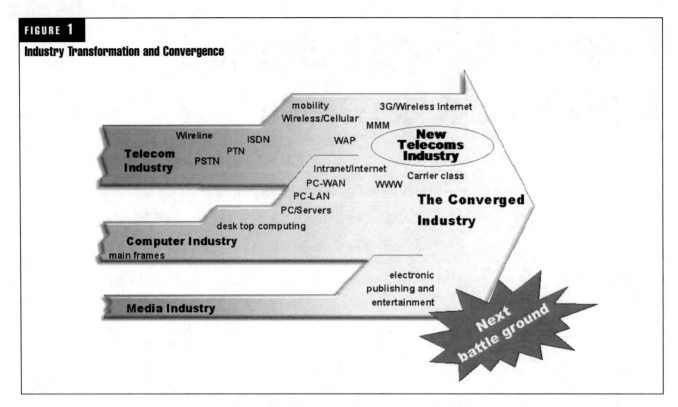

FIGURE 1

Industry Transformation and Convergence

tively immature technology meant providing convergent services in an efficient way to end users was very complex. Furthermore, it was obvious that users were not ready for converged services.

Network operators now re-evaluate convergence and see it as an opportunity to create stickiness and provide more value for their subscribers. Examples of convergence-related services and implementations available today are as follows:

User services:
- Bundling of fixed, mobile, and broadband subscriptions
- Triple play (telephony, Internet, and IP television [IPTV] via broadband)
- Single phone number
- Single mailbox

Devices:
- Seamless WLAN/2G/3G connection
- Multi-access mobile phone with licensed and unlicensed mobile access
- Media and PC functionality in mobile devices
- Networks
- Layered architecture with IMS

Putting the End User in Focus

In this section, the term "end user" refers to consumer and enterprise business users. The different end-user behaviors are described to get an understanding of what influences the users now and in the future. This is an input to which services the operator should offer. The relation to convergence is described later in End-User Expectations.

Changing End-User Behavior

The communication landscape is richer now than it has ever been. This has far-reaching implications on how the end user will be looked at in the future. Research has shown that the following trends are relevant from a user-service perspective:

- Globalization results in trends becoming very similar globally, for example, among young people. Globalization and the impact of mass media make trends diffuse and spread much faster on a global scale.

- The user places increased value on individuality at the same time that we have a boom in the supply of products, services, and information. This leads to new choices that were not previously available. Customers demand more customized and tailored offerings. With this comes the requirement for personal control of services and information. This increased individualism is also true for former collective environments, such as the home.

- Tribalism emerges as a counterforce to the individualization trend. People are self-organizing into communities and "tribes" of shared values and interests. A tribe can be local or spread across the world. So what we see is a collective individualism. The tribe plays a major role in the individualization process.

- New patterns in social networking are both enabled and amplified by technology. Already, scheduled social interactions are increasingly rare among young people and there is a growing expectation and preference for ad hoc social interactions.

- A nomadic lifestyle, where the borders between private and business use are fading and the user expects (or is expected) to have access to the services anywhere and at anytime, is also evolving. Communication capabilities that support context switching and multi-tasking in all aspects of life will be critical.

End-User Expectations

The user needs based on the changing end-user behaviors described, and today's basic user expectations, lead to the following three areas for further consideration, all of them closely related to convergence:

- *Convenience and ease of use*: Users expect similar user interfaces for most services without having to consider which network is used. Services should be adapted to the device and access characteristics being used, including simplified processes for identification and payments, as well as the ability to control cost.

- *Always best connected*: Users expect to be able to connect anytime, anywhere—even when on the move—by their device of choice. Users also expect to be able to specify in each situation whether "best" is defined by price or capability.

- *Reliability and security*: Users expect reliability in all transactions, independent of access, and guaranteed connection quality. From a security point of view, the user expects no viruses, no worms, no fraud, no breach of privacy, and the ability to know who requests a communication session.

Thus, one of the most profound changes in the way we look at convergence today, as compared to a decade ago, is the increased end-user focus as a driver of convergence. Earlier, the focus was far more on operator and network efficiencies. These advantages, however, still remain and have gained increased importance because of the new user and service behavior together with maturing technology.

Opportunities and Challenges for Operators

There are different approaches to the changing business environment for different operator categories. However, it is clear that convergence will play an important role for all categories independent of selected business direction.

Existing Operator Categories and New Players

In most markets, competition for telephony is intensifying, both for fixed and mobile network operators. New players, such as cable-TV companies and dedicated voice over Internet protocol (VoIP) application vendors, are also entering the telephony segment. To grow, most network operators and other players on the market start to address areas outside their traditional offering, which means that new offerings are created by combining previously disparate offerings and traditional operating boundaries are fading.

Mobile-only network operators are striving to offer more advanced services to mobile customers to provide a competitive alternative to fixed network services and in that way increase market share. Others focus on offering convergent services to the enterprise segment, including mobility.

Fixed-only network operators are searching for ways to add mobility to their broadband networks by becoming mobile virtual network operators or exploring mobile license opportunities to create a bundled offer. They are building WLANs through hot spots using unlicensed spectrum. Fixed broadband operators also offer triple-play services (telephony, internet and TV) to utilize their bandwidth advantage. Furthermore, they sell carrier capacity to other operators.

Combined mobile and fixed network operators utilize their combined customer base through bundles and value-added offerings. They can consolidate their networks to cut costs. They can, of course, also target the same areas as fixed-only and mobile-only operators.

Cable-TV network operators are striving to enter the broadband and telephony segments. It is anticipated they will address mobility in a similar fashion as fixed-only network operators as well as the wireless hot-spot access provisioning. Cable network operators tend to focus on entertainment services where they have strong partner relations.

VoIP providers are striving to dramatically increase their telephony market share to capitalize on their interconnect revenues. They are exploiting the rapid growth of broadband IP data access.

It is not possible to specify a single path forward for each operator category, as each operator must take into consideration much more than just the current business operation when defining its strategy. However, it is obvious that convergence will be an important aspect for all operator categories.

Challenges

Some important challenges related to the current and future business environment and network evolution are outlined below.

In a changing business environment, each operator must explore ways to expand its business and decrease costs in the long term. This implies finding ways to attractively bundle services and subscriptions to create stickiness, hence reducing churn as well as subscriber retention and acquisition costs. Further alignment and convergence of the bundled services will enhance the offering.

There are also different regulatory aspects related to convergence that could affect an operator's business plans, for instance, bundling restrictions because of perceived competition limitations.

Network operators must act fast in a dynamic market and adapt quickly to the new business environment. Operators also need strong marketing efforts to leverage the potential opportunity for growth and cost savings. A service delivery platform that enables fast deployment of new converged service offerings is a prerequisite to succeed.

Operator Categories and Business Direction

An important part of the operators' strategy to address market opportunities is to choose in which direction to drive their business. There are three main scenarios, all plausible, for how operator business can be developed: walled garden, bit-pipe, and channel provider, illustrated in *Figure 2*.

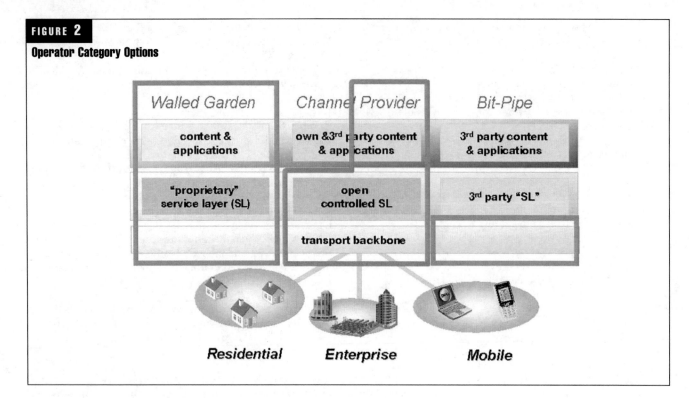

FIGURE 2

Operator Category Options

The walled garden approach means that the operator strives to have an exclusive relationship with the end user, from the device to the service.

For bit-pipe operators, the main objective is to focus on operational excellence and maximum efficiency in the network. There is little focus on service provision.

The channel provider operator could be seen as a combination of the walled garden and bit-pipe scenarios, where the operator supports content and application providers with key network characteristics and services. Identification, authentication, roaming, and presence management are examples of characteristics. This channel provider option drives convergence in all layers.

Enabling Technologies

While technology was a limiting factor a decade ago, it has today reached a level of maturity that enables convergent services. One advantage that now makes convergence a reality is the evolution towards one common network, which is IP–based. Furthermore, high-speed broadband connections, both fixed and wireless, make it possible to offer converged multimedia services, independent of access type.

Evolution toward All–IP

The trend is clear: IP paradigm is, or will be, used in almost all areas of communication.

A common IP–based network enables a multitude of common functions and therefore reduces costs in the form of planning and operation. The potential savings for operators are substantial and one of the most important drivers of network convergence.

And to save the best for last; when the underlying structure is more structured and standardized, other areas have more

room for variation. This means that customized services—which, of course, still can be convergent according to the definition—can be provided efficiently.

Access Technology Development

The fixed broadband penetration is increasing with double digits every second year. Home networks are becoming the norm. The PC penetration is still rising. Communication standards within the home have evolved around Ethernet and IP, enabling high capacity connectivity between the home gateway and the service provider. Already today the technology exists to build high capacity and low cost Ethernet solutions for the mass market, which enable triple play and other innovative services.

The development of wireless access technologies has reached an important breakpoint. Today, it is possible to transfer packet data with high bit rates also over a mobile radio interface. This will lead to a new generation of services such as broadband Internet access, Video over IP and IP telephony.

This rapid development of radio technology is an enabler for true converged services, meaning that the same end-user service can be reached via both fixed and mobile access via the same user interface. *Figure 3* shows the capabilities of some existing and emerging radio technologies.

There is already support in the radio access standards (for instance, Wideband code division multiple access [WCDMA] and CDMA2000) for VoIP, including support for mobility.

The appropriate standardization bodies are addressing updates in radio access standards so that optimized transport of telecom-grade IP telephony services includes full support for mobility.

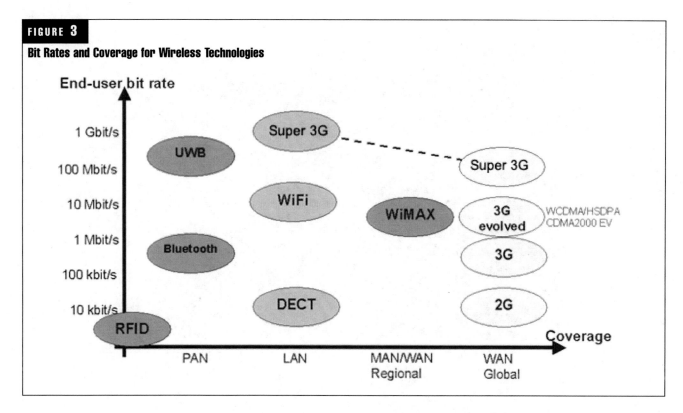

FIGURE 3

Bit Rates and Coverage for Wireless Technologies

From Theory to Practice

Short-Term Convergence Implementation
Network operators already offer convergent services. In general, user services can be offered relatively independent of the evolution of networks and devices. Services such as "one number" and "follow-me" are available. These services, together with bundling of subscriptions (fixed, mobile, broadband), can be an opportunity to gain new revenue streams and decrease churn.

Enterprise solutions giving fixed and mobile users a single numbering plan with access to the same private branch exchange (PBX) services are also offered. Video telephony can today be offered from personal computers (PCs) and mobiles. Services such as Push to Talk and combinational services (such as combining telephony and multimedia sessions) are offered now.

It is important to ensure that new services are introduced in a future-proof fashion. Standards must be followed to allow interoperability and to ensure that the solution can be upgraded. IMS is a cornerstone for converged service offerings for fixed, mobile, and enterprise segments.

As described earlier (see Convergence Is Back), there are also other important new converged services available, such as triple-play and WLAN plus second-generation/third-generation (2G/3G) bundled subscriptions.

Service Convergence
Convergence of services and applications implies that the same service can be accessed from different types of terminals, for example, sending messages from a mobile user to a PC, or browsing the Internet from a handheld mobile phone, and different types of networks—cable TV, mobile, or fixed.

IMS, which is a standardized solution for applications based on session initiation protocol (SIP) for multiple accesses, is a key component for delivering converged services with telecom-grade quality of service. IMS makes it possible to increase network efficiency and makes the introduction of new services faster and easier. The common service execution environment of IMS supports user applications that are available over multiple accesses (access-aware service platforms). There will be one common user and services management function, a common charging system, and a common (based on a subscriber identity module [SIM] card) identification and authorization system.

Flexible authentication and identity mechanisms are crucial in a converged environment. The nomadic lifestyle of the end user will require a convenient solution. The use of authentication based on SIM card will be expanded into new domains. One solution is wireless communication between a personal SIM-card holder and the device being used. The SIM-card could be carried as a personal item, always being with you, as exemplified in *Figure 4*.

The presence information is a key component for many IMS–based services. Presence-aware communication allows a user to see recipient information before connecting (e.g. availability, geographical position). Presence enables the user to see possible communication alternatives based on device and network capabilities. The presence information will be available from any device (mobile and PC). Presence enables a paradigm shift in person-to-person communication.

The majority of communication sessions using converged services, such as voice calls, video calls, chat sessions, file transfer, on-line games and whiteboard sessions, are typically initiated via the active phone book. The active phone book is one application that uses the presence information

FIGURE 4

A Personal SIM Card for Universal Authentication

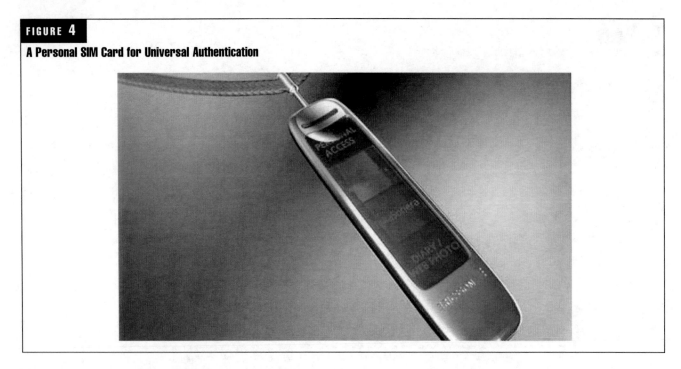

from IMS. An example of an active phone book is shown in *Figure 5*.

Device Convergence

Device evolution can be seen as a mirror of the network and access evolution. It is in the device that the new applications will be available and the identification and security mechanisms are implemented; therefore, it is also where access capabilities need to exist. Another possible functionality for mobile device convergence, in the short/medium term, is adding support for unlicensed mobile access (UMA). This allows a mobile phone to use WLAN/Bluetooth to access the local fixed broadband to connect to the GSM core network. Furthermore, in the future, IP telephony needs to be implemented in the communications protocol of the device so that future conver-

gence opportunities can be fully utilized.

The pace of evolution in this device convergence area will increase in coming years as new access technologies are introduced. It can be expected that additional access technologies will lead to extra cost for the device. Overall device and network economy must therefore be considered when introducing new accesses. Storage requirements and capabilities will dramatically increase in both devices and networks. *Figure 6* shows the evolution of the capabilities of the mobile phone.

Some key network characteristics must also be available to devices connected to the fixed network. The evolution of connected home networks with a multitude of applications, including interactive TV, games, music downloads, shop-

FIGURE 5

Converged Active Phone Book

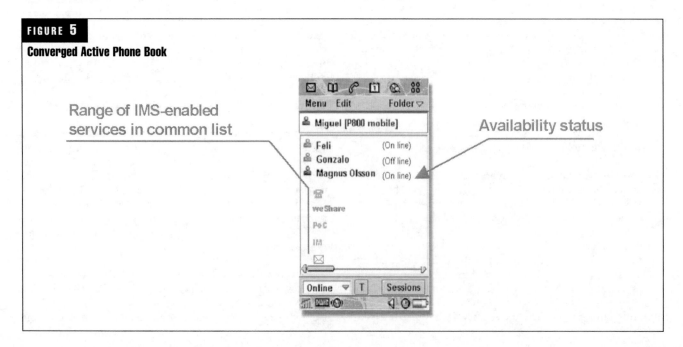

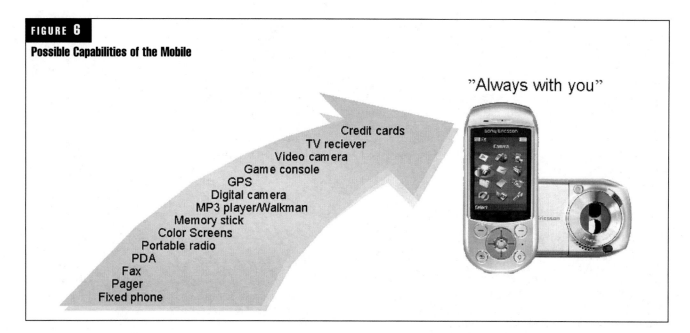

FIGURE 6

Possible Capabilities of the Mobile

"Always with you"

Credit cards
TV reciever
Video camera
Game console
GPS
Digital camera
MP3 player/Walkman
Memory stick
Color Screens
Portable radio
PDA
Fax
Pager
Fixed phone

ping, and home security, will require some of the characteristics that are delivered by IMS (e.g. quality of service, user management, charging, and security) also in the fixed network and to stationary devices. *Figure 7* shows some examples of devices connected to the home via a broadband connection that could benefit from network information, just like the mobile devices.

Telecommunications players have an advantageous opportunity to address this home networking need with business model solutions originating from the mobile market, e.g., the authentication and security related to the SIM card.

Extended use of the SIM-related functionality will increase the operator's value to the content providers as well as to the users.

Network Convergence
Until recently, networks for wireless, wireline, data, and cable-TV services have existed in isolation. The next-generation solutions represent a more efficient way to build networks using a common multiservice layered architecture. The networks will have a layered structure with a service layer, a control layer, a backbone layer and access networks. Having one converged network for all access types is a sig-

FIGURE 7

Digital Infrastructure at Home

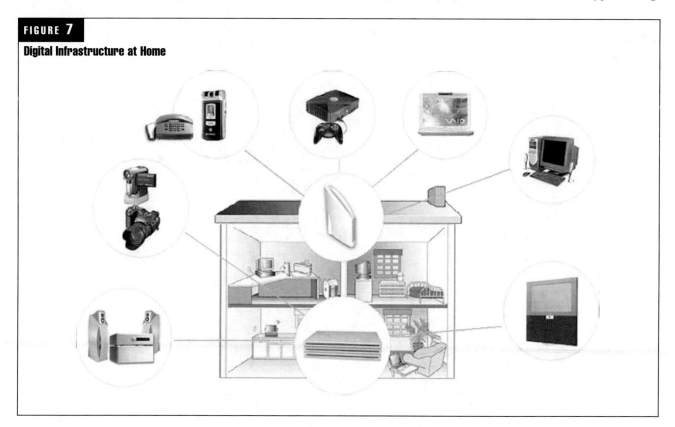

nificant benefit of layered architecture. This can improve service quality and allow the efficient introduction of new multimedia services based on IMS.

Operators can increase network efficiency using optimized transport and coding solutions and will not need the overcapacity required when the networks are separated. Significant cost savings can arise from having one network with fewer nodes and lower operating costs. From an investment perspective, it is possible to optimize the use of control and media processing resources, hence reducing the need to replace technologies and the cost of network updates.

IMS is a key component of multiservice layered architecture. IMS is a subsystem supporting multimedia sessions, standardized by third-generation partnership project (3GPP) and using the SIP from the Internet Engineering Task Force (IETF). IMS is a common foundation for fixed, mobile, and enterprise services, delivering services over multiple accesses such as CDMA2000, WCDMA, GSM, fixed broadband and WLAN. Thus being a cornerstone in a converged solution. *Figure 8* shows a target network architecture based on IMS.

A converged network using IMS allows the following resources to be shared, regardless of service or access type:

Charging	Provisioning
Presence	Media handling
Directory	Session control
Group and list functions	Operation and management

In addition to making converged user services faster and easier to introduce, as described in Service Convergence, the common shared resources also increase operational effi-

ciency in the network.

The network evolution path is unique for each operator and depends on many factors, including the business environment, cultural heritage, regulations, end-user behavior, and PC and mobile penetration rates. The transformation is usually done step by step toward the target network with an all–IP solution based on IMS.

Key Operator Aspects

There are a number of decisions and actions to be considered when defining the strategy. Both the business and network convergence aspects must be considered.

Successful operators will do the following:

- Through market research, considering global user trends and local market situations, create winning propositions toward end-user segments, bearing in mind users' expectations of converged services.

- Select solutions that follow standards to ensure service, network, and device interoperability, and partner with vendors who are driving and adapting the agreed standards. Full interoperability is a foundation for converged services.

- Select vendors and partners that can provide a product portfolio covering all solutions within the areas of fixed, mobile, and enterprise. This is needed to secure a homogenous end-to-end solution covering the services, device, and network domains, and to consider convergence from all aspects.

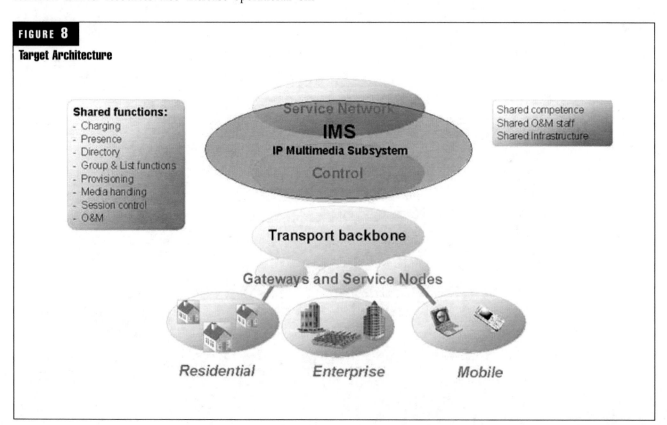

FIGURE 8

Target Architecture

- Consider short- and medium-range radio technologies (WLAN, WiMAX) and how they can be integrated as an extension to fixed broadband access and how this would affect the possibilities to offer converged services.

Conclusion

For years, the telecom industry discussed convergence in its many forms. Even though the theory was right, it did not become a reality. However, today's trend is clear—convergence has returned, and this time with a solid user, technological, and service motivation. Examples of convergence-related services and implementations available already today are bundling, triple play, seamless WLAN/2G/3G connection, multi-access mobile devices, layered architecture, and IMS.

The communication market is evolving rapidly. New dynamic players are emerging, and competition is increasing. New partnerships are formed and old boundaries for conducting business are fading. In this changing business environment, operators are exploring different ways to find new revenue streams, reduce operating costs, and provide solutions that create stickiness and reduce churn. To offer convergent services is one way to achieve this.

Operators have network assets that they want to continue to leverage during the evolution toward next-generation networks. This requires substantial planning and typically a staged approach. It is not possible to specify a single path forward for each network operator category, as each operator is unique. However, it is obvious that convergence will be an important aspect for all network operator categories.

Technology that enables converged services exists. IP and the Internet paradigm are being introduced in all areas of communication. Rapid development of radio technology leading to increased bit rates and support for mobility enables true converged services, meaning the same end-user service can be reached by both mobile and fixed access via the same user interface.

The successful operator will provide a multitude of new services. Many of them will be available by mobile and fixed access. Others will represent a combination of converged TV, Internet, and telephony services. Operators that act now and adapt their strategic business plan considering the changing environment, with an early introduction of converged services, will gain a competitive edge.

IMS is a cornerstone for efficient converged service offerings. It is a key to deliver multimedia services with telecom-grade quality of service. IMS makes it possible to increase network efficiency and makes the introduction of new converged services faster and easier.

References

Ericsson white papers:

Key Business Issues in the Service Layer: www.ericsson.com/products/white_papers_pdf/service_wp_layer.pdf
Mobile Multimedia: www.ericsson.com/products/white_papers_pdf/mobile_multimedia.pdf
Enterprise Communication Trends, Needs, and Opportunities: www.ericsson.com/products/white_papers_pdf/enterprise.pdf
IMS—IP Multimedia Subsystem: www.ericsson.com/products/white_papers_pdf/ims_ip_multimedia_subsystem.pdf

The Seventh Strategy for Fixed-Line Carriers

Outsourcing Network Maintenance and Operations

Bruce Marshall

Principal, Communications and Media Practice
A. T. Kearney, Inc.

Lawton "Mitch" Mitchell

Vice President, Communications and Media Practice
A. T. Kearney, Inc.

No matter who you are or where you are in today's telecom industry, the goal is the same: to return to the high financial returns and profitability that prevailed before 2001 and the onset of the well-known "nuclear winter." Most carriers have tried to recapture those heady days by pursuing any one of six strategies directed at growing the top line, driving down the cost structure, or both (see *Figure 1*). While these strategies have helped carriers improve their business performance—some more than others—the analysts and markets continue to demand more. So the search for the next line of attack continues, with most fixed-line carriers looking for the "seventh strategy" that could increase operating profits by an average of 15 to 20 percent.

In our experience, this strategy already exists and is being implemented by a handful of carriers. The approach? Outsourcing elements of network and maintenance operations. Brazil-based Telemar, Telecom New Zealand, and TeliaSonera are all examples of carriers that have implemented or are in the process of implementing network outsourcing strategies and realizing significant returns as a result. They are cutting costs and expenses by 20 to 30 percent, reducing head count by 40 percent, improving lines per employee by 60 percent, and increasing revenues per employee by upward of 50 percent. Combined with improved cash flows, reduction in capital expenses, and freeing up management to focus on customer-centric issues, the impact on shareholder value has been extremely positive.

That's not all. By outsourcing network maintenance and operations, carriers are doing the following:

- *Reducing resources and operational expenditures (OPEX) and capital expenditures (CAPEX)*: Rather than the carrier, the outsourcer takes on the costs and responsibilities for process improvements; cross-training; investments in new automation; and most light assets such as tools, vehicles, and personal computers (PCs). The result is that the carrier doesn't have to make a capital investment or wait two to three years to achieve results.

- *Improving customer focus*: By outsourcing network operations, carriers increase the number of employees directly involved in customer activities, thus changing the organization to one that is customer-centric.

- *Creating a variable rather than a fixed cost structure*: Outsourcing provides for a variable cost structure that allows carriers to be more flexible and better align their costs with actual market demand.

- *Meeting the demands of a fluid market*: With new competitors on their heels, carriers have to stay a step ahead. They must be able to influence product rollouts such as digital subscriber line (DSL) and create special offers or bundles that address changing market conditions.

- *Increasing cash flow*: Slashing operating expenses combined with the reduced need to purchase light assets quickly improves cash flow.

By outsourcing network maintenance and operations, these carriers are supporting a customer-centric strategy while setting the bar for future competitive advantage—perhaps even securing industry leadership for the rest of the decade and beyond. Hutchinson Australia and Cesky Telecom are solid examples of companies that used outsourcing to significantly increase shareholder value. Hutchinson increased share price by 30 percent after announcing its intention to outsource third-generation (3G) managed services to Ericsson. Cesky Telecom increased its share price by 9 percent by signing a nonbinding memorandum of understanding on outsourcing.

Today, if you're the chief executive officer of a fixed-line carrier, you should be outsourcing your network maintenance

FIGURE 1

Fixed-Line Carriers Use Six Popular Cost and Growth Strategies

Cost-improvement strategies		
Process reengineering and automation	Enterprisewide strategic sourcing	Next generation cost reduction
Streamlining and integrating systems, networks and functions are key tactics. AT&T's Concept One and Concept of Zero are examples.	Reducing external expenditures can offer major cost-reduction opportunities.	Reviewing the total cost structure and operating model can result in multi-year initiatives that can radically reduce costs.

Growth strategies		
Customer-centric business models	M&A strategies	Disruptive technology

and operations. Not only is it an available option, it is an intelligent one.

The Truth about Outsourcing

For years, companies and industries have turned over some non-core functions to outside providers—from information technology, building maintenance, security, and fleet management to human resources, call centers, and credit and collections. The practice has been particularly vigorous in the airline and automotive industries, where companies have spun off functions they once considered core, such as production, to strategic partners. The telecommunications industry is also not a stranger to outsourcing. Carriers have been generally comfortable handing over much of the network deployment function to close technology partners, allowing them to design, build, and install switches. Third-level maintenance support is also outsourced under the umbrella of network maintenance contracts that cover spare parts and technical support for major problems or outages. And most carriers outsource the construction function to subcontractors.

But, if outsourcing is nothing new to carriers, why is it suddenly considered the seventh strategy? The answer has to do with what is being outsourced. The differences between outsourcing construction and outsourcing network operations are like night and day. Construction is not core to the business, so it is mostly risk-free, while network operations are still considered a core competence, which means there are more potential risks that require a more strategic and longer-term transformation plan.

Deciding what makes the most sense to include in an outsourcing initiative often depends on how a carrier views its core versus non-core, or strategic versus non-strategic, functions. For example, how critical is it to continue to own the network switches? The network planning function? Both could be viewed as strategic, or they could be considered non-strategic and therefore candidates for outsourcing. The decision depends on the operator, its corporate strategies, and its explicit labor and vendor situations.

In Network Operations, What Is Core, What Is Not?

In recent years, the definitions of core and non-core functions have expanded as technologies have become more sophisticated. For example, the provisioning and assurance functions are critical from a customer satisfaction level and have long been considered core functions. Today, however, some carriers are cutting costs by outsourcing these functions to suppliers that are more able to deploy innovative technologies and cross-train technicians. The business impact of doing so can be significant, and the implementation effort is more manageable and relatively fast (see *Figure 2*).

Figure 3 illustrates the basic processes of a fixed-line operator. The following is a rundown of common functions and describes to what degree each is appropriate for outsourcing:

Network Planning and Development
This function typically incorporates the network planning and design activities, which most carriers view as strategic and are therefore not willing to consider for outsourcing. However, a few carriers are beginning to change their minds, believing that certain technology providers can perform this function more effectively if they openly agree to a best-of-breed design criteria.

Network Engineering
Carriers are split as to whether or not the actual engineering of the network must be retained in-house, with some carriers expressing concerns about the security of the network if it is outsourced. Given the issues of national and international security, these concerns are understandable and legitimate. However, as companies become more adept at ensuring security and protecting network integrity, network engineering may become an outsourcing option.

Construction
Most carriers already outsource network construction to subcontractors but continue to manage it to control quality

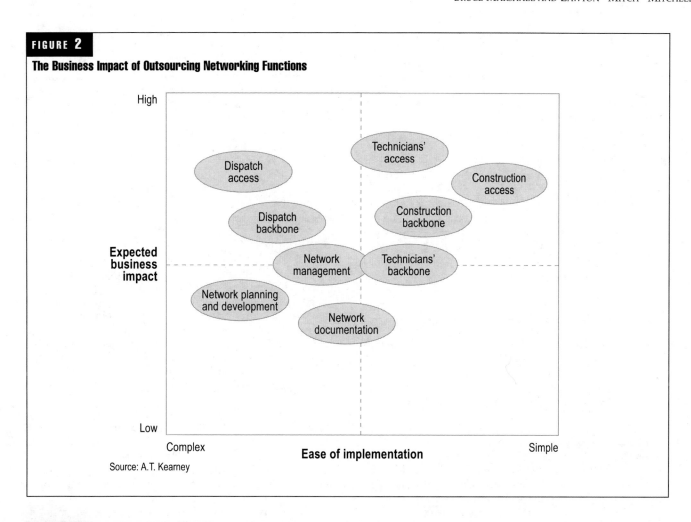

FIGURE 2

The Business Impact of Outsourcing Networking Functions

Source: A.T. Kearney

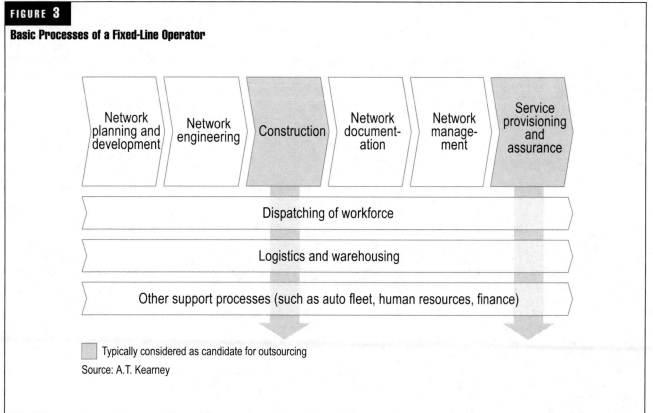

FIGURE 3

Basic Processes of a Fixed-Line Operator

Typically considered as candidate for outsourcing

Source: A.T. Kearney

and performance. These same carriers also prefer to retain the materials-management function to ensure quality and to help negotiate lower prices. Yet as network infrastructure providers gain scale, they are becoming more effective project managers, and many are negotiating lower prices for materials—at times obtaining prices even lower than the carriers obtain.

Network Documentation

Many carriers still rely heavily on paper drawings to document the location and engineering details of their outside plants. Those that upgrade this to an electronic format and outsource it to a qualified provider are realizing significant rewards. For instance, by digitizing the elements of the network, the providers help carriers improve service provisioning, repair and maintenance key performance indicators (KPIs). However, carriers that outsource this function should maintain ownership of the database, or the value is lost.

Network Management

Today, most carriers oversee routing traffic, handling peak surges, establishing network priorities, and restoring service as strategic functions. The job requires a limited number of highly skilled employees and is generally centralized into one or two centers. As a result, most carriers consider network management as a "second-phase" outsourcing opportunity.

Service Provisioning and Assurance

The heart of any service is activation. Thousands of work orders are issued, each of which require special skills and expertise to complete. By necessity, service provisioning is decentralized and comprises the largest segment of the network workforce. Carriers that outsource this function do so because they recognize that an outside provider can increase efficiencies by automating processes, improving records management, and cross-training the workforce.

Service assurance is divided into on-demand and non-demand activities. On-demand activities are the normal request-for-repair calls that have a direct impact on customer satisfaction, while non-demand activities are preventive-maintenance calls, which have a direct impact on the demand activity and can extend the life of network components. As with service provisioning, this is a decentralized function that encompasses a significant part of the workforce.

Dispatch

The daily scheduling of customer provisioning and repairs is integrally linked to a carrier's workforce management system. Dispatch is typically integrated with service provisioning and assurance and has an effect on the overall network KPIs. Therefore, a carrier that decides to outsource service provisioning and assurance should outsource dispatch to the same provider.

Logistics and Warehousing

Operating a network requires a significant logistics capability. Maintaining inventory levels requires considerable amounts of capital, which only creates value once it is activated. Warehouse operations are expensive and do not provide much in the way of customer value. Lost, inactive, or underused inventory is not an easily identified expense, and few carriers are able to track assets at the item level. Carriers that outsource this function to network infrastructure suppliers achieve more efficient logistics and warehouse operations and maintain lower levels of inventory that could potentially be used across multiple customers.

These are the major elements of a network. But given the number of full-time employees in the network unit of a carrier, other support functions are also necessary. Activities such as information technology (IT), human resources, and finance, which are dedicated to the network unit, need to be identified, isolated, and incorporated into any network outsourcing evaluation.

Building the Business Case

An initial assessment provides a basis for building a strong outsourcing business case. This should incorporate all "in-scope" activities such as identifying all costs (including support costs), outlining forecasts and the potential rate of return, and valuing all assets. Other functions will also logically fall into the initiative. For example, construction management, logistics, and materials procurement are all complementary to the network and would help provide a holistic, end-to-end approach.

A well-defined program management plan will ensure that the business case is developed around actual costs that can be offset by network outsourcing. It will also identify potential risks. These mostly involve process and governance issues but can also include rushing in without first devising a rigorous evaluation process for each outsourcing provider.

Also, the business case should call for internal improvements before outsourcing. This point is important because employees might initially reject the new strategy, claiming the benefits can be achieved without outsourcing. Therefore, ask managers to incorporate the proposed benefits of outsourcing into their current or next-year business plans. This exercise can generate some internal improvements not previously considered, and if the decision is made not to outsource, the internal organization will have a plan to execute to achieve the targeted results.

Finally, the plan should address the need for a formal performance evaluation process. This will alleviate confusion brought on by outsourcing suppliers that have slightly different approaches to network outsourcing and will try to steer carriers accordingly.

Who's Who Among Providers

Finding a good network provider can be a challenge. There are fewer major carriers in each regional market than there were five years ago, and the convergence of technologies across wireless and wireline segments means the market for traditional providers of network infrastructure equipment and products is much tighter.

As a result, network infrastructure original equipment manufacturers are expanding their service offerings into new areas—including becoming third-party network infrastructure providers. Indeed, major players such as Alcatel,

Ericsson, Flextronics, Lucent, Motorola, NEC, Nortel, and Siemens are developing their service businesses to support the evolving network outsourcing market. As network infrastructure providers, these companies offer distinct advantages: They can employ international best practices, cross-train the workforce, deploy newly automated tools and, most important, use their process improvements across geographic regions and customers.

However, before making a deal, carriers should rigorously evaluate which network provider is best positioned to meet their needs. Each provider will have a slightly different approach to network outsourcing and will try to steer the carrier accordingly. This can be avoided by using a multi-step evaluation process in which "go" and "no-go" decisions help ensure fairness and thoroughness (see *Figure 4*).

During this process, the carrier should ensure that the potential provider can meet the required responsibilities, can be held accountable for its performance as measured by defined service-level agreements (SLAs) or key performance indicators, and has the scale necessary to achieve the expected cost reductions. Our experience at one European telecom company demonstrates that an outsourcing provider can achieve, and in some cases exceed, aggressive SLA and KPI targets within 90 days.

Also, carriers should pay special attention to the commercial aspects of the outsourcing provisions. For example, it is critical to split the scope of work into distinct services, identify a clear scope of work and define responsibilities for each service. This avoids conflicts as to what services the provider gives for the agreed-upon price. It is also a good idea to calculate the current internal prices per service so that you can properly compare price quotes of potential providers. And if you are signing a multi-year contract, make it "modular" to provide the flexibility necessary to accommodate changes in market conditions, technology, and services.

For compensation, it is best to establish a fixed or variable scheme. This provides the flexibility that fluctuating markets require and the month-to-month predictability a carrier needs. Also, firms that establish specific KPIs that properly reflect the targets of their sales and marketing organizations have a better chance of meeting their goals. Strict forecasting requirements will ensure that providers deliver the expected services with the targeted KPIs and SLAs, while penalties for falling short should be incorporated into the agreement.

In addition, if tangible rewards can be linked to the provider's overall performance, put them into the agreement as well. All of the commercial elements should be incorporated into a master agreement with supporting service documents that can be modified as conditions for a specific service. Step-in rights and governance along with exit, liability, renewal, and key operating provisions should be addressed in the master frame agreement.

Finally, given the complexities of most outsourcing agreements, a strong legal team, supported by experienced outside legal counsel, is generally required. The commercial terms should be agreed upon during the evaluation process and nailed down in a detailed memorandum of understanding early in the process.

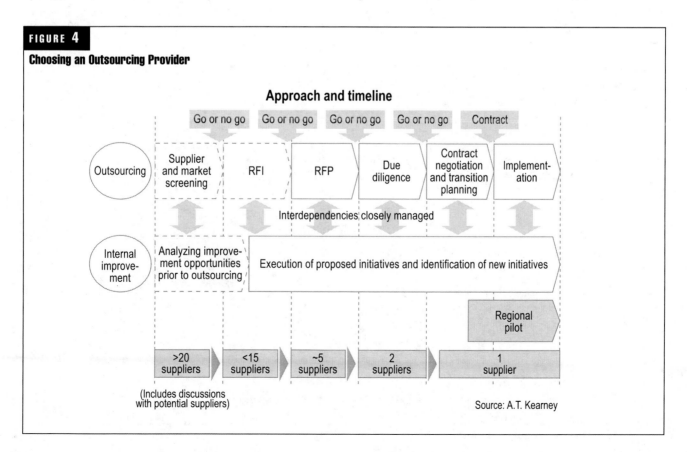

FIGURE 4

Choosing an Outsourcing Provider

Source: A.T. Kearney

Success Breeds Success in Network Outsourcing

The most successful initiatives are those based on a solid foundation of pre-outsourcing strategies. For example, we often recommend cutting costs by consolidating functions into larger regional centers before outsourcing. We also suggest automating processes to improve efficiencies and reduce redundancies, and aligning the workforce to match current and projected market conditions. Other pre-outsourcing tactics include rationalizing networks and developing a multi-skilled workforce.

Combined, these steps help drive down costs and increase returns before outsourcing. It also forces the new provider to use its own skills, competencies, and technologies to achieve the next level of savings. At one carrier, pre-outsourcing tactics reduced total head count by 29 percent. The outsourcing initiative then identified an additional 35 percent reduction in head count.

In our experience, to achieve successful outsourcing initiatives, take the following key actions:

Ensure Executive-Level Involvement
Top executives must be involved from the beginning of the network outsourcing project. The board of directors and, if appropriate, supervisory boards need to be advised and updated throughout the process. Depending on the final scope and scale, this is likely to be one of the most significant projects for a carrier, which means all business units must be informed and engaged.

Additionally, be prepared to communicate with the financial investment community and media outlets. When they learn of the project, they will want to discuss the financial and operating implications of network outsourcing with senior managers. In countries that require public disclosure, regulatory approvals, or union coordination, carriers must coordinate strategies and hammer out the details. Early and open disclosure is the best approach to communicate the process.

Clearly Define the Operating Model
The scope of activities to be outsourced and the interfaces that will be required are central elements in the network outsourcing operating model. In-scope activities should be based on processes—not technologies—that can then be aligned with specific service offerings such as DSL provisioning or assurance. A process approach will help simplify and reduce the number of required interfaces. The operating model should be designed to give the provider the flexibility to make it work and to ensure financial success. *Figure 5* depicts the key success factors in the operating model.

Design the Outsourcing Process
The right design has a major impact on the overall benefit to the carrier even if the project is stopped at one of the go or no-go decision points. The four main outsourcing process steps are as follows:

- Optimize relevant processes before outsourcing
- Define distinct phases and make go or no-go decisions at the end of each phase
- Assess the outsourcing offers against the internal improvement options
- Plan implementation on a phased basis to test the operational quality of the selected provider

Create a Life-Cycle Process
Finally, the outsourcing process has to be understood as a life cycle that requires reassessments over time based on changing conditions. Management needs to continually

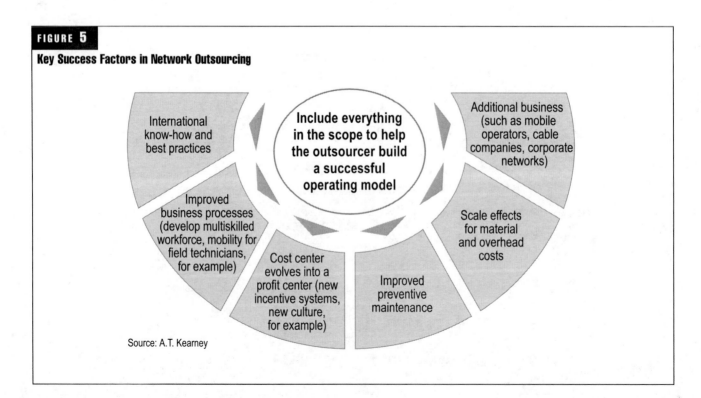

FIGURE 5

Key Success Factors in Network Outsourcing

International know-how and best practices

Additional business (such as mobile operators, cable companies, corporate networks)

Include everything in the scope to help the outsourcer build a successful operating model

Improved business processes (develop multiskilled workforce, mobility for field technicians, for example)

Scale effects for material and overhead costs

Cost center evolves into a profit center (new incentive systems, new culture, for example)

Improved preventive maintenance

Source: A.T. Kearney

monitor the outsourcing agreement, the supplier, and the demand organization to ensure it is achieving maximum benefits.

Get Smart

Fixed-line carriers are witnessing the rapid erosion of their traditional revenue streams as new competitors force prices down, wireless and fixed-voice services converge, and new services increase the substitution effect. From now on, success for carriers will depend on their ability to significantly reduce their cost structures to free up investments for future service offerings that really add value for the customer—including fiber-to-the-curb, voice over Internet protocol (VoIP), and other content-based offerings.

Ultimately, fixed-line carriers will likely follow the path of technology manufacturers. When faced with high cost structures and collapsing markets, the manufacturers were smart enough to outsource significant parts of their operations. Fixed-line carriers that adopt the seventh strategy will be well ahead of the game.

Note

1. The evaluation process and pre-implementation process requires approximately 12 months, and savings can be achieved in the first year of implementation.

Transforming Access Networks for Success and Profitability in the 21st Century

John Mellis
Chief Technology Officer
Evolved Networks

Matthew Edwards
Chief Marketing Officer
Evolved Networks

Summary

This paper reviews the hidden factors driving the rapidly increasing importance of the access network (AN) and identifies the major challenges in implementing network transformations and upgrades, which are required to deliver next-generation services such as triple play and Internet protocol television (IPTV). In particular, the critical role of high-quality network data in supporting intelligent, cost-effective network operations, ubiquitous access to networks, and profitable new services will be emphasized. Innovative methods are described that regenerate network data with high quality, and their use in a major case study is referenced. A newly developed dynamic model of AN operations is used to illustrate the dependencies of business processes on network data quality and to quantify the enormous business benefits that can be gained by creating and maintaining accurate network data.

Business Drivers in the AN

The AN—once the Cinderella of the telecom industry—is now assuming vital importance as the key enabler for the profitable delivery of next-generation services. The increasing demands for broadband, along with the mounting business pressures on incumbent service providers, mean that improved efficiency in AN operations is essential for the delivery of high-quality customer service and the maximization of profitable new-service revenues.

ANs (see *Figure 1*) are of huge scope and extent, and are by definition the service provider's greatest asset in reaching its customers. They are also the service provider's greatest challenge in providing new services and reducing operational overheads. With prices for traditional services falling rapidly and fierce competitive pressure on the profit margins for new services, operators must look again for operational efficiencies and next-generation revenue opportunities. To ensure survival, success, and profitability in the future, network operators must reduce customer churn,

increase customer satisfaction and average revenue per user, and lower costs across the entire expanse of AN operations. As a specific example of the enormous scale of the AN's challenge, the BT copper local-loop network, the United Kingdom's national network operator, comprises some 46 million lines, 4.3 million distribution points (DP), 5,600 local exchanges (central offices), and around 115 billion meters of copper wire pairs.[i]

Such copper-based local-loop networks are being rapidly augmented by optical fiber cables, which, a decade ago, were expected to render the copper local loop completely obsolete. Now, however, few people believe that the copper local loop will die soon. Its life extension has been guaranteed, courtesy of asymmetric digital subscriber line (ADSL) technology, and the next generations of generic digital subscriber line (xDSL) systems and triple-play services will utilize a pragmatic hybrid of copper, optical, and wireless network platforms.

The general market conditions that are driving network operators around the world to adopt various access transformation strategies include the following:

- Shrinking revenues because of product substitution and price erosion
- Growing competition and battles for market share
- Demands for improved customer delivery and operational efficiencies
- Regulatory pressures, e.g., to allow defined or "equivalent" access to network facilities
- Market demographics

Specifically, today's incumbent communications network providers are faced with the following three broad transformational challenges, which must be met simultaneously:

- Providing new broadband xDSL and IP services, efficiently and profitably, over a huge copper local-loop network that was "written off" more than 10 years ago

FIGURE 1

A Physical View of the Multi-Technology AN

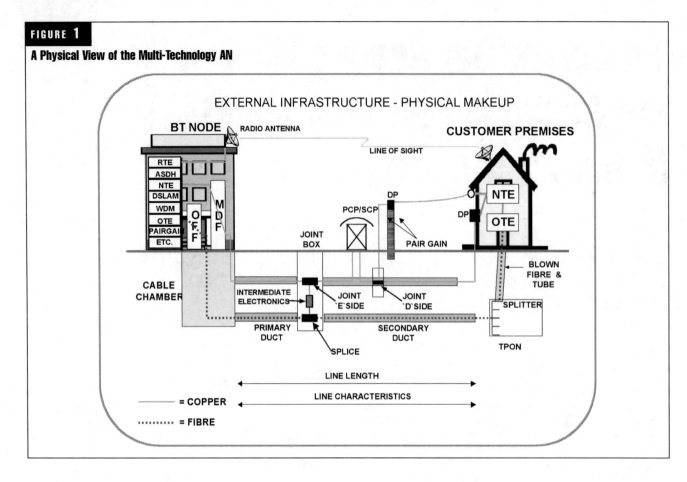

and has been subsequently neglected with respect to data recording, network maintenance, and operations support system (OSS) investment

- Managing eroding monopolies and the increasing penetration of other licensed operators (OLOs) and service providers while maintaining return on capital expenditures (CAPEX) and operational expenditures (OPEX) and moving to lean operations

- Introducing next-generation transmission technologies, including very-high-data-rate digital subscriber loop (VDSL), fiber-in-the-loop, and wireless access, to provide further improvements to customers' connection bandwidth and mobility and enable the triple-play services of voice, video, and high-speed Internet

New entrants in the market are faced with the additional challenge of competing with established networks with widespread geographical footprints. As a result, the next 10 years will see a renewed focus on AN operations, by incumbents and OLOs alike, along with a continuing intensification of AN market competition.

Figure 2 illustrates the pressures exerted by market competition in the AN, using the familiar five forces analysis. Market conditions around the world vary significantly, and this variation is causing operators to adopt different competitive strategies. For example, local-loop unbundling (LLU) is a well-established trend in some European countries—notably in France—but it has yet to gather momentum in others.

Intense competition among cable TV network operators to provide triple-play services is a feature of the North American market, and some European incumbent operators, as in Germany and the Benelux nations, also face similar pressures. In general, the response in these markets has been to develop the following two-stage strategy:

- Win or retain market share by accelerating the deployment of fiber-to-the-cabinet networks, which enable the fast delivery of wider bandwidth and more content-rich services (including high-definition TV) to the majority of customers through xDSL

- Then, utilize the fiber-to-the-cabinet deployments as a platform to enable the cost-effective rollout of fiber-to-the-premises for selected customers or demographic market segments

In East Asia, regulatory pressure and customer demand has led to a more aggressive strategy, resulting in a far higher penetration of optical fiber in the AN, than is typical in the rest of the world. Around 1.2 million customer premises were already connected by fiber in Japan by early 2004, increasing at a rate of some 100,000 new customer connections per month.[ii] Demand for *true broadband* and wireless services combined with regulatory pressure may also accelerate AN transformation in the rest of the world. In the United Kingdom, in an as-yet-unique innovation, regulatory pressure has resulted in the creation of BT *Openreach* in January 2006. Openreach is a new, independent division focused exclusively on managing the national AN and on establishing *equivalence* (i.e., equal access to network facili-

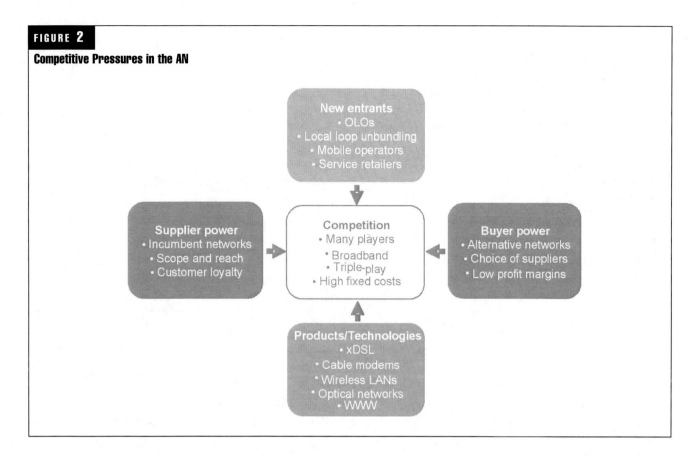

FIGURE 2

Competitive Pressures in the AN

ties and infrastructure, across the industry, including BT's Retail division, value-added resellers, and OLO). This development of a new U.K. market model and of BT's 21st Century Network (21CN) transformation program will be closely observed over the next few years by commentators, regulators, and service providers worldwide.

A Common Denominator: Network Data Quality

In implementing any of these strategies to modernize and transform the dynamics of the AN, one common practical problem is immediately encountered—the inventory data that describes the available capacity, location, and configuration of existing network facilities (e.g., ducts, cables, and cross-connection nodes) is frequently out-of-date or otherwise in error. Unless this critical information is updated and contained in new, accessible data models, hopes of efficient network transformation, equal access, and *lean operations* in the local loop are badly compromised and the prospect of continuing high levels of human network intervention, manually intensive service provisions, and operational overheads is faced.

Improving the quality of AN data has an immediate impact on improving the efficiency of existing operational processes and provides a platform for further future operational streamlining through intelligent automation. The key steps required are as follows:

- Generate and maintain up-to-date physical cable network data from sources, including network maps, logical inventory records, and customer service systems,

and maintain high levels of accuracy in network data through effective in-life processes

- Use both logical and physical data to provide optimized broadband connections across the AN, in response to service connection requests (i.e., broadband network service provisioning)

- Design and plan wholly new networks, and extensions to existing networks, to provide broadband connections to new customers with minimum design time and CAPEX

These three operational processes form a *virtuous circle*. For example, improvements in data accuracy lead to improved speed and quality in the provisioning processes and flow through to the automatic design of network solutions, which can be fully documented and then accurately recorded in network inventory systems.

The data-improvement process is complicated by the fact that, traditionally, two types of data have been held, in separate systems, to fully describe the network. Physical data describing the location, type, age, and construction of equipment is required to support network repair and maintenance activities, and this map-based data is often used to direct the work of field technicians. Logical data describes how the network elements are interconnected from exchange/central office to customers, and how this information is often stored in customer-relationship management (CRM) systems rather than inventory databases. It has been shown that the typical usage of logical and physical data in the AN, and the business impact of this separation of network data into different database systems.[iii] The conclusion is that automation of access operations requires a combined view of both logical and physical data in the AN.

BT's PIPeR Leading the Way

PIPeR is a $50 million program initiated by BT to address the physical inventory, planning, and e-records needs of the future, as defined in the company's Network Engineering Journey (NEJ). The overall aim of the NEJ is to support BT's 21CN program with world-class inventory management systems, and to transform the end-to-end planning and network recording process so the speed of network provisioning and repair can be radically improved.

The NEJ is well under way at BT and involves dozens of major commercial supply companies, including Evolved Networks. The PIPeR systems will hold all external line records to provide a single view of all BT's physical inventory, including external physical cable and duct routes, as well as the Ordnance Survey (i.e., U.K. national) map base. The system will support sophisticated and highly automated network planning tools, and enable easy, and securely controlled, access to critical network infrastructure data from anywhere in the United Kingdom.

The PIPeR program is one of the most ambitious and visionary programs of its kind in the world[iv], and it will do the following:

- Change the way thousands of people do their job in 64 U.K. locations
- Transform the network data into an intelligent format for 60 m network assets, including some 40 million copper pairs and 400,000 optical fibers
- Develop new data-generation methods and tools
- Deal with the backlog of data awaiting recording to the database systems
- Implement major new enterprise systems with specific functionality for the e-records and network planning communities
- Implement new business processes
- Manage a major program with 10 supplier companies, including Evolved Networks, 300 people onshore, and 3,000 people offshore (in India and the United States)
- Fully integrate with other programs in BT's NEJ
- Deliver the significant business and cross-program benefits

These benefits must be delivered while, of course, maintaining business-as-usual quality of network operations and service to customers.

Challenges and Solutions

The principal requirement for the successful interworking of PIPeR and the other NEJ systems is that the network records have to be as complete and as accurate as possible, and that network assets have to be interconnected and represented consistently with the differing data models of the enterprise systems in the NEJ suite. Before the initiation of the NEJ, network information was held in BT's systems in *silos* according to assets types, leading to missing, inaccurate, and inconsistent data sets, as well as to significant backlogs of network data awaiting validation and loading into the central inventory recording systems.

To meet these requirements, a new approach to data validation and migration has been conceived, specifically to do the following:

- Validate data by comparing records across several source systems
- Identify and detail instances of missing and inconsistent data
- Produce complete connected data sets using information from various BT systems
- Deduce the best possible solution to fill missing data based by intelligent application of network planning rules
- Save time and money compared to traditional labor-intensive methods of data capture and migration

The result is an automated process to regenerate and migrate BT's network data *en masse* to new inventory systems. The key features of the process are shown in *Figure 3*. First, data is copied in batches from the legacy information systems, most important from the customer service system (CSS) and the geographical information system, which holds the location of the external plant and equipment against a background of Ordnance Survey maps. Since, in this project, the source geographical maps are held as raster bit-map images, the maps and duct-schematics are digitized and converted from the raster images to digital form. This data comes, as XML files, to the critical next step of data validation. Here, the data sets are automatically cross-correlated, exceptions and inconsistencies are resolved according to agreed rules, and data gaps are filled by inference from the most reliable data sets. They are usually derived from the customer service systems, which are current and up to date to support accurate billing for payment.

As part of this step, entire layers of the network data are generated—for example, the connectivity of copper wire pairs through the cable network is deduced from the other data sets—and this automated deduction and inference engine dramatically reduces the time required to generate complete validated and labeled data sets, in formats ready for upload to the new inventory systems.

Process Innovations and Realized Benefits

Several features of this automated data transformation process constitute radical improvements on the traditional manually intensive methods used until now. The process uses BT's most reliable and up-to-date customer data records, so that the generated network data is also up-to-date, and automatically captures the backlog of data previously awaiting recording in the physical inventory system. Gaps in the data record are filled using inferences that apply network planning rules to *reverse engineer* the data describing missing network elements. Data validation and migration is greatly accelerated. Data describing a complete central office/exchange area network is generated on a time scale ranging from a few minutes to a few hours, compared to the many days required by conventional labor-intensive methods.

Specific demonstrated benefits of the process include the following:

- Automatic validation of network data—including connectivity and containership of equipment compo-

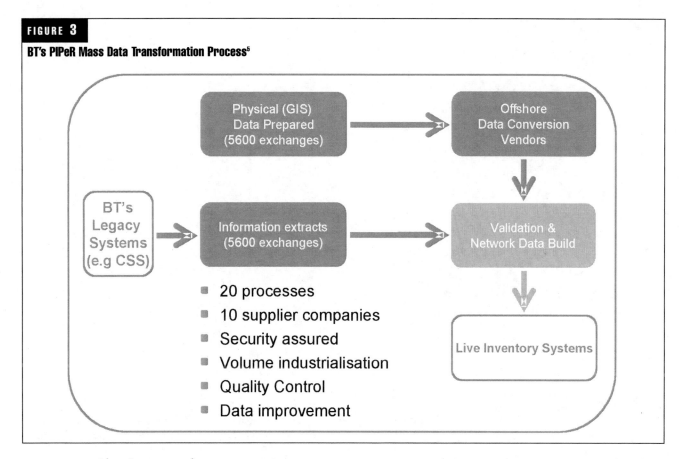

FIGURE 3

BT's PIPeR Mass Data Transformation Process[5]

nents—with reference to the most current customer data

- Consistent application of quality assurance (QA) standards and the production of QA reports covering 100 percent of data
- Improved project management information and simplified process control
- Increased data security

The net results of this innovative approach are improved data quality, radically faster data migration to new systems, and a reduction of overall costs in the migration process through the elimination of much manual effort in map digitization, data checking and validation, re-working, and data loading. Production rates of up to 30 exchange areas per day have been demonstrated in the BT process. The achievement of such speeds means that it will become feasible to validate and migrate the data of even the largest incumbent telecom networks in projects of less than 12 months total duration.

Business Impacts of Data Quality

In evaluating the business benefits of a major systems and data upgrade like BT's PIPeR program, the following key questions must be addressed:

- What metrics can we use to define the quality of network data?
- How does investment in improving the quality of network data pay back in business benefits?

These questions have been addressed by using a dynamic model of AN operations, mostly developed to quantify the

impact of inventory data quality on the efficiency of the three principal network operational processes: service provisioning, network repair, and proactive network maintenance. *Figure 4* illustrates the structure of the model. Data quality is defined as the percentage of AN connections that are correctly recorded, in both the physical and logical inventory databases, and end-to-end (i.e., from the line-card in the local switch to the customer premises). The three job flows are defined in terms of customer orders (for service provisions) or network job volume (for repair and maintenance). In typical AN operations, completion of many of these jobs requires an engineering visit to the network and some intervention in the network, for example, to repair or reconfigure it appropriately. The success rate of provisioning, repair, and maintenance visits depends sensitively on the accuracy of the network inventory data. For example, the field technician may arrive at an AN node only to discover that the bearer (copper pair or optical fiber) which he has been assigned to in order to connect to the customer is already in use by another customer. In this case, the field visit may be wasted and re-work will be incurred both in the back office (in network planning) and in making another *truck roll*. In another example, the repair technician assigned to repair a broken cable joint may experience a wasted journey if the network node location is different from that recorded in the network records. Therefore, the quality of the network data has a direct impact on the flow-through success rates of the three work streams.

Data quality is also linked to the efficiency of network operations by the network records generation process employed by the workforce, i.e., do they accurately record the changes to the network configuration they implement? If so, and the

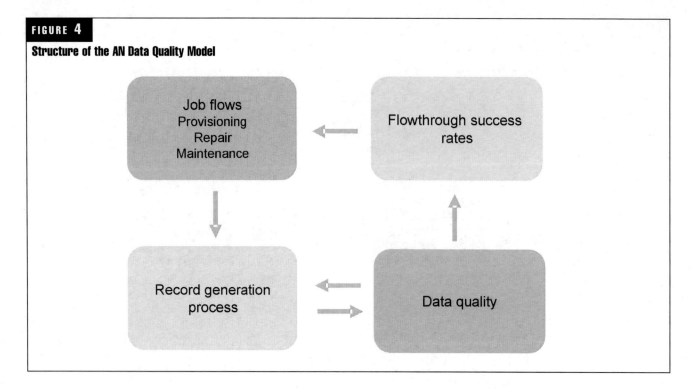

FIGURE 4

Structure of the AN Data Quality Model

record generation process is of high quality, then data quality will tend to improve over time. If not, data quality will quickly degrade.

The model was used to quantify the effect of improving data quality on the business operations of a hypothetical AN operator, and the results are shown in *Figure 5*. In this case,

the assumed size of the AN is 10 m customer connections, with 200,000 provisioning, repair and maintenance visits to the network occurring every month. Each network visit is assumed to incur an operational cost of approximately $225.

The modeled results show how important the impact of network data quality can be. In the base case scenario, an

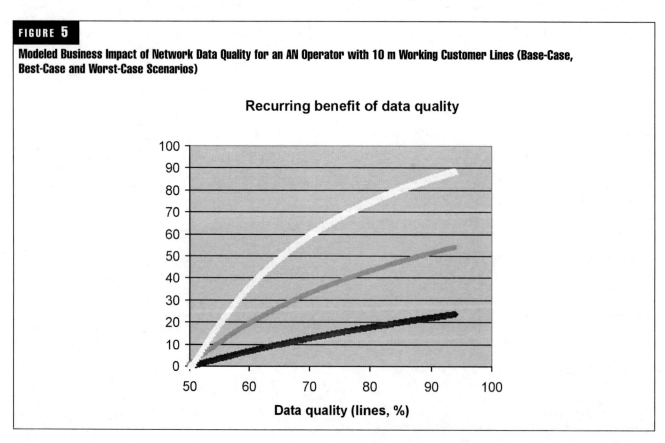

FIGURE 5

Modeled Business Impact of Network Data Quality for an AN Operator with 10 m Working Customer Lines (Base-Case, Best-Case and Worst-Case Scenarios)

improvement in the network data quality metric from a starting point of 50 percent to above 90 percent results in an annualized benefit to the business of the order of $60 million. This benefit consists of the avoidance of re-work and wasted network visits, accelerated customer revenues through faster fulfillment of new customer orders, and reduced customer churn through improved customer satisfaction. Best- and worst-case scenarios have also been modeled, where the sensitivity of job success to data quality is higher or lower, respectively. In these scenarios, the range of realized benefits for this service provider ranges between $30 million and $100 million per year. Such potentially large financial benefits present a compelling argument for operator investments in new OSSs and network data regeneration programs.

Conclusions

The AN is resurgent and is the focus of intensifying pressures of worldwide competition and capital investments in regeneration. A variety of upgrade and expansion strategies are being urgently pursued by network operators in the race to provide profitable next-generation broadband services. New business models are emerging, driven by regulatory and market pressures, as exemplified by the creation of BT Openreach in the United Kingdom.

An important practical and specific obstacle to AN regeneration has been emphasized: a historical under-investment in systems and processes has led to a widespread problem of network data inaccuracy, which impedes progress to efficient network operations. New approaches have been developed to solve this problem cost-effectively. These have been pioneered by the PIPeR program in BT, which is now set to deliver major business benefits. The nature and scale of the business impact of the data quality issue have been explored and illustrated using a dynamic model of AN operations, which predicts the financial benefits achievable from network data quality improvement.

Great improvements in AN infrastructures and network operations have been made in the past few years, but much remains to be done. This analysis shows that operators can achieve cost-efficiency benefits of tens or hundreds of millions of dollars per year through fundamental improvements in AN data quality, operational process streamlining, and the acceleration and retention of customer revenues through faster service fulfillment. The next few years will be a time of unprecedented transformation in the AN, and the most effective AN "transformers" will have equipped themselves with a key advantage in the forthcoming battles for 21st-century market

Notes

1. *BT's Revolutionary Approach to Data Migration in Access Networks*, Colin Noakes (BT) and John Mellis (Evolved Networks), Presentation at Tele-Management World, Nice (May 2005).

2. *Trend of Optical Access Networks – Driving Factors and Technologies*, Yukio Nakano (Hitachi Communications Technologies), Ninth European Conference on Networks and Optimal Communications (NOC 2004), Eindhoven, Netherlands (2004).

3. *Revolution or Evolution? Transforming access networks with new operations software systems*, John Mellis and Mathew Edwards (Evolved Networks), to be published in IEC report "Evolving the Access Network" (2006).

4. *BT's Revolutionary Approach to Data Migration in Access Networks*, Colin Noakes (BT) and John Mellis (Evolved Networks), Presentation at Tele-Management World, Nice (May 2005).

Czech DWDM National Research Network: A Case Study

Václav Novák

CESNET2 NREN Research and Deployment Engineer
CESNET, Association of Legal Entities

Karel Slavicek

CESNET 2 NREN Research and Deployment Engineer
Masaryk University, Czech Republic

Rita Puzmanova

Independent Networking Specialist

Introduction

The driving force behind optical communications and optical networking research has been a continuous demand to improve transmission capacity, configuration capabilities, and flexibility of networks based on optical fibers while reducing operational costs.

Meeting those goals has required a continuing series of innovations as fiber optics moved from single-channel use of a fiber in synchronous optical networks/synchronous digital hierarchy (SONET/SDH) to wavelength division multiplexing (WDM) on point-to-point links to dense WDM (DWDM) ring networks with multiple add/drop points. Technologies such as reconfigurable optical add/drop multiplexers (ROADMs), photonic cross-connects (PXCs), and wavelength tunable lasers have all been developed to meet these needs.

The most recent optical technologies (r)evolution and advances in optical networking may be best seen in modern research networks that serve as a test bed for new technologies as well as a reliable tool for academic communication and research performance. The advanced research network backbones have been based on the optical technologies for some time. But, similar to commercial optical transport networks, they have gone a long way from old SONET/SDH to scalable WDM. Currently, most of the research networks run over dark fiber that is very cost-efficient and allows for full management and dynamic administration of the network.

DWDM technology allows large capacity not only for bandwidth-intensive research applications (medical, space, or quantum physics), but also for running separate networks independently on the same fiber (production and experimental). DWDM has recently evolved one step further with the development of ROADM for dynamic reconfiguration of wavelength paths (lambdas) and will shortly allow for multipath ROADMs. Provision of lambda services is important for bandwidth-intensive applications running for a limited time end-to-end between the same parties and requiring high reliability and sustainable quality of service (QoS). These applications are not satisfied with basic packet-switched services provided by IP networks and would benefit from end-to-end optical circuits provided directly by the underlying optical infrastructure.

Thus, an old-fashioned circuit switching suddenly revitalizes as an additional requirement for pure IP networks. Due to the optical transport nature, a combination of the apparently mutually exclusive principles is possible. The modern optical networks will become hybrid networks, providing both IP packet services and lambda services, based on an IP–routed network over an optical switching infrastructure.

The following paper introduces the recent optical development of one of the top national research and education networks in the world, Czech CESNET2, and related advances in the optical communication, developed, implemented, and tested in the network. The objective of CESNET2 continues to be the same—to satisfy users in the academic and research communities. As the demands on this special group of customers grow exponentially—not only in terms of capacity, availability, and reliability, but also in terms of service types—the network has to keep developing continuously (or rather incrementally) while at the same time testing new technology based on research results.

Brief Network History

The national multigigabit optical CESNET2 network, which serves research and educational purposes (and is the next generation of the Czech Republic's National Research and Education Network), is developed and administered by CESNET, an association of Universities and the Academy of Sciences of the Czech Republic. The nonprofit CESNET association, which celebrates its 10th anniversary this year, is currently financed mainly from the resources of the governmental Committee for Research and Education and the resources of the members of the association. The objective of the association is to perform research and development in the area of information and communication technologies.

Although CESNET2 does not have a long history due to the communist regime strictly limiting any international networking (both social and technical), since the so-called velvet revolution, it has evolved very quickly to become one of the top European national networks, not in terms of scale (the country has 10 million inhabitants), but also capacity, advanced technologies, and services. It may be useful to put the past 10 years of technological network development in perspective: the Czech Republic, represented by CESNET, was the only country from the former communist bloc that participated in the TEN–34 (Trans-European Network) research project in 1996 and also in its follow-up TEN–155 project, which meant a significant improvement in the backbone network capacity, which is based on asynchronous transfer mode (ATM).

The fast development of the optical CESNET network led to the first 2.5 Gbps backbone line using the packet over SONET (PoS) technology put into operation in 2000, when transport lines of such capacity dedicated entirely to Internet protocol (IP) traffic were rare even in the most developed countries. At the same time, it became clear that it would not be possible to implement a gigabit network based on leased transport services within the available budget. A more economical method of achieving the goals was sought: the solution was to rent dark fibers (the network has run entirely on dark fibers for the past two years). The customer-empowered fiber (CEF) networks approach is nowadays widely used, but at the beginning of the 21st century, CESNET was one of its pioneers worldwide.

In 2001, a trans-European network called GÉANT (TEN–155's successor) was launched, and the Czech Republic became part of its core with three 2.5/10 Gbps international connections. Related to this development, a renamed CESNET2 network was officially launched. In the comparative study of European scientific and research networks carried out by the TERENA association, CESNET2 was evaluated as the third largest network in Europe at that time.

Once WDM technology, which allowed the transfer of several independent signals over a single fiber, was developed, the evolution of the network continued. The new generic network structure was designed in 2002 based on the backbone consisting of DWDM rings. As the CEF principle imposed limitations on optical line transfer range, CESNET developed methods called nothing in-line (NIL), which are based on utilization of various types of amplifiers and their combinations. The original NIL-supporting modular amplifier, tested in CESNET2 on distances exceeding 300 km, is called Czech line amplifier (CLA) and is considerably cheaper than commercial amplifiers.

The first DWDM line was deployed in CESNET2 in 2004, providing several independent lines with transfer speeds of 10/1 Gbps. During 2005, the line was upgraded to the 10 Gbps DWDM ring (Prague[1]—Brno—Olomouc—Hradec Králové—Prague) based on 32-channel ROADM technology.

CESNET2 is connected to pan-European GÉANT2 network by a 10 Gbps optical line. The association has participated in numerous other European Union networking research projects such as DataGrid, EGEE, 6NET, SCAMPI, LOBSTER, and SEEFIRE. CESNET has been an international partner of Internet2 since 1999 and is also a founding member of the Global Lambda Integrated Facility (GLIF) initiative, in which lightpaths (based on individual wavelengths) may be provided end-to-end on demand.

Parallel to CESNET2, the association has developed CzechLight, a purely experimental optical test bed connected to the international GLIF infrastructure. This test bed is designed for experimenting with optical transfer technologies and on-demand circuit provision and is utilized for applications with extreme bandwidth demands.

CESNET2 is technologically comparable to NRENs in developed countries. External connectivity capacity increased a hundredfold from 300 Mbps to 35.3 Gbps between 2001 and 2005. Contrary to other new EU member states' NRENs, CESNET2 is not just a sink of external data, as it exports 1.7 times more data than enters the network.

Optical (R)evolution

The migration of CESNET2 to dark fiber lines was completed at the beginning of 2004. The optical lines were designed as point-to-point circuits between points of presence (PoPs). Optical fibers over distances of about 120 km were deployed using the erbium-doped fiber amplifier (EDFA). The technical solution of optical paths using primarily EDFA amplifiers with single-channel transport (a so-called gray solution) did not allow for transmission capacity increase on a single optical fiber. It also represented a limiting factor for end-to-end service provisioning at the level of optical transmission channels. For deployment of optical lines, CESNET2 is in transition to the improved solution: NIL with optical EDFA amplifiers only at the paths' ends. This variant gradually replaces the protocol-dependent Layer 2 (L2) switches used as path repeaters.

There are several technologies deployed for optical lines. For shorter gigabit lines (approximately up to 120 km), there is no need for regeneration or amplification. Therefore, pluggable optics transceivers at router and switch interfaces coarse WDM (CWDM)–1550 and DWDM gigabit interface converter (GBIC) are used with channel spacing of 100 GHz, according to the International Telecommunication Union Telecommunication Standardization Sector (ITU–T) (see *Figure 1*).

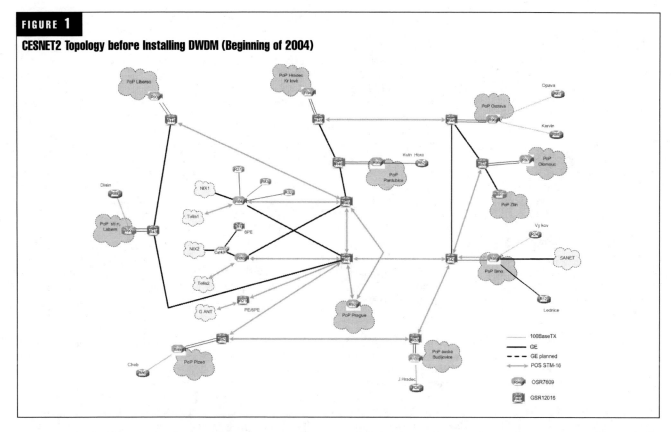

FIGURE 1

CESNET2 Topology before Installing DWDM (Beginning of 2004)

For longer optical lines with higher attenuation (above 32 dB), optical EDFA amplifiers (pre-amplifiers and booster amplifiers) from a well-established vendor are deployed. However, the amplifiers in use show several problematic characteristics such as necessary manual activation after a power failure. Amplifiers currently available on the market have already solved these problems, but the upgrade or modification of the previous generation would be costly.

In 2005, CESNET2 successfully tested self-developed optical amplifier CLA PB01 at the 1 Gbps path between Prague and Hradec Králové (150 km, 35.7 dB, G.652). The one-side amplification (OSA) method was used with a dual amplifier (booster and preamplifier in one) in the Prague node only. The testing showed that the originally developed amplifier is more suitable and cost-efficient than the currently used commercial EDFA amplifiers.

In February 2006, an international 200 km path Brno—Bratislava (interconnecting CESNET2–Sanet, Slovak NRENs) was upgraded to offer 10 Gbps speed. The deployed path uses four-channel multiplexers/demultiplexers at minimum (typically an eight-channel solution) to offer the 10 Gbps optical channel transport capacity. Chromatic dispersion compensation is performed by Bragg grating, which is more cost-efficient in comparison to the classical compensators (dispersion compensating units [DCUs]). The client connections (routers) will be deployed with the DWDM Xenpak or GBIC of the relevant wavelength. Other paths are planned (e.g., Brno—Vienna, Austria).Optical transport layer with core ring topology was created by the extension of optical transport DWDM technology to further nodes and its integration with existing path Prague—Brno. The deployed 32-channel ROADM

technology (with 10 Gbps individual channel capacity) offers the software-managed optical channel provisioning on demand. This capability is available between any nodes of the whole system.

In the next development stage, the implemented optical transport DWDM network will allow for integration with the existing IP network and for transition to the hybrid IP/optical backbone–based network controlled by protocols such as generalized multiprotocol label switching (GMPLS), including the implementation of optical switches for dynamic optical transmission route switching.

From OADM to ROADM

A key element of the first-generation WDM networks was the optical add/drop multiplexer (OADM), allowing the evolution from point-to-point to ring networks. New WDM networks are based on OADM improvement into a form of ROADM. ROADM is an optical network element that supports a remote (re)configuration. The software-reconfigurable ROADM enables the flexible establishment of optical transmission channels over a DWDM network as well as their branching off or insertion at required network nodes, thus providing bandwidth on demand, facilitating increasingly frequent unforeseen data transfers and the related bandwidth requests.

In cases where termination is necessary, the ROADM hands off the optical wavelength, keeping it in the optical domain without any prior electrical conversion to the native DWDM interface of the router, where the electrical conversion is used only for IP processing. Pure optical transmission is inherently more tolerant to bit-rate variations where moves to higher rates and new protocols may still be required in the future, and hence more robust because photonic processing is intrinsically insensitive to protocol changes, unlike typical electrical processing elements. When the filling ratio of wavelength is high enough, ROADMs allow transit at the lambda level, which results in transit traffic bypassing the IP/MPLS routers, making it possible to decrease the router size (thereby reducing capital expenses [CAPEX]).

ROADMs facilitate network design, commissioning, setup, and provisioning, both at first installation and at each upgrade. ROADMs eliminate optical-electrical-optical (OEO) conversion and provide greater reliability and service flexibility. The transmission systems that can be provisioned automatically provide flexible adjustment of bandwidth and restoration and self-healing at the network level. Additionally, the ROADM has enabled mesh networking that combines electronic grooming with transparent wavelength management in national scale networks. The combination of new integrated switching technologies, feature-rich automation, and management tools provide efficient network monitoring and troubleshooting.

As a result, WDM becomes an easily manageable convergence transport layer. The DWDM system itself is designed and constructed to provide up to 32 channels with capacity up to 10 Gbps in C-band with the 100 GHz spacing as well as enable transparent transmission of an independent "colored signal" (i.e., a signal created outside the DWDM system). This system is ready for the 50 GHz spacing to support 64 channels and C-band + L-band transmission. It is qualified for 40 Gbps as well.

The DWDM technology and implementation of ROADM represents a significant shift in the development of the optical network toward provision of bandwidth on-demand and creation of logically independent networks on common optical fiber infrastructure. The added benefit is the capability to satisfy demanding and potentially aggressive applications with specific requests on dynamic response of the network (e.g., low latency) and bandwidth—such as grids—without affecting services provided to traditional applications.

GMPLS

Evolution of the photonic layer is the first step in the overall maturation of the carrier network. In general, photonic remote reconfigurability opens up the possibility of highly automated network operations. Today, networks are managed at their individual layers of hierarchy. A photonic network is provisioned, monitored, and protected without consideration of the subtended IP/switching network.

The ideal, long-term vision behind GMPLS is to operate the composite network in a more seamless and holistic way. GMPLS provides mechanisms that support operations between the traditional infrastructures. In GMPLS, a customer may in the future request a service and then have the back office provision and commission that service across IP routers, Ethernet switches, and optical terminals on an end-to-end basis.

One of the critical integration components of the IP/DWDM strategy is that of control integration between the control planes of the two networks, IP and DWDM. Manual provisioning methods at the transport layer can incur high operational expenses (OPEX) and slow service activation that spans weeks. The control plane intelligence offered by IP/MPLS to the transport networks within service provider networks may be satisfied by GMPLS. GMPLS mirrors the principles of the MPLS mechanism widely deployed in packet-switched networks onto an optical network environment.

GMPLS in an optical control plane will allow for seamless cooperation of all components of an optical network. A flow that potentially starts on an IP network, is transported by the optical network, and then switched through a specific wavelength on a specific physical fiber by the intermediate optical nodes that are GMPLS–capable can be controlled by the overall intelligence of the network. It provides a path for optical elements within the transport network today to become peers of the router elements in the IP network, and it also provides the capability to auto-provision wavelengths directed by the IP control plane.

The benefits to the service provider are significant savings in OPEX across the network through the capability to enable fault correlation between the networks in real time while improving the overall speed to service from end to end. For operators, GMPLS will offer value in the control of the various layers of the transport network. GMPLS will provide unified network auto-discovery, network auto-inventory functions, provisioning through hierarchical layers of the network, and protection services.

Hybrid Networks

A hybrid network combines a classic IP–routed network with an intelligent optical switching over the same optical infrastructure. The network may therefore provide either packet IP service (connectionless) or lambda services (connection-oriented), a connectivity at a level of optical fiber wavelengths.

The rationale for this combination lies in the fact that the prevailing principle of best-effort transport service provided by IP–routed networks is not sufficient for all. Many users require a specific circuit through the network with a fixed set of characteristics, such as high capacity and low latency. Thanks to the optical transport networks, this requirement (however contradictory it may seem for IP packet networks at first glance) may be fulfilled: the user may request an end-to-end optical circuit (lightpath). These lightpaths are then switched directly at the optical layer without incurring the delay in the intermediate nodes.

Hybrid networks may transfer an optical signal over analog optical systems and utilize the combination of circuit and packet switching. General traffic is processed in the packet mode (in the routed network), but for selected traffic (especially for those transferring a large amount of data), independent circuits are established. The applications requiring the special treatment belong to the bandwidth-intensive, aggressive ones (including grid) where communicating partners are mostly the same and require a guaranteed end-to-end QoS. A separate circuit established for a limited time may prove more effective for such application traffic, compared to adding this high-volume data to the regular network traffic and overloading the interconnecting nodes.

CESNET2 Optical Network

For DWDM deployment, CESNET put up a tender for technology and implementation providers. The main requirements for the DWDM technology were as follows:

- Support of at least 32 traffic channels with 100 GHz spacing, according to the ITU–T G.694.1 recommendation
- Possibility to connect "colored" input signals (i.e., signals with a wavelength according to the ITU?T G.694.1 recommendation) without using a transponder
- Guaranteed bit-error rate (BER) better then $10^{?15}$ for all channels
- Automatic laser shutdown by optical fiber cutoff, according to the ITU–T G.694 recommendation
- Add/drop or rerouting of any optical channel in all terminal nodes
- Add/remove or rerouting of optical channels not causing traffic outage
- Automatic power control of optical amplifiers, including automatic reaction to optical channel add/remove and slow changes in optical parameters (e.g., fiber aging)
- Multi-rate transponder supporting networking protocols used or planed in CESNET2
- Suitable management system for full control over nodes and network as well as out-of-band management

The public tender winner was Cisco Systems with a solution based on DWDM system ONS 15454 MSTP[2]. The DWDM network design and installation were performed by Intercoms Systems, one of Cisco Systems' certified partners in the Czech Republic. ONS 15454 MSTP boasts a ROADM feature that allows for software configuration of zero up to 32 channels of pass-through or add/drop in every ROADM node. The schema of ROADM utilization is depicted in *Figure 2*.

ROADM nodes that support software-managed optical path add/drop are based at nodes in Prague, Brno, Olomouc, and Hradec Králové. ROADM itself consists of an optical switch based on planar lightwave circuit (PLC) technology.

The amplifiers and dispersion compensators are located in the intermediate nodes. They are needed for ensuring the proper function of DWDM system, required number of transmission channels (32) and guarantee of optical channel quality (BER value). The amplifiers are also based on the ONS 15454 platform.

The following are the two options for end-system (user) connection to DWDM:

- Using transponders that convert the standard "gray" 1310 or 1550 nm signal into a "colored" DWDM channel according to the ITU–T (OEO conversion) and perform reshaping retiming regeneration/amplification (3R) functionality. Transponders are an integral part of DWDM systems and are software-tunable to more wavelengths (across all 32 channels in C-band with the latest hardware versions).

- Transmitting directly to the "colored" DWDM channel. End equipment must transmit at a wavelength specified in the ITU–T G.694.1 recommendation (typically pluggable optics such as Xenpak and SFP). This

---

FIGURE 2

ONS 15454 MSTP ROADM Node Configuration

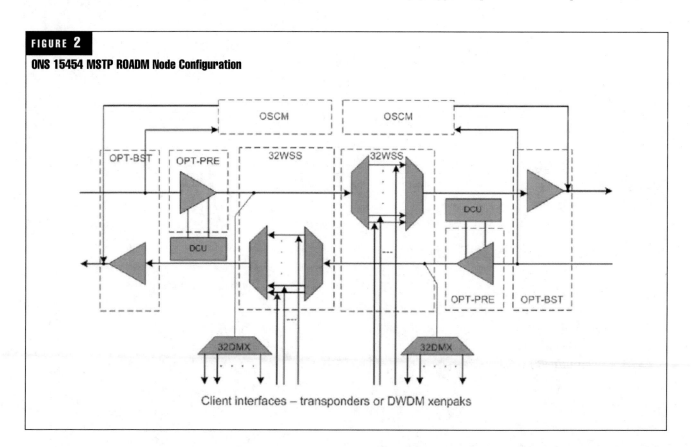

Client interfaces – transponders or DWDM xenpaks

option is cheaper in comparison to the transponder but has a number of limitations (mostly caused by the absence of FEC functionality) and can be used for shorter distances only.

Topology

The introduction of DWDM into the CESNET2 backbone was performed in two stages. The first stage occurred in 2004 when the line from Prague to Brno was deployed. Its detailed topology is shown in *Figure 3*.

The optical line from Prague to Brno is the most complicated one in the CESNET2 backbone network. It consists of three types of fiber optics. Local loops in Prague and Brno are legacy single-mode fibers as specified by ITU–T G.652. For the rest,

fiber-optics cable Draka LT072 (66 SM+ 3TW+/3TW?) is used, composed of fiber-optic lines with either positive or negative chromatic dispersions. Sections with positive and negative dispersions are combined in an effort to achieve a line with minimal chromatic dispersion. Because the provider of this dark-fiber line was unable to provide detailed documentation with complete parameters of the whole line, it was necessary to measure these optic lines, including chromatic dispersion. Measurement results are shown in *Table 1*.

Based on the measurement, the DWDM network plan was developed. It included two terminal nodes (Prague and Brno) and three optical line amplifier (OLA) nodes on the path.

On terminal nodes, multiplexers and demultiplexers upgradeable to ROADM were used. The line was compen-

---

FIGURE 3

DWDM Network Topology: Phase 1

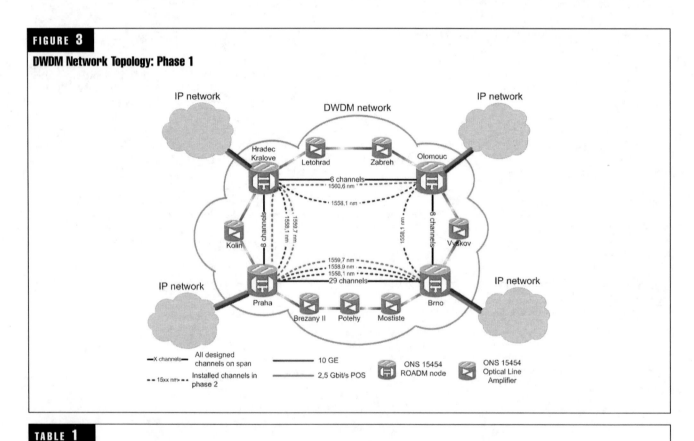

TABLE 1

Optical Parameters of Fibers in Prague–Brno Line

Total Distance Prague–Brno: 299 km

FO	G.652	G.655	G.655	G.655	G.655	G.652
Attenuation dB at 1550 nm:	-6.50	-10.70	-15.20	-21.82	-20.73	-1.50
	-6.50	-11.00	-14.92	-22.10	-20.56	-1.50
Avarage CD "C" band ps/km	504		-10	12	230	
Distance (km)	23.79	39.67	57.14	78.95	94.95	4.45

---

sated via dispersion compensation fiber to near zero chromatic dispersion. In the first step, only two channels were used: one 10 gigabit Ethernet (GE) and one STM–16 POS. In the second stage, which took place in 2005, several more channels were added and terminal nodes in both Prague and Brno were upgraded to full ROADM. Next, ROADM nodes Hradec Králové and Olomouc were added, resulting into a DWDM network of a ring topology with optical path protection capabilities. The final topology of this DWDM deployment stage is depicted in *Figure 4*.

All new fiber-optic lines used in the second stage were legacy single-mode fibers as specified by ITU–T G.652, and all of them were measured. Suspension cables hung on trolley lines or high-voltage lines are used on a portion of these new lines. Because the lines may be influenced by mechanical stress, the polarization mode dispersion (PMD) was measured on most fiber-optic sections. The placement of OLA nodes resulted from these measurements.

The existing DWDM network can be used (without changes in current equipment) on 32 channels. However, only a few channels are used today. The whole network is calculated not to exceed the required BER (10^{-15}). Though transponders used in the network provide a rather high dispersion tolerance, the chromatic dispersion of all lines is compensated. The resulting chromatic dispersion is near zero. The main benefit of this approach is an easy addition or reconfiguration of any number of channels up to the size of multiplexer/demultiplexer matrix, easier scaling of DWDM network on new nodes, and easier utilization of "colored" signals. In all these cases, we do not need to recalculate and check the chromatic dispersion of the whole system.

In 2006, the expansion to new locations is planned. New spans Prague–Pilsen and Olomouc–Ostrava will be interconnected with the main DWDM ring at the optical level to enable optical channel provisioning between any CESNET2 optical PoPs. The technical solution for these interconnections has to be determined. Optical switches or three-way ROADM (based on wavelength cross-connects [WXCs]) may be used. The planned DWDM span Olomouc–Ostrava is important for the CESNET2 DWDM network extension to the Polish town of Cieszyn. There is a Czech CESNET2 and Poland Pionier DWDM system interconnection planned as a part of the cross-border fiber (CBF) project that builds cross-border optical NREN interconnections. The resulting topology is depicted in *Figure 5*.

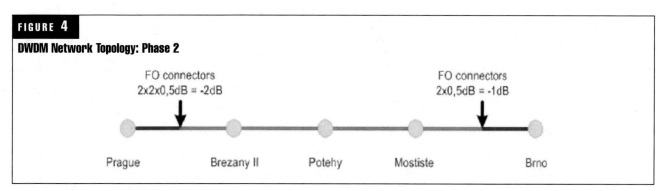

FIGURE 4

DWDM Network Topology: Phase 2

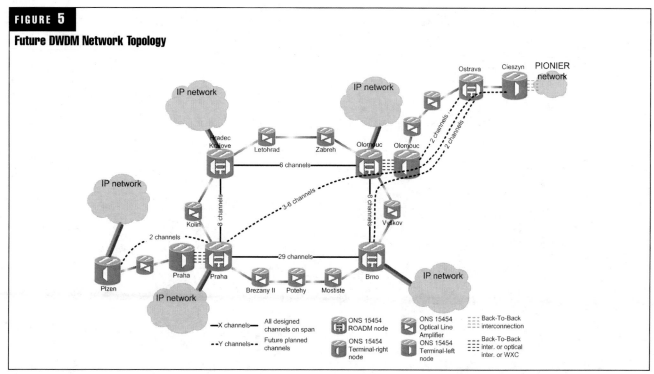

FIGURE 5

Future DWDM Network Topology

Case Study

In long-haul WDM systems, a proper design of DCUs is necessary to mitigate the combined impact of linear and nonlinear fiber transmission effects that would otherwise introduce excessive signal distortion.

Linear distortion is mainly introduced by chromatic dispersion (CD), while the main sources of nonlinear impairments are self-phase modulation (SPM), cross-phase modulation (XPM), and four-wave mixing (FWM).

These effects strongly depend on launched power, fiber type, channel spacing, and chromatic dispersion management. In the following figures, various nonlinear (split-step) simulations, both at 10 Gbps and 40 Gbps per channel, are shown for two fiber plants: one using a standard single-mode fiber (SMF) and the other using alternating sections of non-zero dispersion shifted fibers (NZDSF) with opposite signs of CD, namely true wave plus (TW+) and true wave minus (TW?).

DCUs have been optimized based on the preliminary extensive simulations to minimize the combined impact of all transmission effects.

Figures 6 to *9* show nonlinear simulation results for 10 Gbps (*Figures 6* and *7*) and 40 Gbps (*Figures 8* and *9*) over both TW+/TW? and SMF for the Brno?Prague link. Launched power is between 0 dBm and 2 dBm per channel in all cases.

All figures show at the top the penalty introduced for each channel by each nonlinear effect (SPM, XPM, and FWM) and, in red, the cumulative penalty. The bottom portion of the figures details the level of cross-talk introduced by XPM and FWM separately. These are the values that have been used to evaluate the penalties on the top portion of the figure.

MF can support both 10 and 40 Gbps over the line Prague?Brno with less than 1 dB of transmission penalty. DCUs for SMF accurately compensate chromatic dispersion slope; therefore, the penalties are almost the same for all channels.

Penalties mainly originate from single-channel effects (CD/SPM). TW+/TW? can support both 10 and 40 Gbps over the line Prague?Brno with less than 2 dB of transmission penalty. The main source of penalty at 40 Gbps is from single-channel effects (CD/SPM) due to the inability of TW+/TW? to compensate for dispersion slope.

This has been mitigated in the design by using "special" enhanced large effective area fiber (ELEAF) DCUs for the short SMF sections present in the link.

Conclusion

The CESNET2 optical network layer development, based on originally developed DWDM technology with ROADM, is the key requirement for new services introduction. It

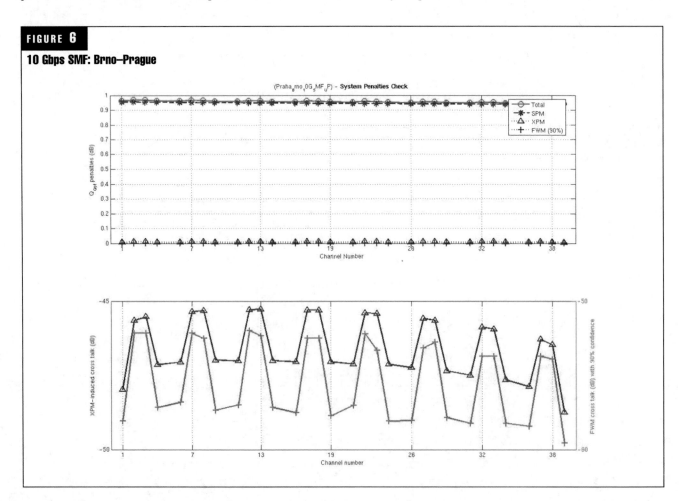

FIGURE 6

10 Gbps SMF: Brno–Prague

FIGURE 7

10 Gbps TW+/TW: Brno—Prague

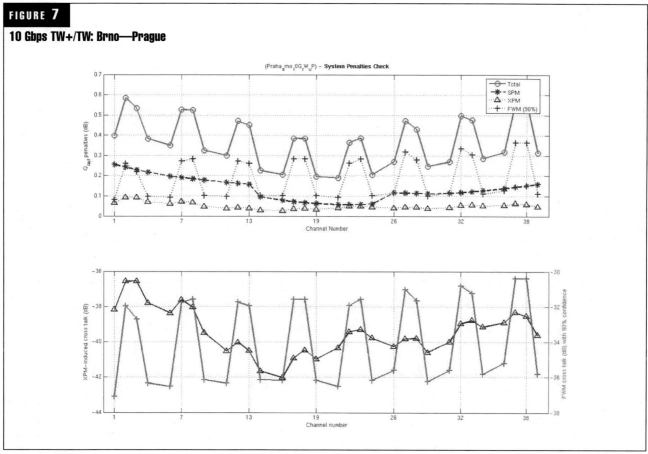

FIGURE 8

40 Gbps SMF: Brno–Prague

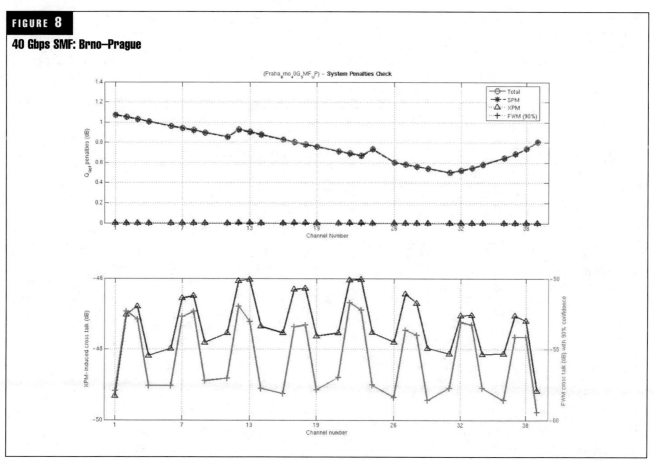

FIGURE 9

40 Gbps TW+/TW: Brno—Prague

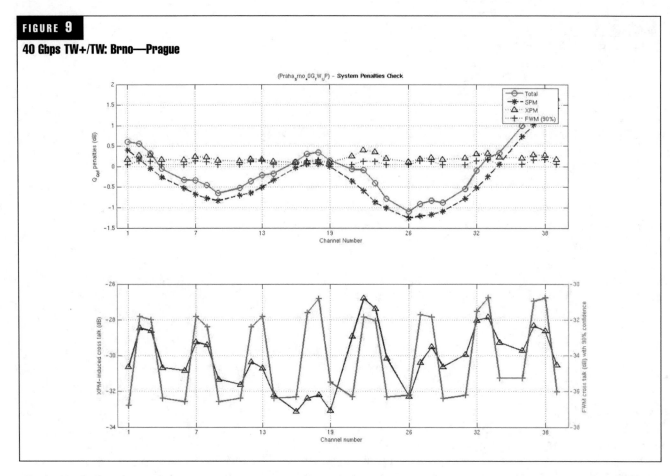

enables the support of on-demand end-to-end services on L1 and L2 network layers between the optical PoPs. It also provides an easy capacity scaling of the dark-fiber optical lines and optical channel provisioning. The objective of the DWDM deployment is an integration with an existing IP network and a step forward to a hybrid IP/optical network controlled by GMPLS or a similar protocol, including deployment of optical cross-connects for dynamic optical path switching.

The used DWDM technology could be upgraded to 40 Gbps and higher capacity support with the 50 GHz spacing at both L and C bands. The DWDM optical transport level could also be completed by SDH/Ethernet technology to achieve better scaling and flexibility if needed.

The optical DWDM network must be designed with respect to the real optical fiber parameters (attenuation, PMD, and chromatic dispersion) to guarantee reliable and stable services from the optical transport network. The current net-

work design reflects 10 Gbps transmission per optical channel. The simulation of 40 Gbps over the channel satisfied the requirements for accuracy design on the most complicated fiber line, Prague–Brno. Based on the performed simulation, the migration to 40 Gbps is feasible for CESNET2.

References

CESNET: www.cesnet.cz
GÉANT2: www.geant2.net
TERENA Compendium 2005: www.terena.nl/activities/compendium/2005/toc.html

Notes

1. For Prague, the capital of Czech Republic, the original name is used in figures – *Praha.*
2. See detailed description of the equipment at www.cisco.com/en/US/products/hw/optical/ps2006/products_data_sheet09186a00801849e7.html.

Important Mechanisms Available Today for Bridging the Digital Divide by Bringing Connectivity to Underserved Areas of the World

Shahla Riaz

Telecom Regulatory Analyst
Great Bear International Services, Pvt. Ltd.

Ensuring that consumers have a right to choose from the innovative technological facilities at the most cost-effective, affordable price is the main objective for bridging the digital divide. Mechanisms involving comprehensive strategies such as the following are required to address the issue of digital divide, particularly in underserved areas of the world:

- Open regulatory framework and universal access can spark up the digital connectivity by allowing innovative communication technology to be used for providing connectivity between communities, particularly in rural areas. Communication infrastructure, including telecommunications, Internet connections, and broadcasting, should reach not only urban high-income class but also the semi-urban as well as rural areas.

- Content industries should be developed to provide relevant information that actually matches to the local people's needs. This way, access to the international market for local products and services can be provided by the innovative usage of technology. However, availability of good and relevant content is a prerequisite for the development of the information society.

- Creating an e-business in developing countries, particularly in underserved areas of the world, requires a precise road map. Technical support industries can play a crucial role in producing, distributing, and maintaining personal computers (PCs), software, e-financing facilities, networking, and other solution industries that are essential to the digital revolution and to bridging the digital divide. In rural areas, where the digital divide is wide, certain factors need to be addressed by adopting a strong consumer gratification

mechanism for imparting technology. Technology just for the sake of technology never works and cannot succeed in any domain; until and unless its objective is focused on benefiting consumers at certain affordable prices. Some of the impacting factors in this domain are as follows:

- Putting up certain solutions for problem avenues in underserved areas using technology as a tool in a way that involves consumers' interest in adopting that solution (e.g., providing farmers certain guidelines or solutions for agriculture or involving e-commerce at local shopkeepers where people could be given access to the shops via certain cheap technology such as wireless for shopping rather than coming to the market). Railway stations, bus stations, schools, etc., could act as telecenters.

- Initial revenues may be small in this scenario, or even nonexistent in certain cases, while initially making consumers used to certain technology and eradicating their fears of using technology. For that time, industry can get help from the Universal Service Fund (USF) for bridging this digital divide.

- When the consumers get used to certain technology-based solutions and start getting certain benefits from it, many would be happy to pay for certain services that benefit their life routines.

The consumers exercise the choice of the services as well as service providers. Now their choices definitely confine them with respect to their living area (i.e., urban or rural) and hinge upon their income status as well as their need for a certain service, particularly with the invasion of technologies arrayed against one another through competing services in rural areas of developing countries where agriculture is the cornerstone of the economy. It is quite a cumbersome job to introduce technology-based needs as preferred requirements in a place where customers' decisions are vital. Therefore, consumers' needs and behaviors directly impose certain disciplines in the industry.

The industry plays a pivotal role as the center of gravity. Service implementation and timely delivery that is compatible to the needs of the consumer market (based, of course, on the financial affordability) is the main concern particularly in rural areas.

- To gain popularity in the underserved areas of the world, technology infrastructure needs to be cost-effective in terms of deployment as well as maintenance. For example, where the issue is providing connectivity to a larger area, wireless local loop (WLL) provides unparalleled large coverage in a wireless realm, allowing networks to be built with far fewer cell sites than is possible with other wireless technologies. Fewer cell sites covering larger areas ensure reduced operating expenses, which results in savings for both operators and consumers. These intrinsic advantages of WLL allow fast penetration levels along with expeditious network deployments.

- A migration from service-specific licensing regime to unified licensing regime is proposed. In the unified access licensing regime, both basic and cellular mobile service providers can offer basic and/or cellular mobile service using any technology. Unified licensees shall optimize their resources by offering all kinds of services using any technology as per the prescribed terms and conditions.

- Villages that are not included in operators' present plans for cellular mobile coverage, as well as several other villages, could be covered through the "niche operator concept." These villages will be those that have less than 1 percent teledensity, as defined for the niche operator concept in the unified licensing draft recommendations.

  ○ Niche operators shall be mandated to operate in only those areas where rural teledensity is less than

1 percent. They will be levied a license fee of a certain percentage of their average revenue on a quarterly or annual basis. Niche operators may also get support from the USF.

- A model of a largely self-sustainable business for the rural areas is needed. An increase in the rural teledensity of the country would lead to an increase in the gross domestic product (GDP) per capita, which would ultimately be the index of prosperity in the countries where about 70 percent of the population resides in rural areas.

- Private organizations and governments need to collaborate on more project initiatives that ultimately result in the access of digital technology–based facilities at cost-effective rates that are affordable to the local inhabitants.

- Rather than going for the establishment of the new infrastructure; leasing the already established infrastructure to secure lower interconnect offers needs to be addressed. Several countries are using the infrastructure-sharing model to reduce the cost of offering telecom services. Two examples are the white-zone concept in France and the sharing of infrastructure between two mobile operators in Australia to offer third-generation (3G) services.

- Finally, to bridge digital divide and utilize information and communications technology (ICT) for socially beneficial purposes, social change agent roles are essential. To bridge up the two ends of the rope where technology-based solutions reside at one end and consumers of the underserved areas are at the other end; consumers need to go through a thorough process of change in their lifestyles. Social change agents may work in various sectors using ICT. It is human nature to fear change. Certain technology-based U-turns can be made to conform to each community's needs and lead to enhanced connectivity with the rest of world on a massive scale. For that sake, change agents may even be chosen from the local community, but they will be using innovative models to bring about a change in the social process. Education, health, environment, and governance are the typical areas where ICT will create real change.

Building Networks that Create Value

Stef van Aarle

Vice President of Marketing
Lucent Worldwide Services

"Is more really better?" That's a question more than a few consumers are asking themselves. Some cannot wait for anytime, anywhere services and are willing to pay for them. Others are juggling handfuls of cell phones, laptops, personal digital assistants (PDAs), and other devices, all choked with applications and cringe at the thought of more choice, complexity, and unfiltered content. This is without mentioning those consumers who have yet to translate the technology already in hand into their first ring tone, text message, or buddy list—or minutes and bytes on a network.

Nowadays, as the boundaries among telecom, Internet, and the media and entertainment worlds continue to blur, nearly every service provider faces a rapidly changing competitive landscape rife with opportunities and challenges. They know they need to reduce churn, create brand loyalty, boost network traffic, cut operating costs, speed new services to market, and expand the total revenues spent in the telecom market instead of fighting with other carriers over the same slice of pie.

While each service provider has or will have a unique approach to the challenges it faces driven by its unique set of competitive advantages and legacy skills, the end goal, the sweet spot for the industry will be a truly converged communication ecosystem capable of delivering blended services to consumers. This end state is called blended services, and a three-phase, strategic path for service providers to achieve it has been identified: from virtual bundling to network IP. It's an end state that can be enabled and powered by the IP multimedia system (IMS) network architecture. Once unlocked and unleashed, this IMS opportunity and environment can be a competitive advantage.

In the telecom industry, people have talked for years about killer applications. None have truly surfaced. But an IMS–based network could be a killer ecosystem capable of rapidly trialing and launching salvos of high-margin applications that not only differentiate the service provider from other competitors, but also deliver the unified and personalized services that business people and consumers alike crave to restore simplicity in their lives. Blended services are centered on that simplicity. IMS is the engine that drives it and makes it all possible with a menu of services that are continually revamped, refreshed, and blended—and easy to access and use.

Today, if a friend of mine changes his or her telephone number or e-mail address, that causes me problems. I have to sift through devices to update buddy lists and databases. Wouldn't it be nice to have a trusted service provider hosting all my buddy lists so I only have to make that change once? Or maybe that service provider could notify me when someone on one of my many lists makes a change in his or her information.

I recently traveled from my New Jersey office to the west coast. My cell phone rang at 5 a.m. Why? Someone on the East Coast thought I was in my New Jersey office. I wasn't. The caller didn't know, and neither did the network. The ringing cell phone was an unwanted wake-up call.

Our lives are becoming more complicated, even blurred, 24 hours a day and seven days a week. No longer can it be taken for granted that at 3 p.m. on any weekday that a person is in an office, at a desk just working. Things such as children in day care who get sick and need to be picked up complicate our lives. It has happened to me. As a result, people have multiple devices, doing many different things, for all sorts of needs, for family and friends, customers and companies.

"Prosumer" is a buzzword in our industry today for those end users whose professional and personal lives overlap, and they "toggle" between the two. But I think that is an oversimplification because "prosumer" assumes only two states—professional and consumer. For teenagers, families, and businesspeople, life is actually more complicated. Each individual already is capable of shifting among many different states.

In my work life, sometimes I'm the manager of a large team, sometimes I'm a spokesperson, and sometimes I'm a sales person talking with a client. Depending on what role I'm playing, what state I'm in, my communications needs are very different. It's the same in private life. Tonight, I might be playing in a band. Tomorrow I'll be at home putting kids to bed, shopping, or at a sporting event. Prosumer and toggle imply on or off, one or the other. But every day, at any given time, our lives intersect with people, places, and events, all requiring different communications needs.

Our research shows that teenagers, families, and the professional consumer are all interested in services that allow them to share information with multiple people, across multiple networks in ways and at times of their own choosing. Teenagers want universal messaging. They want to know when a friend is available. They want to instantly share video and data files with multiple people within their own communities of contacts and interests. Families want shared schedules that are automatically updated, and alerts when schedule changes are logged. Professionals want to instantly establish conference calls and share files.

Consumers want blended lifestyle services as their lives become more and more dependent on networks. They want access to their networks, their applications, their content, contacts and communities in a way that follows them easily and flexibly—services that are personalized, content-rich, reliable and secure, from anywhere, anytime video and voice services, to Internet and data services.

It's a tall order. It's the challenge ahead for traditional wireless and wireline service providers who not only face a phalanx of new competitors eager to siphon away revenues but also a market where there are few barriers to keep consumers from jumping to the next best communications deal. A business model grounded in an IMS network that delivers blended lifestyle services can be the glue that keeps them from straying.

Figure 1 illustrates the three phases to move service providers from existing networks to next-generation networks: virtual bundling, network IP, and blended services. These stages are not exclusively mutual. In fact, a service provider could be working all three stages in parallel.

With virtual bundling, service providers can address the consumer's demand for one-stop shopping by consolidating telephone, television, and Internet services on the same

bill and giving a discount. It provides a converged experience on legacy networks, but it's not really a sustainable model. A service provider can achieve top-line growth and extend its name and brand in the market and into new service areas, as well as achieve some back-office integration in terms of billing and operations. However, long-term virtual bundling basically destroys value because with these bundles the service provider is delivering more service for less revenue. The bundles don't create any new dollars. The dollars are simply being taken away from some other competitor. In terms of the industry, consumers are actually spending less. It creates a price war among competitors as well as the continual migration of consumers. Yet, as an interim step, virtual bundling is a viable strategy considering the capability of today's networks. It works. It does draw customers. But for service providers to create value, they need to take additional steps.

The network IP stage is one of those steps in which service providers focus on driving costs out of their networks by converging their disparate networks—and each has many—into a single network, a federation based on IP, where all the network elements can easily interact with each other. The primary objective is to achieve operational efficiencies through physical changes to the network, specifically by deploying IP/MPLS solutions to the core and implementing next-generation optical solutions to transport, access, and share data. However, network IP still maintains the old paradigm of point-to-point communications and does not create an environment for seamless blended lifestyle services. A service provider could stop here and even introduce new services, in all likelihood, one silo or smokestack at a time.

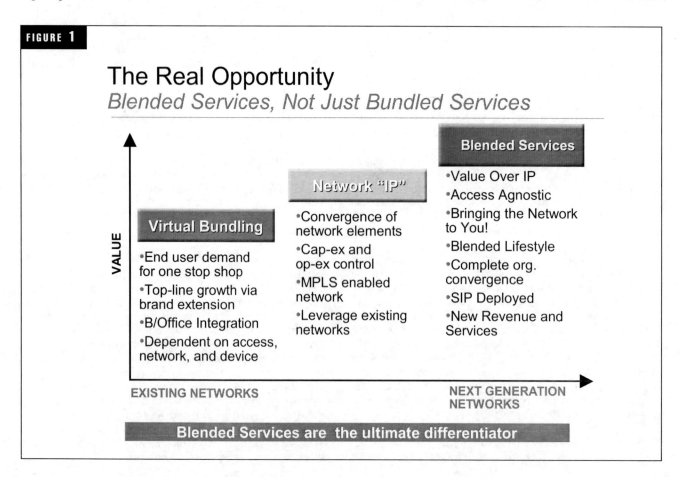

FIGURE 1

The Real Opportunity
Blended Services, Not Just Bundled Services

Blended Services
- Value Over IP
- Access Agnostic
- Bringing the Network to You!
- Blended Lifestyle
- Complete org. convergence
- SIP Deployed
- New Revenue and Services

Network "IP"
- Convergence of network elements
- Cap-ex and op-ex control
- MPLS enabled network
- Leverage existing networks

Virtual Bundling
- End user demand for one stop shop
- Top-line growth via brand extension
- B/Office Integration
- Dependent on access, network, and device

VALUE

EXISTING NETWORKS

NEXT GENERATION NETWORKS

Blended Services are the ultimate differentiator

But we believe that the next step—blended services—will be the ultimate differentiator and the wellspring of value creation in our industry.

That is one reason why IMS has gained so much momentum in our industry and why we have invested so much in the IMS network architecture as the enabler of convergence and as the industry standard that will both speed the adoption of next-generation services and boost network traffic and revenues. IMS offers plenty of opportunities and it addresses our customers' challenges, along with meeting their needs. If, as a consumer, I want to roam across networks and devices with my personalized services and content, that anytime, anywhere device experience with any community is not going to happen unless there is a fairly standardized, technical way to do that easily. IMS is the closest standard that we have to make that possible, to create that world of universal and unified services. IMS is as real as it gets today.

From the 1.3 billion people who use the Internet—including 19 million bloggers—to the 3.2 billion people tethered to the public switched telephone network and the 2.6 billion using mobile phones, market forces are colliding.

To couple these numbers with advances in real-time collaboration technologies, consumer and enterprise market dynamics (such as growing broadband penetration and high-speed mobile data), new, standard approaches for operations and architectures, and finally with blended, personalized, anytime, anywhere services and solutions, would result in the future market for next-generation services being far greater than the sum of its present-day parts.

Carrying all that surging traffic—all those commoditized bits—might turn out to be a good business for some large players, the big prize, or the industry sweet spot, will go to the service providers who hold relationships with the end users, whether they are consumers, business professionals or some as yet unknown and undefined digitally connected and savvy hybrid.

By unlocking and unleashing the opportunity in their networks that an IMS–based architecture offers, service providers can win customer loyalty and remain at the cutting edge of innovation and competitiveness against all corners. For end users, the future of their communications is more about brand, personality, community, ease of use, and fun. This is exactly where the battle for customers is going to be won or lost. IMS is a technical solution and it is an architecture only limited by a service provider's imagination.

Since IMS is an architecture that allows any kind of access—wireless or fixed—for any kind of media from multiple devices and endpoints, it provides an ideal environment to quickly deploy new applications and blended lifestyle services, either as innovations or to counter competitive threats. Because it is a standard architecture with open interfaces among applications, network layers and back-office systems, service providers can also tap the resources of third-party applications developers. IMS–based blended services also will spur service providers to look at different ways and different business models to deploy blended services.

For example, service providers might decide to develop their own applications in-house, or sometimes outsource the entire process to a third party who would also host the service. Under this scenario, a service provider could potentially increase speed to market by avoiding the internal churn and talent acquisition necessary for in-house development. By outsourcing the entire process, a service provider could quickly trial a service. If it catches on with the market, they could bring it back in-house and build it out to full deployment—or leave it with the hosting vendor until the service runs its course. If the service does not fly, the service provider could quickly drop it and move on to the next opportunity. This is just one business model that IMS can enable.

An IMS–based architecture also counters an emerging uncertainty in today's market, and that is the unpredictability of what services consumers or business people want to buy. In the past, Bell Labs invented a service and the regional phone companies deployed it to people who wanted to buy it which truly turned into a push model for services. Today, however, services are very "fashion sensitive." Some applications just catch on, such as text messaging. They fill a void or create a need. As a result, the industry is seething with companies focused on innovation and the desire to mine technology for the next popular service, refining the mother lode into revenue. The network capability to deploy that service, to reduce the cycle to get it to market, is an inherent strength of the IMS architecture. It's a rapid-deployment machine.

IMS–based blended services also will enable service providers to build their own community with blended lifestyle services that win customer loyalty and establish new revenue streams. Research shows that a 5 percent improvement in customer retention can increase a company's profits by 70 to 80 percent. Service providers that can create a personalized and customized experience for consumers are service providers that can create a community, reduce churn and increase the stickiness of their relationship with customers. IMS is the invisible glue. It is the seamless current that consumers unknowingly float on as the network churns away beneath the surface, moving them to their destination of choice.

In addition, it is important to remember that their destination isn't always a foregone conclusion. It is not always known when translated into a blended lifestyle service. As a result of people not knowing what they do not know, unpredictability exists in our industry—what sells and what doesn't—and it has certainly become an issue today. It is risky business. The next great thing, or small thing, is hard to predict and that is part of the beauty and functionality of IMS–based blended services. It is a blended services ecosystem where applications can quickly be rolled out and trialed. Service providers can play into what works and easily change course when something doesn't.

There are some things that are known, too. Service providers can build value and loyalty through network-based quality. Quality of service, security, scalability, personalization, presence, and location capabilities are all part of the equation in winning customer loyalty. People are not going to trust their buddy lists with just anyone. In addition, they want to gather where their communities are accessible—and that service provider's network will allow them to interact with the people on their buddy lists in the best way possible with the greatest ease. The future for them is all about contacts, content, and communities.

With IMS–based blended services, providers can own the customer relationship with a rich, personal customer experience which creates trust in the brand and in the company. Once a service provider has earned that trust, it will be hard to break. IMS and the blended services it can deliver is a game changer. Similar to any game, those who adapt, hone their skills, and innovate will have an edge. However, a service provider needs to be in the game, have the right equipment, and be on the right playing field to win. Finally, they need a playbook full of blended services.

Next-Generation Services

Road Maps for Next-Generation Data Transport Networks

Lawrence Jacobowitz

Development Manager
IBM Thomas J. Watson Research Center

Casimer DeCusatis

Distinguished Engineer
IBM Corporation

Introduction

The data communication marketplace is divided into a high-end enterprise market and a low-end commodity-driven market. Most networking innovation has originated in the enterprise market, where data communication networks continue to grow and evolve at a rapid pace. This continues to be a hybrid, multivendor environment; protocols such as Fibre Channel, fiber connection (FICON), and Ethernet are used in high volumes at peak data rates of 2 to 4 Gbps, while enterprise users still consume significant amounts of legacy channels (enterprise systems connectivity [ESCON], asynchronous transfer mode [ATM], low-speed Ethernet) at speeds of less than 1 Gbps. Emerging technologies such as InfiniBand promise to increase the aggregate data rate to 30 to 60 Gbps using parallel communication buses in the next few years. In this article, we will describe the industry trends and directions for supporting data communication protocols across metropolitan-area networks (MANs) and wide-area networks (WANs). After a brief review of the existing telecom and datacom environments, we will discuss industry road maps for next-generation Internet and synchronous optical network (SONET) networking, including competing standards proposals for the extension of both technologies.

At the same time that enterprises are adopting new, innovative technologies, they continue to rely heavily on traditional services that are well understood and widely deployed throughout telecom and datacom networks. These may include frame relay, ATM, or private line services. However, future growth will occur with the installation of advanced packet switching services, including metro Ethernet, IP virtual private networks (VPNs), Layer-2 VPNs, and voice over IP (VoIP). Advanced services offer benefits such as lower cost per gigabit of data transferred, simplified network topologies with greater flexibility, and higher aggregate bandwidths. However, the migration to these new services is happening slowly, partially because of the prolonged equipment sell cycles in the enterprise and partly because customers are unwilling to give up key benefits of the legacy services. Emerging technologies are not yet mature enough to provide service-level agreements (SLAs) that guarantee carrier grade availability of up to 99.999 percent per year (no more than five minutes downtime). By contrast, until recently, most implementations of metro Ethernet were "best effort" delivery only.

Changes in customer demand are causing carriers and service providers to adopt new service-oriented architectures, lower capital expenditures (CAPEX) and operational expenditures (OPEX), and bridge from circuit-switched voice traffic to packet-switched data traffic. Worldwide revenue from voice traffic is expected to stay flat over the next five years; in areas where competition is heavy and services are commoditized, network providers will be forced to compete strictly on price with corresponding decreases in revenue and margins. By contrast, data revenue is expected to grow between 10 and 15 percent per year. For example, metro Ethernet services reached $333 million in 2003 and are forecast to exceed $1.3 billion by 2008 with a 31 percent compound annual growth rate (CAGR). Similarly, there are strong market drivers favoring growing in disaster recovery and long-distance data communications, including a recent series of natural and man-made disasters affecting major corporations. As outsourcing drives many business operations across the globe, the economic impact of lost or temporarily unavailable data is steadily growing. In addition, this market is being driven by a host of new data integrity regulations that carry heavy financial penalties for noncompliance, including the Sarbanes-Oxley Act, the Basel Capital Accord, the Health Insurance Portability and Accountability Act (HIPAA), the Gramm-Leach-Bliley Act (GLBA), and California State Bill 1386. All of these trends favor growth in storage and data communication over WANs.

Microelectronics development is becoming an increasingly critical enabler for next-generation network (NGN) features. Available chipsets do not necessarily meet market performance requirements. For example, the largest commercially available generic frame procedure (GFP) chips today handle about 128 virtual concatenation (VCAT) groups, which is not enough to meet current network demands. For example, a 10 Gbps optical backbone providing subscriber connections at twice the rate of a telecom T-1 line may require over

1,600 VCAT groups. Higher-layer network management functions such as traffic queuing, shaping, grooming, forwarding, and classification are usually handled at Layers 4 to 7 by custom-designed application-specific integrated circuits (ASICs). More recently, this approach has become increasingly difficult as designers are faced with rising non-recurring engineering expenses and longer time-to-market design cycles. As the semiconductor industry moves toward smaller geometries (0.13 micron and 0.09 micron) coupled with larger wafers (300 mm), the higher volume per wafer output will limit the number of manufacturers able to achieve break-even volumes required to offset their up-front engineering costs. Following the telecom downturn in 2001 and 2002, many network equipment providers have migrated to programmable designs built around a new class of off-the-shelf network processors. These designs provide the flexibility to scale data rates from less than 200 Mbps to more than 10 Gbps, and in some cases to use field programmable gate arrays (FPGAs) to create customized processors that can respond quickly to changing industry standards such as InfiniBand. The development of cost-effective ASIC functions may even drive new trends in enterprise storage networking. For example, high-end storage-area networks (SANs) are based on FICON and Fibre Channel, but as single-chip implementations of Internet small computer system interface (iSCSI) become available in the next few years, 10 Gigabit Ethernet becomes a more plausible storage fabric for these applications. Likewise, application-specific functions such as data encryption will become more pervasive as they are integrated within larger ASICs as part of network equipment control planes.

Next-Generation SONET Road Maps

SONET (and the closely related synchronous digital hierarchy [SDH] standard for international telecommunications) was developed to optimize the transport of large volumes of data over WANs. SONET traffic is carried in fixed bandwidth increments and by itself does not provide the ability to dynamically adjust bandwidth in response to user requirements (data streams smaller than the minimal increments can be transported, but they typically waste the unused bandwidth and are not directly monitored by SONET network management). Since this constant bit rate approach does not adapt well to packet-based bursts of traffic characteristic of data communication systems, extensions to the standard collectively known as next-generation SONET have been proposed. For example, VCAT allows protocols such as Gigabit Ethernet and Fibre Channel to be efficiently transported over a combination of SONET frames, even though their basic data rate does not match SONET bandwidth increments. An extension to VCAT is the link capacity adjustment scheme (LCAS), which makes it possible to vary bandwidth distribution based on factors such as time of day or other customer usage requirements. In addition, LCAS makes for a more robust network by providing the ability to recover from failures of one or more VCAT elements in the network. Without LCAS, the failure of any one member in a VCAT group will cause the entire circuit to fail; with LCAS enabled, the circuit can continue to operate using a lower bandwidth provided by the remaining available members of the group. Bandwidth allocation can also be done manually to route traffic around failures or recover from disasters.

Generic Frame Procedure

Another emerging next generation SONET feature is GFP, defined in International Telecommunication Union (ITU) standard G.7041. This describes a set of methods for framing different Layer 1 and 2 traffic types into a common structure compatible with SONET equipment. Essentially, GFP provides a protocol-agnostic frame container that allows for the encapsulation of many standard datacom protocols into SONET compatible frames. Any 8-bit/10-bit encoded data can be remapped into SONET using this approach. The basic GFP procedure for datacom protocol transport involves decoding each 10-bit character of an 8-bit/10-bit data sequence and mapping the result into either an 8-bit data character or a recognized control character. This data is then re-encoded as a 64-bit/65-bit data sequence, with control characters mapped into a pre-determined set of 64-bit/65-bit control characters. In GFP terminology, the resulting data sequences or control characters are known as words (this differs from the server definition of a word, which is usually taken as either a 4-byte quantity or a 40-bit string of four 8-bit/10-bit characters. We will use the GFP terminology for consistency throughout the remainder of this discussion). A group of eight such words is assembled into an octet, which is provided with additional control and error flags (note that this differs from the server definition of an octet, which is usually taken as an 8-bit byte). An octet is then assembled into a "superblock," scrambled, and a CRC error check field is added. The resulting frames are compliant with routing through a SONET/SDH network flow control, including quality of service and related features; the original 8/10 encoded data is reassembled at the other end of the network. Collectively, the next generation optical transport network (OTN) is defined in ITU standard G.872; the network interface definitions are part of ITU G.709.

G.709 "Digital Wrapper"

Ratified in 2001, G.709 is part of the ITU body of standards. It defines standard interfaces and data rates derived from existing SONET/SDH line rates while taking into account roughly 7 percent overhead due to the addition of forward error correction and other functions [1-3]. Commonly known as "digital wrapper," this standard differs from SONET framing in that it operates at the wavelength layer, adding overhead bits to a 2.5 Gbps or 10 Gbps wavelength, rather than applying SONET framing to multiple low-speed data channels. Digital wrapper also enables performance monitoring and traffic management, as well as forward error correction (FEC), which is required for extended distance links beyond 10 Gbps data rates. FEC in the G.709 standard is analogous to bit interactive parity (BIP)-8 error monitoring in SONET networks; this offers benefits for protocols that do not have built-in error protection schemes such as Ethernet or fiber channel. The standardized G.709 line rates and corresponding SONET rates are given in the table below; in addition, there is a nonstandard interface for 10 Gigabit Ethernet LAN clients that operate at 11.095 Gbps. Digital wrappers are also a key component for future wavelength routed optical networks, which will be discussed in the following section.

MPLS and GMPLS

There are many ways to modulate information onto an optical fiber communication system, including amplitude,

TABLE 1

G.709 Interface	Line Rate (Gbps)	SONET Rate	SONET Rate (Gbps)
OTU-1	2.666	OC-48/STM-16	2.488
OTU-2	10.709	OC-192/STM-16	9.953
OTU-3	43.018	OC-768/STM-256	39.813

phase, and frequency modulation. Recent advances in dense wavelength division multiplexing (DWDM) fiber-optic networks have led to interest in developing the means to switch optical wavelengths (or lambdas) directly, rather than the conventional methods, which require performing optical-to-electrical conversion, switching the electrical data, then converting back to the optical domain. Lambda switching technology transmits many high-speed (2.5 to 10 Gbps or higher) data streams over a common fiber-optic cable using different wavelengths of light, but unlike conventional DWDM, lambda switching also incorporates a set of emerging industry protocols to manage higher-level functions such as switching and routing. These functions provide the ability to engineer traffic (specify how, when, and where data flows), ease network management, increase fault tolerance and reliability by providing redundant lambda paths, and implement new types of network topologies such as ring mesh. The key to lambda switching is the ability to automatically connect the endpoints in an optical network. Conventional networks require tedious, expensive configuration of each device, fiber, and protocol in the network; this approach is prone to errors that may leave critical data paths unprotected or exposed to potential single points of failure. Lambda switching automatically reconfigures the network by integrating the switching functions with higher-level protocols; thus network designs that would have previously been impossible to manage can be constructed. There are many devices in a lambda switched network, including optical cross-connects (OXCs), DWDM devices, and others; the key to making all of these devices work together is the ability to share information about the network topology among all attached devices. This requires some form of in-band communication path within the lambda switched network.

Conventional lambda switching requires a separate optical wavelength channel to be reserved for advertising network information to all attached devices. This information may include which devices on the network can be reached by which paths, available bandwidth, quality of service, and extensions of current Internet routing protocols such as open shortest path first (OSPF) to determine the optimal paths for data flowing through the network. Using this information, lambda switching devices can each construct a network topology map (or traffic engineering table) as a basis for subsequent wavelength switching operations. This is a self-constructed topology map, which changes over time without requiring input from end users. When a connection over the network is required, the ingress OXCs can transmit setup messages requesting that the downstream OXCs allocate one or more wavelength channels for the data. This signaling will likely be accomplished using emerging stan-

dards such as multiprotocol label switching (MPLS) standardized by the Internet Engineering Task Force (IETF), or adaptations of the resource reservation protocol (RSVP) employed on Internet connections. To engineer the traffic flow, extended MPLS or RSVP messages must flow over the network, using paths set up by the traffic engineering tables. Alternately, traffic flow can be handled by routers that manage the network against some connection criteria, such as bandwidth use. Fault tolerance is achieved by requesting backup light paths, and services such as VPNs can be implemented. After processing the setup messages, the OXCs would signal successful light path resource allocation by passing MPLS or RSVP messages to upstream neighbor devices. The optical connection is completed and ready to transmit data from the edge of the network devices when each OXC between the two endpoints has assigned a label to the light paths.

A related application for tunable filters is reconfigurable optical add/drop multiplexers (ROADMs), which may also be based on micro-electromechanical system (MEMS) technology. These devices determine the path of multiplexed wavelengths traversing an optical network node and have the ability to switch paths remotely rather than requiring manual reconfiguration of a fixed filter system. The ROADM can perform its function on single wavelengths without affecting other traffic in an optical network. Recently, there have been many requests for proposals on ROADM technology from leading service providers interested in automating MANs. By transferring the flexibility and ease of control associated with SONET networks to the optical layer, ROADMs can significantly lower operating costs. They also enable more scalable optical network architectures and features such as wavelength-dependent protection switching.

An extension of the MPLS standard, the generalized MPLS standard (or GMPLS) includes not only packet layer equipment, but also WDM, TDM, and other types of optical switches. GMPLS is a control plane standard for provisioning bandwidth in optical networks, including both signaling and routing features. As a common control plane interface for equipment residing in different networks, equipment from different providers, or configurations with different classes of service, GMPLS is intended to unify carrier networks and enable end-to-end bandwidth provisioning. Further, proposed features of this standard enable dynamic, autonomic operation of network management tasks that were previously labor-intensive manual operations (for example, provisioning capacity on demand). For example, the time required for an experienced service provider to install new services on an existing WAN can vary from five to 45 days, which is significantly longer than customer requirements. This translates directly into cost reduction and faster revenue recognition and is seen as a first step toward intelligent optical networks.

Emerging Standards: IMS and GENI

The emergence of market trends that make the end user the focal point of service delivery, rather than the network providers, has led to conflicting proposals for future data transport standards [4]. These standards could determine how closely future subscribers are tied to specific service providers or carriers and may force changes in the traditional carrier business model. The two major standards proposals in this area come from the ITU (whose membership tends to favor telecommunications carriers) and the IETF (which tends to drive data communication priorities); both proposals attempt to address the convergence effects of datacom, telecom, video, message-based routing, and peer-to-peer applications. The ITU has proposed an NGN standard that allows service providers more control over the network, called the IMS architecture [5]. Initially conceived by the third-generation partnership project (3GPP)—a collaboration of telecom standards bodies initially chartered to define specifications for 3G mobile wireless systems, but which has since been extended to include wired communication networks [6]—IMS essentially replaces the control infrastructure in the traditional circuit-switched telephone network, separating services from the underlying networks that carry them. IMS enables services, such as text messaging, voice mail and file sharing, to reside on application servers anywhere and be delivered by multiple wired and wireless service providers. For example, one system could hold information about the preferences and access rights of each user, which could be made available on many other systems to handle mobile users. By defining how IP networks should handle voice calls and data sessions, IMS essentially takes the place of the control infrastructure in the traditional circuit-switched telephone network, with the key difference that it separates services from the underlying networks that carry them. IMS enables the migration of the legacy telecom networks to IP while maintaining telephony-borne features such as emergency services, wiretaps, call handoff, and billing.

Although the IMS model is complex to implement, it can be broken down roughly into three basic components. These are the home subscriber server, which stores information about each user; application servers, which run the services subscribers use; and the call session control function, which regulates how each session works and is merged with other sessions. At the core of IMS is the session initiation protocol (SIP), a signaling system for setting up and handling calls and data sessions, which already is the standard for VoIP products. SIP brings together the three major elements of IMS. The traditional circuit-switched telephone network can tie into IMS via gateways. Using enterprise communications equipment or peer-to-peer network software, SIP enables corporations and consumers to manage their own voice or data sessions. If this happens, future carriers may be rele-

gated to the role of providing pure high-speed IP connections at commodity prices. More likely, carriers will take advantage of the rich feature set and complex implementation of IMS to more tightly control quality of service with their customers. IMS compliant solution offerings are scheduled to become available from major vendors such as Nortel and Cisco by early 2006. Interoperability among different vendors' IMS platforms is perhaps a year or two away.

New bandwidth-intensive applications continue to put strain on the public telecom infrastructure. As an alternative to the conventional telecom network, business customers are beginning to substitute public Internet connectivity for frame relay, ATM and private lines as the performance of the Internet improves. An alternate proposal is the global environment for network investigations (GENI), a model for facilitating change in the current Internet, in particular to improve security and reliability. GENI, a component of the National Science Foundation's Future Internet Design Initiative (FIND), is a prototype network for deploying research innovations conceived by FIND participants. By helping to bridge small-scale lab experiments and commercial deployment, GENI might facilitate changes to the Internet's basic design and support experimental validation of new architectures. To simulate real-world deployment, GENI supports virtualization and user opt-in. Virtualization allows a GENI to be partitioned to run a given service or architecture independently of others in adjacent partitions. Users can opt in to various services and applications on a per-user, per-application basis, which makes it possible to attract users necessary to validate new designs.

Conclusions

As the industry debates the various options for NGNs, the business models of network operators from both legacy telecom and the Internet hang in the balance. The disruptive impact of IP, high-speed data communication, and digital multimedia has brought these issues to the forefront of next generation transport network design. Previously separate markets such as wireless/wireline, voice/data, and commodity/enterprise are beginning to consolidate. Customer service–oriented architectures are emerging within the network infrastructure, which will provide the differentiation and added value of next generation transport networks.

References

1) G. Simo, Nortel white paper, "Updates to key system acceptance tests for G.709 based long haul systems," Proc. National Fiber Optics Engineering Conf. 2003, see www.nortelnetworks.com/products/01/optera/long_haul/dwdm/collateral/nfoec2003_acceptance.pdf.

2) Next Generation Networks 2005 conference, Washington, D.C., Sept. 26–30, 2005.

3) C. DeCusatis, Editor in Chief, "Handbook of Fiber Optic Data Communications," second edition, Academic Press (2002).

4) J. Duffy, "Next-gen net seen at a crossroads," Network World, Oct. 10, 2005 www.networkworld.com/news/2005/100305-ngn.html.

5) S. Lawson, "What IMS promises enterprises and carriers," Network World, Sept. 26, 2005 www.networkworld.com/news/2005/092005-ims.html.

6) Third-generation partnership project (3GPP): www.3gpp.org.

Information Revolution: The Next Generation

Ashok K. Kapoor

Principal
SemiSolutions

Introduction

Historical growth of information technology (IT) in the past century is synonymous with the history of electronics. The unprecedented growth in the world economy and the financial markets, especially between 1990 and 2000, had its roots in information technology. The productivity gains achieved in this period were enabled by the information revolution. *Figure 1* shows the labor productivity, measured by the worker output per hour, between 1960 and 2004 in the data published by Ed Yardini [1]. The accelerated productivity growth from 1996 on was instrumental in the rapid growth of the financial markets, as the workers were able to produce more goods in a given period.

This article is an attempt to find an order underlying the tremendous growth of IT as it becomes uncovered with time. Each stage in this revolution has brought new concepts, new technology, and a new set of winners and losers. Interestingly, each stage of this revolution coincides with an economic event—mainly a market downturn. This article attempts to build a model that identifies the various stages with the beneficiaries and the losers, and use this model to develop a picture of the next phase of the information revolution.

Phase 0: The Age of Infrastructure Development

The invention of the solid-state transistor was a major milestone that set off the information revolution in 1945. An interesting fact often overlooked is that this invention was made in the research laboratory of a communication company—the Bell Telephone Laboratory—and it was first used to improve the telephone switching circuits. The first portable commercial transistor radio was an instant success in 1954. The computer industry played catch-up in using this new invention; the same year, IBM announced that transistors would replace vacuum tubes in their computers. Gradually, the computer companies embraced the solid-state transistor and prospered with it. The invention of integrated circuits in 1957 enabled a whole new way of building complex functions in a very inexpensive way. Since then, the integrated circuits—or "chips," as we know them—have been getting smaller, faster, and cheaper. Gordon Moore

observed this trend back in 1965, and it is still going after four decades. [2]

The big computer manufacturers such as IBM, Sperry, Burroughs, and CDC grew rapidly in that period. At the dawn of the information age, every machine was based on a proprietary system, from integrated circuit chip to the operating systems. Data communication between those computing machines was a rarity.

In the years that followed, the data processing machines went through a process of disintegration where the big machine was gradually replaced by smaller computers.

"Minicomputers," made by Digital Computers, started in 1970.

"Workstations"—made by Daisy, Apollo, ValidLogic, Silicon Graphics, Mentor Graphics, SUN Microsystems, etc.—started in the 1980s.

Unfortunately, very few of them exist today in their original business model.

In parallel, ARPA had been driving a project to connect various computers across the country. They had funded the development of the network that came to be known as ARPANET. ARPANET expanded rapidly from 15 hosts in 1970 to more than 700 in 1980. This was the dawning of the information revolution. One might argue that the information age arrived in 1980 with the arrival of the first computer virus, which shut down nearly 70 percent of all computers.

Phase 1: Tools for Information Creation

The first phase of the information revolution started with the availability of the tools for creating information by the masses, namely the computing hardware and the application software in 1978 by Apple Computer. For instance, it enabled a small-business owner to buy a personal computer, load an application such as Lotus 1-2-3 spreadsheet and start using the computer without having any knowledge of the programming languages, with the total price tag of less than $5,000. The personal computers were finding place on

The Rise in Labor Productivity between 1960 and 2004

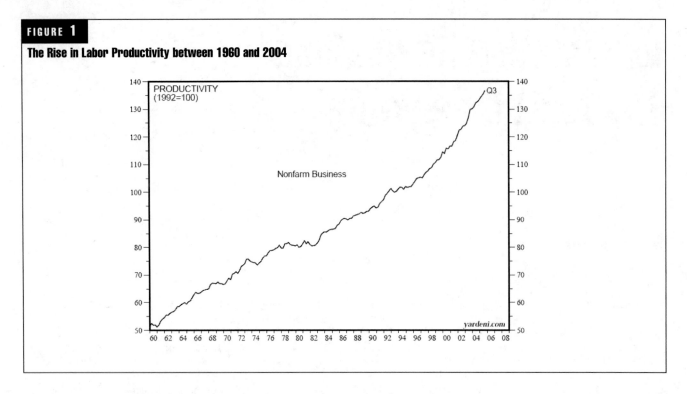

every desktop in small and medium-sized businesses and even in some homes. The data processing was no longer a privilege of the scientific community—it had begun its ascendance to the masses.

First Bottleneck: Data Storage
The early personal computers had very limited functionality. The first IBM personal computer (PC) in 1981 cost $2,945 and was powered by an Intel 8088 processor running at 4.47 MHz; it had 16 kilobytes of memory and a 5.25″ floppy, a monochrome monitor, and a keyboard. Additional memory was expensive; memory boards with 16, 32, and 64 kilobytes cost $190, $325, and $540, respectively. A floppy drive cost

$570. Applications such as Lotus 1-2-3 spreadsheet were limited by the available memory in the computer. In essence, the real bottleneck was the available memory in the desktop machine.

It is no surprise that massive investment went in the development of dynamic random access memory (DRAM) at that time. Starting with the 64kb DRAM made with two-micron lithography in 1978, memory prices have kept declining as predicted by the Moore's Law, as shown in *Figure 2*.

In tandem with the decrease in cost per bit, memory use in a PC grew exponentially, as shown in *Figure 3*.

Memory Price Decline over the Past Two Decades

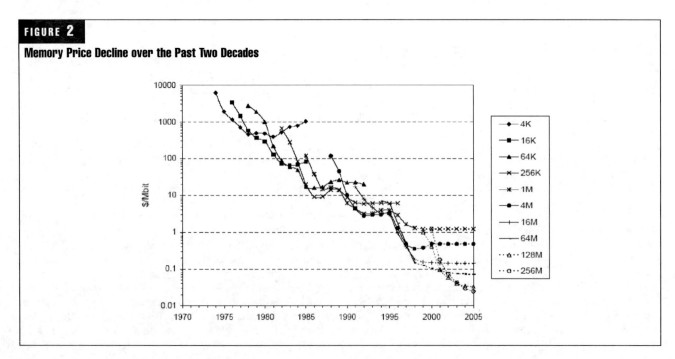

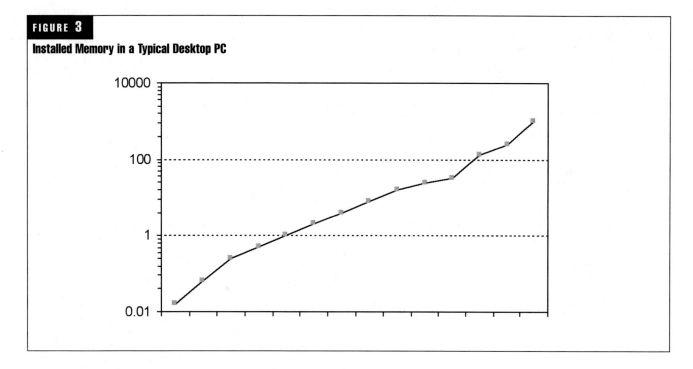

FIGURE 3

Installed Memory in a Typical Desktop PC

Not surprisingly, the DRAM became the technology driver, being the first product to be designed with every shrink of the linewidth. Japanese manufacturers took the lead in DRAM development, driving most of the U.S. manufacturers out of DRAM business gradually. This rise in the semiconductor industry in Japan contributed to its financial markets' meteoric rise, which ended in 1990. The capacity for DRAM manufacturing was being added in excess of demand. A few years later, the decline in the price of DRAM accelerated. Many of the manufacturers of DRAM exited the business, leaving Micron Technology as the only U.S. supplier of DRAMS today. By 1995, DRAM prices had decreased to less than $10 per megabit. It was not limiting factor for the machine performance anymore.

While the price of DRAM had been declining, the disk drive industry was following a similar growth curve and kept on increasing capacity and reducing cost per bit at an even faster pace.

Second Bottleneck: Processing Power
With the cost of memory declining rapidly, the desktop machine became limited by the computing power of the processor. The processors became the drivers of technology in the early 1990s. They drove the innovation in semiconductor technology, taking the lead over from DRAM manufacturers in early 1990s. In addition to X86 processors pioneered by Intel, PowerPC by Motorola and various reduced instruction set computing (RISC) processors from SUN, MIPS, and others saw significant investment and growth.

After taking over the leadership from DRAM, the processor business prospered. The processor clock speed has increased from 4.47 MHz to just below 4 GHz in a period from 1981 and 2004. The new processor and the operating system were introduced in tandem, and they became the twin engines for growth of the desktop machines. The clear winners to that era were the two companies driving the

business: Intel and Microsoft. The processing power of X86 machines over time is shown in *Figure 4* below.

Along with chip companies, software providers also prospered, led by Microsoft. The Microsoft windows operating system slowly established itself as the standard for the desktop machines, along with the application suite. The market value of Microsoft grew even faster than that of Intel, maintaining a similar growth pattern.

The Losers
With the progress of the first phase of the information revolution, the mainframe manufacturers saw their market stagnate, and a gradual exodus from the business started.

The dominance of the desktop machines as the performance benchmark ended in mid-1990s with the focus shifting to networking.

Phase 2: The Age of Information Exchange

By 1990, most of the desktop machines were running at clock speed of greater than 10 MHz, capable of supporting most of the application software for routine tasks in the offices and homes. The next challenge was to exchange the data with peers electronically. The need for connecting desktop computers became a compelling proposition to increase productivity. Various manufacturers started offering local-area networks (LANs) to connect the computers such as IBM's token ring, AppleTalk, Novel's Netware and Unix. In 1989, Windows NT joined the cadre of NetWare providers with the release of version 2.0.

Wide-area networks (WANs) had been evolving since the 1960s in various government-led initiatives. The only problem was that the cost of bandwidth on switched networks had barely changed in a decade. (It had taken 79 years to decline by 50 percent.) The bandwidth of the network now

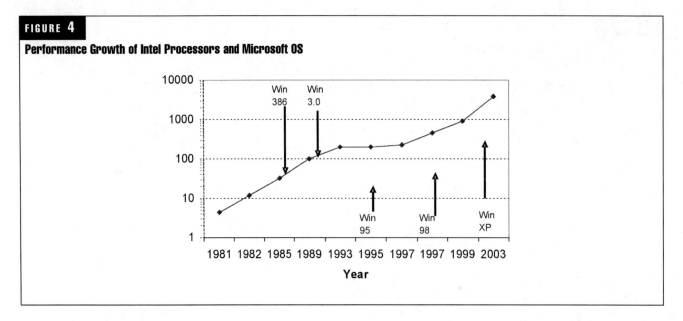

FIGURE 4

Performance Growth of Intel Processors and Microsoft OS

stood as the limiting factor in information revolution. The goal of seamless data communication across the WANs came after the release of the World Wide Web by the Corporation for Research and Educational Networking (CREN). Factors that hastened this phase of the information revolution are explained below.

Reduction in the Cost of Network Bandwidth

In the early 1990s, dense wavelength division multiplexing (DWDM) enabled multiple signals coded in different frequencies to be sent on the same fiber, thereby increasing the bandwidth of the optical fiber–based long-haul networks. Bandwidth explosion using DWDM began with narrow-band WDM in early 1990s, in which two to eight channels spaced at an interval of 400GHz in a 1550 nm window were used. In a few years, DWDM systems with 16 to 40 channels and spacing from 100 to 200 GHz were developed. By the late 1990s, DWDM systems had evolved to the point where

they were capable of 64 to 160 parallel channels, densely packed at 50 or even 25 GHz intervals, resulting in a more than twentyfold increase in bandwidth of the fiber. To illustrate the point, the cost of data communication, measured in dollars per megahertz of bandwidth, is shown in *Figure 5*. [4] OC-3 is the standard used for sending the data at 155 MHz over the optical fiber. T1 is the conventional coaxial copper cable transmitting data at 1.544 MHz.

The increase in data traffic over the WAN was also limited by the switching speed of the routers. The excess router capacity in large cities was exhausted by 1997. From 1998 onward, Internet traffic increased in tandem with router capacity, doubling every six months. Advances in the computer technology were able to support the expansion of the router capacity, and the router manufacturers enjoyed the price premium awarded the pace-setting technology.

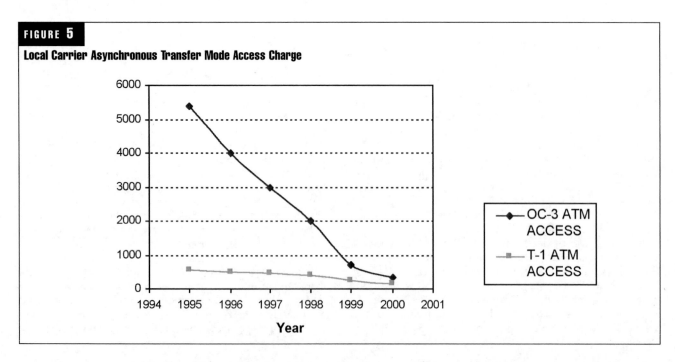

FIGURE 5

Local Carrier Asynchronous Transfer Mode Access Charge

Availability of Web Browsers Supporting Hypertext Translation Protocol (HTTP)
The Mosaic-based Web browsers released by Netscape and Microsoft totally changed the user experience in accessing Web content. The rivalry between the two companies sped up the technology development and made the consumer access to technology a reality.

The annual report of the Federal Reserve Bank of Dallas contained a chilling comparison of the cost reduction achieved in the past decade, as shown in *Table 1*. [5]

The age of information exchange was pioneered by networking companies such as Cisco, Juniper Networks, Lucent, Nortel Networks, Corning, Netscape, and Microsoft. The market value of these companies skyrocketed at a rate much faster than any historical norm. The network had unleashed such a tremendous gain in worker productivity that the users were willing to pay any price to access the technology. The decline in the cost of leasing bandwidth had made it economical to operate the World Wide Web over the long haul.

The anticipated problems with the arrival of the year 2000 caused the demand to increase temporarily, which resulted in large build up of inventories. By the year 2000, the network capacity of the fibers had increased to the point that less than 10 percent of the fibers were being used. It became evident that the demand was not expanding at the same rate as supplies for the network bandwidth, and the prices went into a free fall. The euphoria ended in what is commonly known now as the "dot-com bust." Rates for long-haul line lease declined by 85 to 90 percent in 2001. The premium associated with the data communication eroded, marking the end of the second phase of the information revolution.

Losers
The long list of losers with the second phase of information revolution includes the following:
- Manufacturers of storage devices, both DRAM and hard disks
- Manufacturers of proprietary processors
- Proprietary data networks, especially those based on mainframe computers

Phase 3: Need for Information Management— Here and Now

The age of information exchange ended with the price collapse of desktop machines, network bandwidth, and routing capacity. The result was availability of extremely low-cost

bandwidth and software tools for effective information exchange over the long-haul network. Low-cost bandwidth allowed access to the Web. With inexpensive desktop machines and high-speed network to access the World Wide Web using HTML–based application interfaces, the information available to a typical network user became overwhelming. The impact of information overload was felt in the enterprises gradually. The ease of copying e-mails to any number of users, access to the enterprise-wide calendars, and ability to store and retrieve data from any node within the enterprise firewalls started having an opposite result on worker productivity. The amount of time devoted to addressing the plethora of nonessential e-mails and meetings started taking time away from the workers. The Department of Commerce reports that the labor productivity growth has slowed down to less than 3 percent per year, which is in line with the values seen in 1980s. [6] In this age of information overload, a new business model that supports intelligent management of information is beginning to appear; arrival of the age of information management, the new phase of the information revolution.

Since the early days of Internet, the challenge for the software companies has been their ability to make the users pay for the services they provided. With limited content available and limited capability to sort through the information, the solution offered to the users was inadequate. Initial euphoria with the on-line shopping seemed to suggest that the all the retail stores were to be replaced by on-line stores. This dream also ended with the dot-com bust. Out of the ashes of the bubble emerged a few corporations, including eBay and Yahoo that had built viable businesses managing information exchangeable over the Internet. These companies are prospering in the current environment with more powerful tools at their disposal. The following are a few salient examples, presented here for reference:

- *SAP*: The company has taken the lead in complete information management within the enterprise.

- *EBay*: The Internet auction powerhouse has created a whole new medium for merchandising new and used goods over the Internet. It has continued to grow and prosper despite the economic downturn of 2001 and the dot-com bust.

- *Yahoo*: Many companies attempted to present a portal to the users to meet all their information access needs, but Yahoo is one of the few still in the business today.

- *Google*: This organization's growth is an excellent example that a business can be made in the search

TABLE 1

The Cost Comparison of Processor Speed, Storage, and Data Transmission over Time

	1970	1980	1990	1999
Cost of 1 MHz	$7600.82	$103.40	$25.47	$0.17
Cost of 1 MBit storage	$5256.90	$614.40	$7.85	$0.17
Cot of sending 1 trillion bits	$150000.00	$129166.67	$90.42	$0.12

engines. The challenge of attaching a business proposition with the free services has been addressed by this company very profitably.

- *Apple's iPod and iTunes*: This is an information management device with an associated service to provide a recurring revenue stream. With consumer electronics as the driver of technology, one can expect to see many more applications.

Non-volatile memory has become the semiconductor device of choice in almost all information management devices.

Many more software and hardware companies are bound to emerge as the winners in this phase of the information revolution. Those companies will allow users to maximize productivity while utilizing the tremendous pool of information available over the Internet. One can safely project that the users are becoming aware of the cost involved with the access to information and are willing to pay for the services.

Challenges

The age of information management can be expected to solve many challenges facing users, including the following:

Security

The access to the World Wide Web can be compared to the Wild West two hundred years ago. Without the appropriate gear (antivirus software), one constantly runs the risk of downloading a computer virus while accessing the contents over the Web. Driving a parallel between the unruly outlaws who terrorized the travelers two centuries ago and the traffic on the Internet today, one cannot help imagining a time when the Internet will be formally managed.

Web Search

The search engines today take pride in claiming that tens of thousands of matches are found in a fraction of a second on a given topic. Ironically, the user of the search engine can end up spending a few hours to a few days sorting through the search results. The obvious conclusion is that the search engines are not all that efficient in finding the exact information needed by the user.

Management of information over wireless networks is still in its early stages and totally disorganized. Multiple, totally incompatible standards for data transmission (including CDMA, WiMAX, Wi-Fi, Bluetooth, and Zigbee) present a major challenge and opportunity for the management of information systems. The wireless data management is expected to play a very significant role in management of information.

Losers

Clearly, the information management age has left the winners of the previous ages as trend followers, displaced from their previous leadership roles. The processing power of the desktop machines is sufficient (Intel), the software environment of the machines is stable (Microsoft), the bandwidth of the network is sufficient to support the traffic (Cisco, Juniper, Lucent). These giants' glory days are past them. New leaders are emerging.

Summary

Once the tools for information management have been perfected, the age of information intelligence will arrive. When it will arrive is subject to speculation.

References

[1] www.yardeni.com/globalindicator.aspx?Country=US, accessed on November 30, 2005.

[2] G. E. Moore, "Progress in Digital Integrated Electronics." IEEE International Electron Devices Meeting, Technical Digest 1975. 1975, pp. 11-13.

[3] Nadejda M. Victor and Jesse H. Ausubel, "DRAMs as Model Organisms for Study of Technological Evolution," Technological Forecasting and Social Change in 2002, Volume 69(3):243-262, 2002.

[4] Robert Cohen, "Pricing for Advanced Network Connections," Net@EDU annual meeting, Tempe, Feb. 7, 2000.

[5] Federal Reserve Bank of Dallas Annual Report 1998, page 8. (Also available at www.dallasfed.org/fed/annual/index.html#1999, accessed on November 30, 2005.)

[6] www.bls.gov/news.release/prod2.t02.htm, accessed on November 30, 2005.

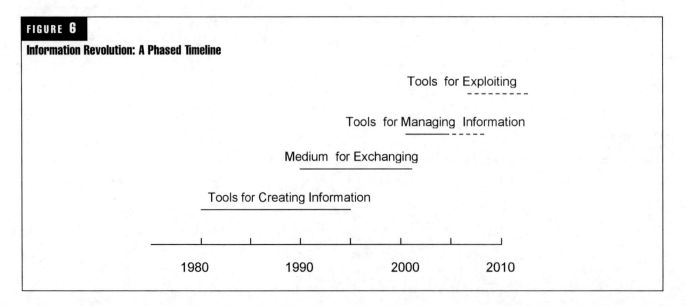

FIGURE 6

Information Revolution: A Phased Timeline

Tools for Exploiting

Tools for Managing Information

Medium for Exchanging

Tools for Creating Information

1980   1990   2000   2010

Advanced Video Encoders in Next-Generation Television

Sean McCarthy

Principal Scientist
Modulus Video

Overview

Television has been evolving for a number of years. Recently, new technologies and economic factors have converged to accelerate evolution, bringing us to the verge of a new species of video telecommunications.

Much has been made in trade journals and popular press about the role of competition as traditional telecommunications companies enter the established purview of cable and satellite video providers. Much also has been made about the new possibilities enabled by the use of Internet protocol (IP) to deliver video. These are important issues, but there are other critical topics that have received scant attention.

The purpose of this paper is to examine one of those critical yet neglected topics—the role of professional broadcast encoders in next-generation television services.

In the past, broadcast encoders were asked to ingest raw video and spew bits in a specified format. The primary objective was compression—deliver consumable video with limited bandwidth. The economic driver was the number of channels that could be delivered across a pipe of prescribed bandwidth. For example, if a new version of broadcast encoder allowed five channels to be sent over a pipe that could only accommodate four channels using an older model of encoder, then it was a fairly straightforward sale. Bandwidth is a costly ongoing expense. A new, more efficient encoder could pay for itself. Moreover, the service provider would be in a position to offer additional services or additional content that could generate additional revenue.

In the next generation of video services, compression efficiency will still be key, but professional encoders will be asked to take on an expanded role. In addition to maximizing channel efficiency, next-generation professional encoders will be smarter network devices that do the following:

- Provide new kinds of services to subscribers at home
- Optimize end-to-end network performance through sophisticated bit-rate management

- Generate new top-line and bottom-line economic opportunities

This paper focuses specifically on several of the most important issues related to professional encoders and next-generation television, including a short discussion of new compression technologies, an overview of the migration from linear broadcast models to switch-delivery models, and a more detailed examination of several new beyond-compression encoder functions that are already becoming critical in the rollout of new television services.

A New Species of Video Compression

The International Telecommunications Union–Telecommunication Standardization Sector (ITU–T), the international standards body behind many of the most important telecommunications advances in modern times, is celebrating its 50th anniversary. As part of the celebration, ITU–T has asked the world to vote for the most influential work of the past half century. Video coding (H.262/MPEG–2 video, H.264/AVC) is currently winning by more than a 2-to-1 margin.

Advanced video coding (AVC) is the latest generation of video compression. AVC, also known as ITU–T H.264/AVC and ISO/IEC MPEG–4 part 10/AVC, was developed in the first few years of the 21st century in response to widespread calls for more efficient video compression. It succeeds MPEG–2 in most application areas and provides tools that enable a wide variety of new and improved applications.

AVC would be better described as a toolkit than as a monolithic compressor-decompressor (CODEC). Compared to its popular predecessor, MPEG–2, AVC provides more video processing versatility. Both MPEG–2 and AVC are block-based transform-based CODECs, but in AVC, the blocks vary in size and the transforms are optimized for real-world processors. AVC also incorporates very efficient prediction modes that do not exist in previous CODECS. Prediction is the way of recycling data that has already been sent to a decoder so it does not have to be sent again, thus saving bandwidth.

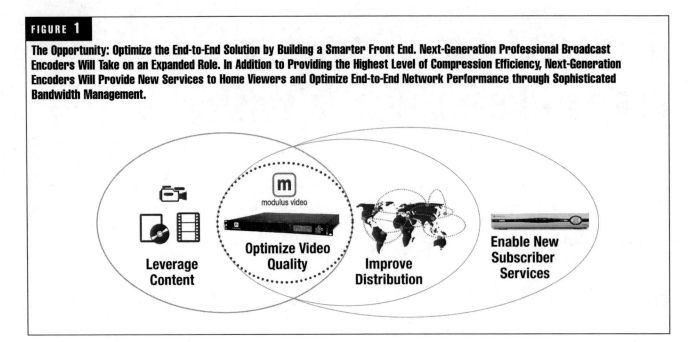

FIGURE 1

The Opportunity: Optimize the End-to-End Solution by Building a Smarter Front End. Next-Generation Professional Broadcast Encoders Will Take on an Expanded Role. In Addition to Providing the Highest Level of Compression Efficiency, Next-Generation Encoders Will Provide New Services to Home Viewers and Optimize End-to-End Network Performance through Sophisticated Bandwidth Management.

Current first-generation professional AVC encoders implement main profile, which is a particular subset of the full AVC toolkit. AVC main profile encompasses both standard-definition (SD) and high-definition (HD) formats. The full AVC standard also has more inclusive profiles—various high profiles—that provide tools for additional kinds of data transformations, data reduction, and color processing.

It was claimed early on that AVC would be about twice as good as MPEG–2 in terms of compression efficiency. Commercially available top-tier AVC–based broadcast encoders meet the twice-as-good mark and arguably do even better. Many IPTV services plan to launch HD television (HDTV) services at 8 Mbps per channel, compared to the 18 Mbps or more that would have been required using MPEG–2. For some content, including movies, top-tier

encoders are flirting with bit rates near 3 Mbps. Future improvements in AVC compression efficiency will come from incorporation of AVC high profile tools into second- and third-generation professional encoders and the use of more sophisticated proprietary video-processing techniques such as noise filters that enhance the efficiency of the core AVC tools. One could imagine AVC HDTV television services being delivered at 5 to 6 Mbps, maximum, in the near future.

A New Species of Television

It may sound like heresy, but the compression efficiencies of AVC alone are not sufficient to revolutionize television. For sure, the improved compression of AVC is worth many hundreds of millions in bandwidth savings, but better compres-

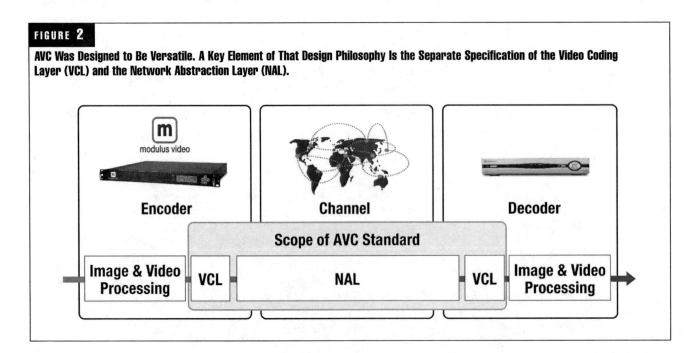

FIGURE 2

AVC Was Designed to Be Versatile. A Key Element of That Design Philosophy Is the Separate Specification of the Video Coding Layer (VCL) and the Network Abstraction Layer (NAL).

sion by itself would not change the way people experience television. AVC is a necessary condition but is not sufficient by itself.

What gives birth to a new species of television is the combination of AVC and Internet protocol (IP) video delivery. AVC with IP–based transport is the magical combination that enables telecommunications companies to compete with traditional broadcast, cable, and satellite companies. It is the trigger that could transform the entire television industry from a linear broadcast model to a switched-delivery model. That transformation is already under way in IPTV from companies such as AT&T and Swisscom. It may soon follow from cable companies in the form of IP–like switched-digital video (SDV) services and in over-the-top services in which viewers bypass traditional television services to obtain content directly from the studios and content producers using the Internet.

In the linear broadcast model, all channels are delivered to all subscribers all the time. The home viewer selects what to watch by changing channels to tune in to an incoming stream. Everyone watches the same programs at the same time. From a network utilization point of view, the pipes into the home must be large enough to carry all available channels at once, whether a home viewer is watching or not.

In the switched-delivery model, only the content that a viewer wants to watch is delivered to the home. In some ways, switched delivery is like a fully evolved video on demand (VoD) service that includes both previously recorded content and live content. A home viewer makes a request through the remote and set-top box (STB), and upstream technology routes the selected content to the viewer. The channel switch happens in the upstream network rather than in the home. From a network utilization

point of view, switched delivery means that the pipe into the home need only be large enough to carry the particular subset of programs requested at any particular time, rather than all the programs that could be requested.

The switched-delivery model is made possible because of the high-compression efficiency of AVC and the customized delivery afforded by IP transport, but the home viewer really does not care about those kinds of things. The home viewer cares about good content, good services, and good prices. It is in these areas that switch-delivery models can really change the television experience. It is also the area in which smarter network-centric encoders become a necessity.

A New Species of Professional Broadcast Encoders

The promise of switched delivery is that the home viewer will have more choices. A viewer can watch first-run and live content by channel surfing, just like now, or ask for reruns or premium theatrical releases using instant VoD. A viewer could recall time-shifted content from a home-based or network-based digital video recorder (DVR) or access specialized niche content that has a select audience. Switched delivery is the way for video providers to let home viewers watch what they want when they want.

With this new model come new challenges. Providing choice can be at odds with providing quality.

In particular, next-generation television will face the following challenges:

- Delivery of a growing number of HDTV programming options

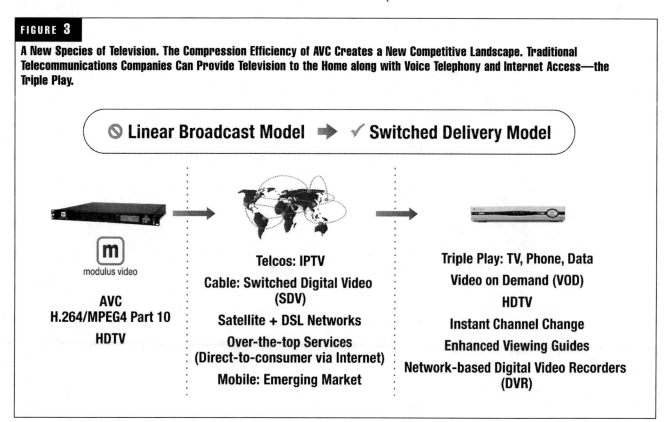

FIGURE 3

A New Species of Television. The Compression Efficiency of AVC Creates a New Competitive Landscape. Traditional Telecommunications Companies Can Provide Television to the Home along with Voice Telephony and Internet Access—the Triple Play.

Linear Broadcast Model ➡ ✓ Switched Delivery Model

m
modulus video

AVC
H.264/MPEG4 Part 10
HDTV

Telcos: IPTV
Cable: Switched Digital Video (SDV)
Satellite + DSL Networks
Over-the-top Services (Direct-to-consumer via Internet)
Mobile: Emerging Market

Triple Play: TV, Phone, Data
Video on Demand (VOD)
HDTV
Instant Channel Change
Enhanced Viewing Guides
Network-based Digital Video Recorders (DVR)

- Joint optimization of triple-play services; i.e., services in which television, voice telephony, and Internet access are bundled and share a common pipe into the home
- The danger of making switched-delivery services too complicated with too many options or with content that is too difficult to find, leaving the viewer bewildered

The answer to how these challenges will be met may be found in two case studies using technologies that were deployed first by Modulus Video and are now being required routinely in next-generation television deployments.

Case Study 1: Optimized End-to-End Network Performance (CF–CBR™)

The encoder defines what the home viewer sees. It sets the upper boundary on video quality. The rest of the network delivers the content to the home.

The problem is that the home is at one end of the lowest-capacity last mile of the network, and the encoder is at high-capacity locations such as super head ends (SHEs) and video hub offices (VHOs). Moreover, the last mile is shared with other services such as voice telephony and Internet traffic.

How can a remote encoder deliver video of specific visual quality to a home user while simultaneously maximizing network performance? The answer is to make the encoder aware of both video quality and bandwidth limits.

Modulus Video developed an encoding mode called constrained-fidelity™ constant bit-rate encoding (CF–CBR). CF–CBR is a set of video processing and rate-control algorithms that find the optimal balance between visual quality and bandwidth utilization. CF–CBR is different from other encoding modes such as variable bit-rate encoding (VBR), constant bit-rate encoding (CBR), and capped VBR in that both bit rate and visual quality are monitored and actively regulated. The result is lower average bit rates than can be achieved by VBR or capped VBR without the visual quality fluctuations that are often observed in CBR mode.

CF–CBR is a preferred feature of professional encoders now being deployed in next-generation applications. One major advantage of CF–CBR to service providers is its ability to jointly optimize triple-play services. Bandwidth freed by CF–CBR is immediately available to be reclaimed by voice and data services. Typical in-service applications of CF–CBR today enable 20 to 30 percent reclamation per video channel without noticeable difference in visual quality. For a single 8 Mbps HD channel, CF–CBR gives 1 to 2 Mbps back to voice and Internet services.

From the point of view of the home viewer, CF–CBR results in better network performance and more reliable picture quality. Voice and data services are more fluid because of reclaimed bandwidth. Video quality is consistent because of the visual quality monitoring and regulation that is part of CF–CBR. On both counts, the home subscriber and the service provider win with smarter next-generation encoders.

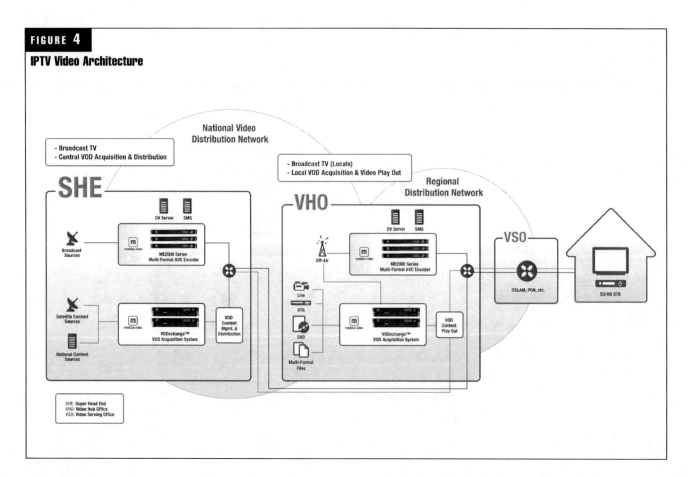

FIGURE 4

IPTV Video Architecture

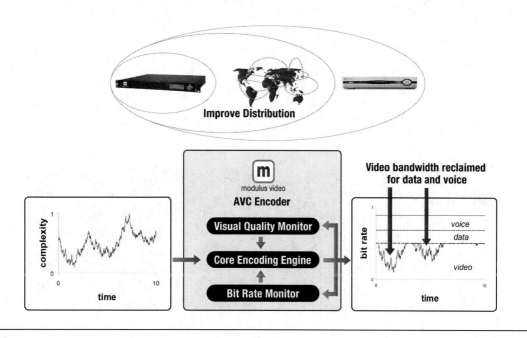

FIGURE 5

Joint Optimization of Triple-Play Services. Advanced Encoding Modes such as CF–CBR Encoding Enable Maximum Last-Mile Performance at Specified Video Quality. Bandwidth Reclaimed from Video Services Helps over All End-to-End Network Performance from the Point of View of Individual Home Viewers and from the Point of View of Entire Neighborhoods.

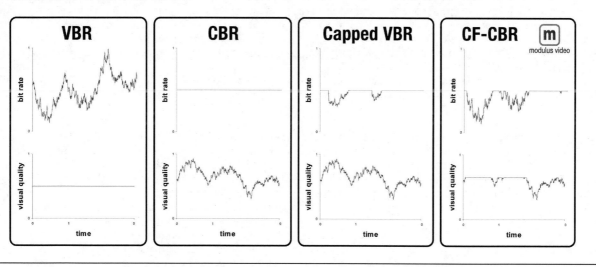

FIGURE 6

CF–CBR Is a New Encoding Mode for Next-Generation Professional Encoders. CF–CBR Enables the Best Features of VBR, CBR, and Capped VBR to Be Used Together to Optimize End-to-End Network Performance. Like VBR, CF–CBR Enables Direct Control of the Visual Quality a Home Viewer Experiences. Like CBR and Capped VBR, It Enables an Upper Bound to Be Enforced for the Video Bandwidth Component of Triple-Play Services.

Case Study 2: New Subscriber Services— Enhanced Viewer Guides and More

The idea that a home viewer can choose what to see and when to see it with almost no restrictions is compelling. But what is the engineering nitty-gritty that can make that happen?

Several industry leaders are building middleware solutions that promise to provide infrastructure that will make televi-

sion viewing a quick and intuitive experience while providing new revenue options for service providers. To a large extent, the design philosophy behind next-generation middleware is motivated by visual preview and selection. That is, small-scale low-resolution video will be used as proxies for full-resolution SD and HD programs. One application for video proxies is enhanced viewer guides that visually let a viewer know instantly what is on. Another application would enable a viewer to watch multiple programs at once.

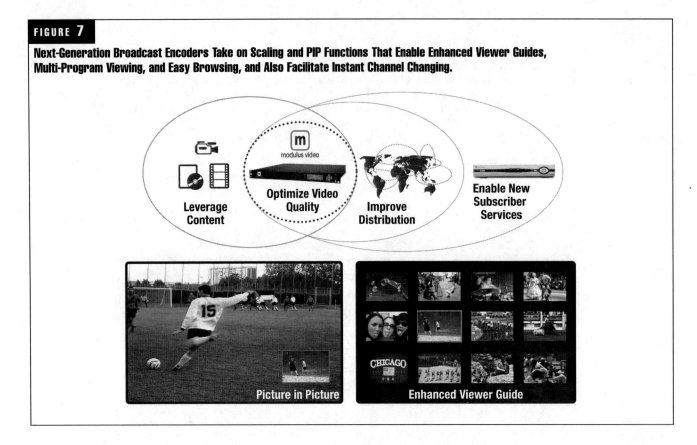

FIGURE 7

Next-Generation Broadcast Encoders Take on Scaling and PIP Functions That Enable Enhanced Viewer Guides, Multi-Program Viewing, and Easy Browsing, and Also Facilitate Instant Channel Changing.

Leverage Content

Optimize Video Quality

Improve Distribution

Enable New Subscriber Services

Picture in Picture

Enhanced Viewer Guide

Yet other applications would serve more prosaic picture-in-picture (PIP) functions such as channel preview.

Creating low-resolution versions of full-resolution video has previously been the domain of home television equipment, but a different scale of processing is needed for next-generation television. The necessary video processing could be included in next-generation STBs, though that would not solve the bandwidth problem created by multiple full-resolution video streams entering the home. Nor would it address the additional cost of STBs loaded with extra video-processing capabilities.

Creating video proxies in the network is a better solution for several reasons. Proxies require less bandwidth than full-resolution video. STBs and other home television equipment can be less expensive because they require less functionality. More important, network-generated proxies enable new revenue service such as targeted advertising and VoD previews.

One possible method of generating proxies in the network would be to aggregate full-resolution video streams in a PIP server, a new class of network equipment. Within the PIP server, video would be decoded, scaled, re-encoded, and repackaged in a distinct IP stream.

A different method of generating PIP streams that avoids the costs and difficulties of creating a stand-alone PIP server has been developed. Instead of a dedicated PIP server, a proxy is created along with its full-resolution counterpart within a single next-generation encoder.

The advantages of encoder-based PIP generation are the same as those that would be achieved with a dedicated PIP server, plus some others. Notably, PIP quality is better because encoder-generated PIP streams are created from the same original source as the full-resolution video. PIP streams created in a stand-alone PIP server would suffer in the process of being decoded then re-encoded. As important, encoder-generated PIP streams leverage the advanced processing functions that already exist within the encoder, including noise filters, high-quality image scaling, multi-format audio processing, closed-captioning, and CF–CBR™. Such functions would need to be engineered *de novo* for PIP servers.

From the standpoint of a home viewer, encoder-generated PIP streams make the television interaction easier. Proxies provide an intuitive means of finding interesting content. For the service provider, encoder-generated proxies are the higher-quality, less expensive future-proof solution. They enable new services and revenue models to be explored without upgrading the home STB or incurring the expense of new network equipment. Simply stated, a smarter PIP–enabled encoder makes for a better television experience for home subscribers and service providers.

Conclusion

This paper examined the expanded role of professional AVC encoders in next-generation television. Professional encoders are moving away from being strictly compression engines. Though compression efficiency will continue to be the most important must-have feature, encoders are becoming more versatile. Next-generation encoders are adding

FIGURE 8

Incorporating PIP Functionality into Next-Generation Broadcast Encoders Is Less Expensive, Easier to Implement, and Provides Better Subscriber Services.

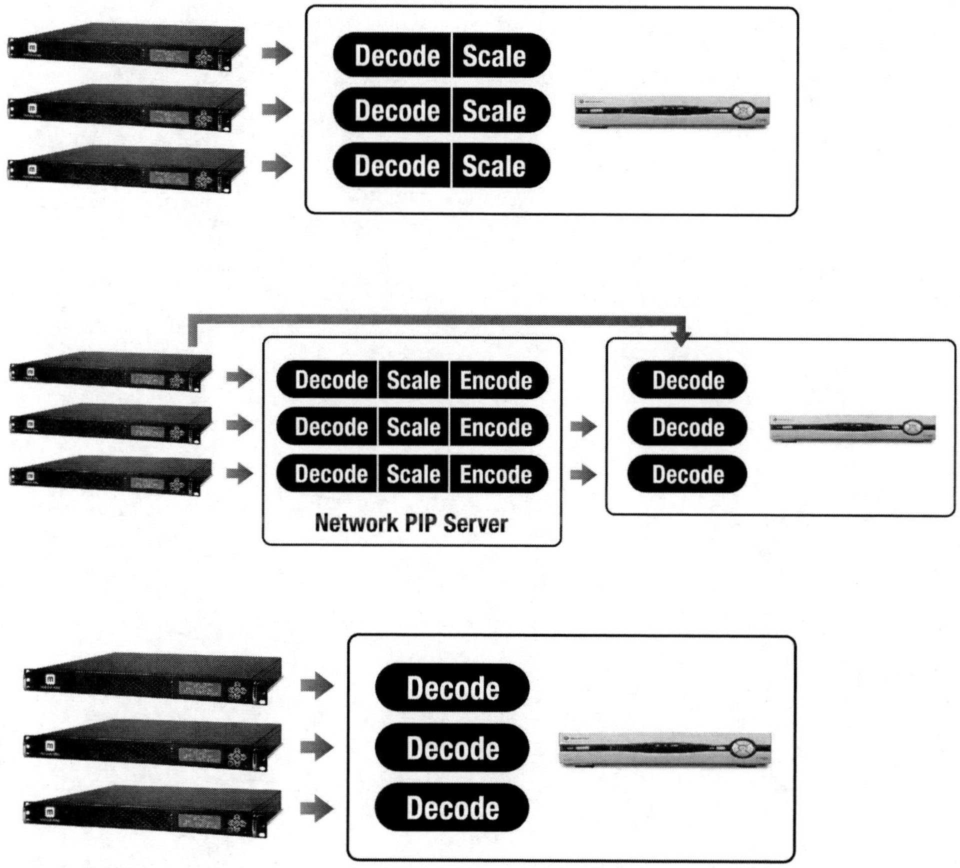

functionality that improves end-to-end network performance by balancing video quality and bandwidth restrictions in new ways. Other new features address the need for service providers to differentiate themselves from competition by offering the home viewer new services. Some new encoder functions facilitate revenue-generating services such as VoD, while other functions, such as joint-optimization of triple-play services, improve bottom-line performance.

Two case studies were presented to illustrate the ways in which challenges of next-generation television can be addressed. Both cases showed that newly developed technology creates a new, smarter class of professional encoder. Both technologies, CF–CBR and encoder-generated PIP, are already being deployed and rapidly becoming necessary components of new television rollouts.

The net impact for the home viewer is more choice, better video quality, and a more intuitive television experience. For the service providers, the next generation of professional AVC encoders enables new revenue-generating services and better ways of managing delivery costs.

Next-Generation Residential Gateway

Avni Wala
Technical Leader
Flextronics Software Systems

Introduction

Consumer electronics (CE) and chip manufacturing companies have envisioned a device that will create networked homes. All Internet, mobile, and CE devices will seamlessly work toward the creation of an interoperable home network, enabling all manufacturers to provide unique value to consumers.

For this customer-premise device to be a success, it has to be low-cost but technically sophisticated. Considering these facts and the engineering sophistication involved in creation of such a device, it is critical for hardware and software manufacturers to consider the design aspects more keenly.

This paper focuses on hardware and software aspects that need to be considered for creation of such a device, called residential or home gateway.

What Is a Residential Gateway?

Imagine a device measuring few inches, which creates your "home network" by connecting all your analog and digital appliances such as set-top boxes, digital video recorders, laptops, home theaters, camcorders, multimedia mobile phones, game consoles, personal digital assistants (PDAs), etc., and is also able to connect to the Internet so that you can manage your home remotely. Such a device is a residential or home gateway.

Thus, a residential gateway is an intelligent customer-premises device that provides the user seamless Internet access and other value-added services by creating a cooperative environment between devices within the home.

How Do You Build a Residential Gateway?

Since this device is consumer equipment, primary concern while designing both hardware and software for the gateway should be the retail cost. You do not want to make it so sophisticated that it gets beyond a retail customer's reach. However, you still want to keep it fairly sophisticated to attract end users' attention. In the next few sections, we will look at the hardware and software designs that will be most suitable for a residential gateway.

Hardware Design
Start point for any hardware design is to identify what the end product is going to do and then select chips that will help you do those activities in best possible way.

Typically any home that is planning to deploy the residential gateway might have one or more of following devices:

- Desktop/laptop computers
- Media servers, digital subscriber line (DSL)/cable modems
- Universal serial bus (USB) devices (e.g., USB hard drive, printer, portable media player)
- High-definition televisions (HDTVs), plasma TVs, digital satellite TVs
- PDAs, camcorders, digital cameras
- Analog telephones

The residential gateway will have to manage and share all of the above devices and maybe more. This should serve as input for the hardware design.

Figure 1 shows a common hardware design that a residential gateway could offer.

Processor
Choosing the right processor is the key to your device performance. Since the processor will be managing a large amount of data and varied devices, choosing a network processor to do the job could be a wise decision. Alternately, any very-high-speed processor will work equally well. The processor, however, should have a peripheral component interconnect (PCI) host interface.

Interfaces and Controllers
- Hard disk drive (HDD) controllers will provide interface to integrated development environment (IDE) devices and CD–ROM/DVD–ROM drives.

FIGURE 1

Hardware Design

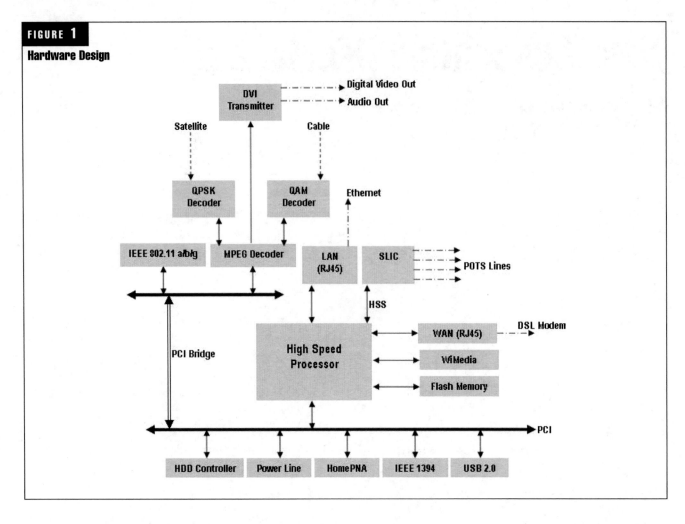

- Power line communication controllers will enable use of existing electricity network cabled throughout the house to drive next-generation home entertainment applications, including HDTV and broadband connectivity.

- HomePNA interfaces can be used to connect the home gateway to existing coax and phone lines, thereby providing data rates of up to 240 Mbps with guaranteed quality of service (QoS) and, most important, no new wires.

- Institute of Electrical and Electronics Engineers (IEEE) 1394, also known as FireWire, is a high-speed audio-video transport interface supporting speeds up to 800 Mbps. More important, a FireWire-based network can link up to 63 pieces of equipment.

- WiMedia provides wireless media connectivity and interoperability between devices in a personal-area network. Wireless USBs can be considered a viable option for WiMedia connectivity.

- USBs, whether wired or wireless, do not require introduction. Today almost all digital storage devices, camcorders, and digital cameras use a USB interface. Although the data rate is not as high as other interfaces introduced earlier, looking at the widespread use of

USBs, it is almost mandatory for our home gateway to provide the interface.

- Digital visual interface (DVI) transmitters will generate digital signals for digital displays such as HDTV and flat-panel TV.

- Quadrature amplitude modulation (QAM) and quaternary phase shift keying (QPSK) decoders will receive audio and data from cable and satellite respectively. Both the decoders comply with digital video broadcast (DVB) common interface format and generate MPEG–2 streams.

- IEEE 802.11 a/b/g controller enables the home gateway to act as a wireless switch to which other devices such as laptops can connect using wireless local-area networks (WLANs).

- Subscriber line interface chips (SLICs) provide audio for enhanced voice quality of voice over Internet protocol (VoIP) applications. It provides direct interface to analog telephones.

- RJ45 interfaces for LANs and wide-area networks (WANs) will enable the home gateway to be used with external DSL modems and Ethernet switches.

- PCI bridges might be required to expand the PCI bus to hold more controllers.

Software Design

Selection of software components will primarily depend on what hardware circuitry you decide upon. However, considering that the hardware design mentioned in previous section, *Figure 2* below shows various software blocks that you might typically need:

The primary objective should not only be to provide as many features as possible, but also to keep the software footprint as small as possible.

Operating System and Board Support Package (BSP)

Considering the fact that the home gateway is going to be heart of a variety of applications with varied speeds and needs, real-time nature of the operating system becomes a necessity. However, since this is a consumer device, considering cost factor, an open source real-time operating system (like Linux) should be the obvious choice. The benefit of open source does not end at cost. You will also have the flexibility to customize—or, rather, "tune"—the operating system kernel as per your needs.

Device Drivers and File System

Of course you will need as many drivers as the devices you want to interact with. But similar to the operating system kernel, you could remove unnecessary features from your driver to make it as "tiny" as possible.

For file systems, you might not get too many choices due to the operating system you might select. Nevertheless, almost any file system will do.

Basic Networking Stack and Bridging

The backbone of any next-generation networking device has to be transmission control protocol/Internet protocol (TCP/IP). The home gateway is no different in this respect. So you will need TCP, user datagram protocol (UDP), IP, address resolution protocol (ARP), and all other basic networking stack components.

Similarly, spanning tree protocol (STP) and virtual LAN (VLAN) implementations will be required for bridging. You can also select later enhancements of STP such as rapid STP (RSTP) or multiple STP (MSTP), but that might not add real value, since the home gateway is not going to work as a pure Layer-2 (L2) switch.

Routing and Streaming

A home gateway should have basic unicast and multicast routing capabilities. Selecting a very complicated routing protocol is not required, as the gateway is not going to be placed in a network similar to a full-fledged router.

Streaming is comparatively important since we are going to use multimedia applications. So having real-time streaming protocol and real-time transport protocol (RTP) will be a good idea.

Security and QoS

A con of any sophisticated device is developing a sophisticated security system. A security breach in a home gateway will result in direct exposure of all the home devices. Firewall, network address translation (NAT), and content filtering are some of the basic components required. You can add as many security features as you want, provided you do not overshoot the budget. Since this is going to be one of the features that will differentiate your gateway and home gate-

---

FIGURE 2

Software Design

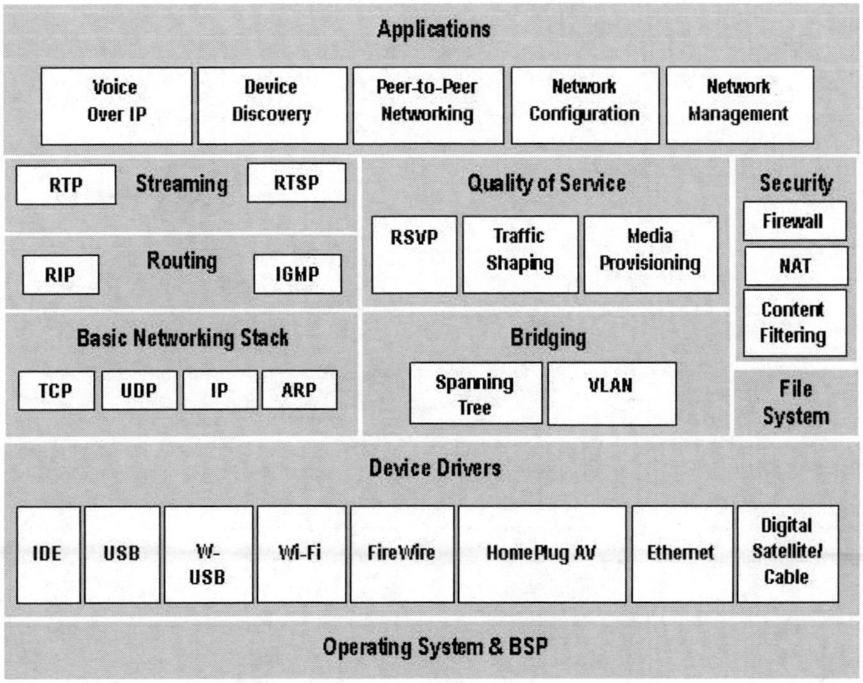

ways of other vendors, it will be a good idea to invest a handsome amount in providing security features.

Similar to security features, QoS features you provide will also be your selling point. A basic need would be providing resource reservation, traffic shaping, and media provisioning.

Applications

For end users, the applications that can run on a home gateway are of primary interest.

However, a major disadvantage of any "next-generation" device, including home gateway, is that there are too many application standards to support. Universal Plug and Play (UPnP), Digital Living Network Alliance (DLNA), open services gateway initiative (OSGI), Home Audio Video Interoperability (HAVi), and JXTA are few of them. It might be a good idea to provide that choice to customers and price the devices accordingly. Since some standards are almost parallel, consumers might want to select only one of them based on what their other home appliances support.

Home Gateway Deployment Scenario

Figure 3 depicts a typical deployment scenario of a residential gateway.

Conclusion

Any next-generation device attracts lot of attention initially. Vendors tend to believe in the philosophy of "If you are late, you do not exist." However, for a next-generation device to succeed in the long run, it has to offer real value to its user. Hence, the process of building such a device has to be well thought out. Remember, "If you are first and worst, you only exist. But if you are sufficiently on time and best, you flourish."

References

www.1394ta.org
www.homepna.org
www.plcforum.org
www.osgi.org
www.intel.com/netcomms/solutions/index.htm

FIGURE 3

Deployment Diagram

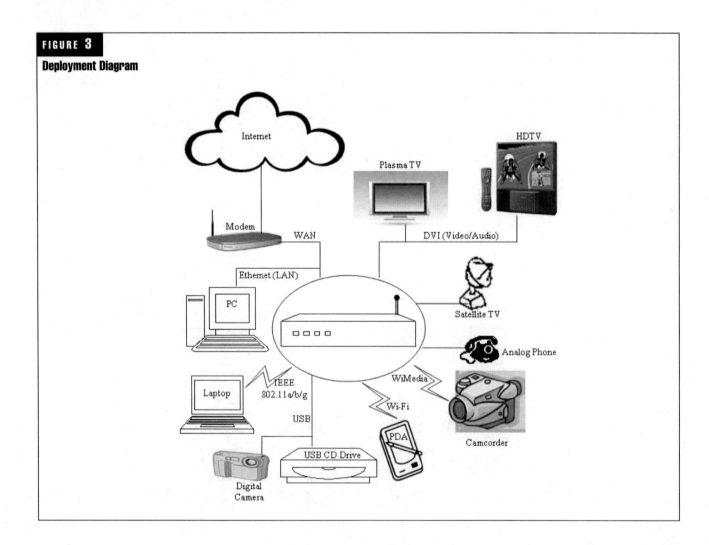

Security

Review and Analysis of WLAN Security Attacks and Solutions

Chibiao Liu

Doctoral Student, School of Computer Science, Telecommunications, and Information Systems
DePaul University

James Yu

Assistant Professor, School of Computer Science, Telecommunications, and Information Systems
DePaul University

Abstract

The growing popularity of 802.11–based wireless local-area networks (WLANs) also increases its risk of security attacks. As a result, there is much research and many publications from academics and industry to address the security issues of WLAN. In addition, there are many standard-based or proprietary products in this area. This paper provides a comprehensive review and analysis of major security attacks on WLAN and classifies them as crypto attacks and non-crypto attacks. Crypto attacks include unauthorized access, man-in-the-middle, masquerading, eavesdropping, session hijack, replay, and tampering, and these attacks can be effectively resolved with cryptographic solutions such as mutual authentications, strong data encryption and integrity protection. Security measures to prevent crypto attacks include wired equivalency piracy (WEP), 802.11i-TKIP, 802.11i-CCMP, and Internet protocol (IP)–based virtual private networks (VPNs). We present the strengths and weaknesses of each approach and propose an integrated approach of VPNs over TKIP/CCMP. We conducted extensive experimental studies to measure WLAN performance, and the results show the integrated approach does not introduce much performance overhead compared to TKIP and VPN. Non-crypto attacks include traffic analysis and denial of service (DoS), and there are no standard or effective solutions to address these attacks. However, we may use sound policies and practices to reduce the effect and risk of such attacks. This paper also provides a direction for further research and design of secured WLAN.

Introduction

Since the standardization of 802.11 in 1999, we see the increase popularity of wireless local-area networks (WLANs) and wide deployment at home, small office/home office (SOHO), campus networks, enterprise networks, and hot spots. The advantages of WLANs (in comparison to other LAN technologies) are flexibility, ease of installation and configuration, high performance, and relatively low cost. However, the popularity of WLANs also encounters a continual increase in security attacks.

We classify WLAN security attacks into two categories: crypto attacks and non-crypto attacks. The crypto attacks include unauthorized access, man in the middle (MITM), masquerading, eavesdropping, replay, tampering and session hijacking, which can be prevented by applying cryptographic solutions such as mutual authentications, strong data encryptions, and data integrity protections. Security approaches used to address these attacks include WEP, VPNs, 802.11i and the integrated approach of VPN over WEP. WEP is the first security technology proposed to protect WLANs. It was defined in 802.11-1999, and was "specified by Institute of Electrical and Electronics Engineers (IEEE) 802.11 to provide data confidentiality that is subjectively equivalent to the confidentiality of a wired local-area network (LAN) medium that does not employ cryptographic techniques to enhance privacy" [1]. However, WEP fails to provide secured data communications between the access point and wireless clients because of several security holes in the WEP security protocol [2]. VPNs were a popular solution to address the WLAN security issues, and there are two common VPN approaches: Internet protocol security (IPSec) with Layer-2 tunneling protocol (L2TP) and point-to-point tunneling protocol (PPTP) [3]. The newly approved standard of 802.11i also addresses many security issues of WLAN, and one solution of 802.11i is temporal key integrity protocol (TKIP) [4]. There was also a proposal of an integrated approach with VPN over WEP [5, 15].

The non-crypto attacks include traffic analysis and denial of service (DoS) and cannot be prevented by cryptographic measures. Traffic analysis is to obtain confidential information by observing the traffic pattern. The DoS attack causes problems with time-sensitive services (e.g., VoIP). Currently, there are no effective approaches to resolve the issues of traffic analysis and DoS attacks.

Because of various WLAN security attacks, it is important to define major attacks on WLANs and provide efficient and effective solutions to protect corporate and personal resources. Recently, many publications and researches have been proposed to address WLAN security issues, and these papers can be classified into three categories. Papers in the first category [6–9] discuss WLAN threats and provide practical guidelines to reduce the exposure of WLAN to those threats. Papers in the second category [10–12] address problems of WEP and present alternative security solutions to protect WLANs. Papers in the third category [13–15] provide a more comprehensive study on WLAN security issues, including threats, WEP problems, and alternative approaches to secure WLANs. To some extents, these papers raise WLAN security issues and help wireless users and wireless network managers secure their WLANs.

The objective of this paper is to provide a comprehensive review of WLAN security attacks and solutions and analyze the pros and cons of different security approaches. One issue with VPN is high performance overhead observed in a wired network [16], and we are interested in the VPN performance overhead in a wireless network. The questions to be investigated in this paper are summarized as follows:

- What are the major attacks on WLANs?
- What are the problems of WEP in protecting WLANs?
- Is 802.11i or VPN alone sufficient to protect WLANs against crypto attacks?
- Does VPN cause significant performance overhead for WLANs as the wired network?
- Which security approach provides the most effective and efficient protection of WLANs against crypto attacks?
- What are possible protection mechanisms to prevent non-crypto attacks of DoS and traffic analysis?

The second section of this paper analyzes major attacks on WLANs. Summary of solutions to tackle the attacks on WLANs is presented in the third section. Sections 4 to 6 review and analyze the security solutions of WEP, 802.11i and the integrated approach of VPN over TKIP along with

detailed discussion of their performances. Section 7 discusses the solutions to resolve traffic analysis and DoS attacks. Conclusions and directions for further research are at the end of this paper.

Attacks on WLANs

The network architecture of common WLANs is illustrated in *Figure 1*. Communications of WLANs can be divided into two parts—wireless-to-wireless communication and wireless-to-wired communication (*Figure 1*). The wireless-to-wireless communication is between wireless stations and the access point (AP) or between two wireless stations using the ad hoc operation mode. The wireless-to-wired communication involves one wireless station and one wired station or server on the wired network. Security attacks could happen on either the wireless part and/or the wired part of the network (*Figure 1*), and those in the wireless part will be discussed in the following sections.

Traffic Analysis
Traffic analysis is the examination of communication patterns to derive significance from otherwise meaningless communication. To launch such an attack, hacker needs only a wireless card operating in promiscuous (i.e., listening) mode and software, such as Netstumbler and commView wireless fidelity (Wi-Fi), to count the number and size of the packets being transmitted. An example (screen dump) of traffic analysis is illustrated in *Figure 2*.

In such an attack, hackers can collect or guess confidential information about the target WLAN, even if the transferred messages are encrypted. The derived information can be used for future attacks. Four cases of information can be collected through traffic analysis (*Table 1*).

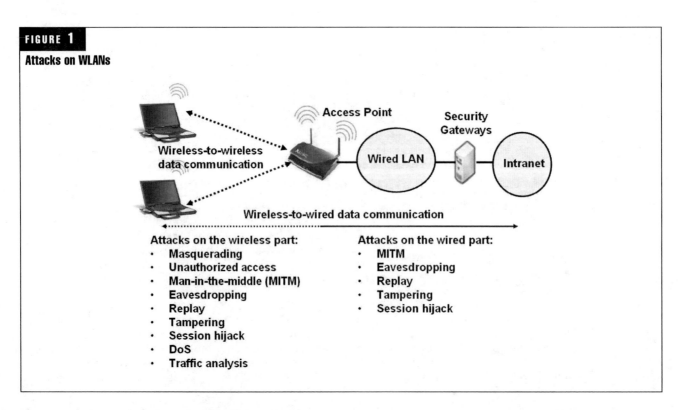

FIGURE 1

Attacks on WLANs

Wireless-to-wireless data communication

Access Point

Security Gateways

Wired LAN

Intranet

Wireless-to-wired data communication

Attacks on the wireless part:
- Masquerading
- Unauthorized access
- Man-in-the-middle (MITM)
- Eavesdropping
- Replay
- Tampering
- Session hijack
- DoS
- Traffic analysis

Attacks on the wired part:
- MITM
- Eavesdropping
- Replay
- Tampering
- Session hijack

FIGURE 2

Example of Traffic Analyses on WLANs

MAC Address	Type	SSID	Encryption	Bytes	Packets
00:12:88:BC:D0:D9	AP	2WIRE565	Yes	338	5
Private:42:7E:3A	STA			2,112	83
00:13:10:A3:70:FB	AP	AdaptiGroup	Yes	10,080	260
Usi:27:AC:26	STA		Yes	17,046	243
Intel:85:D8:58	STA			18,811	354
00:12:88:83:CA:B1	AP	2WIRE289	Yes	27,058	326
00:14:BF:1E:CA:89	AP	Bloom School Wireless		37,608	383
2wire:05:72:D9	AP	2WIRE888	Yes	147,368	2,439
GemtekTech:4F:AE:E4	STA			215,119	3,574
2wire:24:15:F9	AP	2WIRE569	Yes	803,896	24,521

TABLE 1

Collected Information about WLAN via Traffic Analysis

Case #	Collected information
1	Identities of wireless nodes: SSID, MAC addresses
2	Communication protocols used by wireless nodes
3	Number of wireless nodes and location of AP
4	Traffic patterns and the derived confidential information

Because the frame header (including SSID and MAC addresses) of WLANs must be transferred in plain text, it is relatively easy for hackers to collect this information, as shown in the first case. In the second case, hackers may learn the type of protocols being used in data transmission, and then launch an attack based on the protocol characteristics. In the third case, hackers simply count the number of packets of different MAC addresses and use position locator software. In the fourth case, information collected from wireless traffic analyses can be used to derive the activity level of individual wireless clients and access points, timing of communication activities, and mobility/movement of wireless clients.

Eavesdropping

Eavesdropping attacks on WLANs allow a hacker to obtain sensitive information such as passwords, data, and procedures for performing functions without physically tapping into a network. In WLANs, eavesdropping is the most significant attack because hackers can intercept the transmission over the air from as far as 1.5 miles using a suitable transceiver, and it is impossible for the owners of WLANs to find who is sniffing their traffic. The main purpose of the eavesdropping is to find the content of the transmitted messages over air with decoding and decryption if necessary. This is different from traffic analysis attack. Traffic analysis attack derives confidential information through traffic patterns without knowing the content of the sniffed packets.

There are two forms of eavesdropping attacks—passive eavesdropping and active eavesdropping. Passive eavesdropping is to obtain confidential information, for which, a hacker simply listens to weakly encoded or unencrypted packets. In general, strong encryption for WLAN frames can protect such an attack. Active eavesdropping attack is composed of two phases. During the first phase, as shown in *Figure 3a*, Hacker1 sniffs packets from the legitimate user and sends spoofed packets to the access point, which will do the decryption and send those plain-text packets to Hacker2. After receiving those plain-text packets from Hacker2, Hacker1 can compare the encrypted packets with the plain-text packets to crack the encryption key used by the legitimate user and the AP. During the second phase, Hacker1 passively listens to the remaining wireless communications between the legitimate user and the AP and uses the cracked encryption key to decrypt the sniffed packets.

Man in the Middle

Man in the middle (MITM) is a form of active attack in which hackers intercept and selectively modify the transmitted data to masquerade one or more of the entities involved in a wireless communication session. MITM attack against WLAN is much easier to mount than against wired networks, because it does not require physical or close access to the target network. In WLAN environments, hackers can position a rogue AP between two wireless hosts without being connected with them to launch the MITM attack (*Figure 3b*).

There are five phases to launch a MITM attack, which are as follows:

- Collecting information
- Configuring a rogue access point (AP)

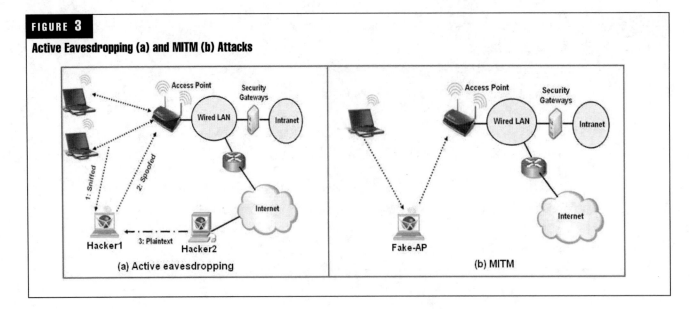

FIGURE 3

Active Eavesdropping (a) and MITM (b) Attacks

- Disassociating target wireless clients from the legitimate AP
- Tricking legitimate wireless users to connect with the rogue AP
- Manipulating target traffic

Information collection can be achieved through traffic analysis and eavesdropping attacks. Configuring a rogue AP can be accomplished using one station with two or more wireless adapters and necessary software (e.g., HostAP). Disassociating target wireless clients from legitimate AP can be done by using physical layer and data-link layer approaches with an appropriate toolset (e.g., AirJack). Tricking legitimate wireless users to connect with the rogue AP has two scenarios. One scenario is that some wireless clients will doubtlessly connect to the rogue AP. Another scenario is that the hacker can force a specific wireless client to connect with the rogue AP. Manipulating target traffic means that the hacker has a full control of the transmitted traffic if it is not properly protected using a higher-layer security protocol such as VPN or secure sockets layer (SSL). The hacker can not only sniff the traffic, but also change its content, insert viruses into downloaded files, change Web pages, and use known vulnerabilities in browser scripting to attack client machines when they visit the fake Web pages [17].

Session Hijack
Session hijack on WLANs is an attack on the existing wireless communication session, which disconnects one legitimate host and takes over the remaining communication session. There are three phases for the session hijack attack: information collection, disassociation of target, and takeover of the remaining session. The first phase can be achieved through traffic analysis and /or eavesdropping, which will collect necessary information for session takeover. The second phase can be accomplished as shown in *Figure 4*. When the legitimate wireless User2 wants to talk with the wireless User1, it sends ARP request to find the MAC address of the wireless client with the destination IP address (*Figure 4, step 1*). After the legitimate user sends back a valid ARP response message (*Figure 4, step 2*), a hacker who sniffs the ARP request, might fake an ARP

response message using tool such as arpspoof and send the faked response ARP message to User2, who will update its ARP table using the new entry of 192.168.1.3 : 66:55:44:33:22:11. From this point, the communication session between User2 and User1 is hijacked, and User2 will send all the traffic with a destination of 192.168.1.3 to the hacker. Since there is no traffic from User2 to User1, User1 just closes the current session. However, User2 continues sending traffic to User1 without knowing that all the traffic is actually sent to the hacker.

Masquerading
Masquerading is the act of a hacker posing as a legitimate user to gain access to the target network. A masquerading attack based on the spoofed MAC address is more serious for WLANs than wired LANs. The reason is that WLANs use broadcast media, and every wireless node within the area can send and receive data, while a wired LAN requires hackers to have physical access to the target network to launch a masquerading attack.

MAC address can be used to track or authenticate wireless clients to get access to a wireless access point. The fake MAC address (*Figure 5a*) cannot only hide the hacker's presence, but also bypass AP's access control list and get access to WLANs. Another form of masquerading attack is called authenticated user impersonation, which has two phases. In the first phase, both the hacker and the target user are authenticated to access WLAN services, which are configured with different priorities and/or permissions based on their MAC addresses. In the second phase, the hacker changes its MAC address to the target MAC address to access restricted services available only to the target user.

Unauthorized Access
If a hacker wants to access a wired network (the context is the Layer-2 access), he or she needs to be physically at the premise and find one unsecured network port. If there are security staff members at the entrance, the hacker cannot even get into the building. Because of the physical nature of WLANs, a hacker can easily gain access to the WLAN, even from a long distance. Unauthorized access might be relatively harmless—for example, a residential user may try his

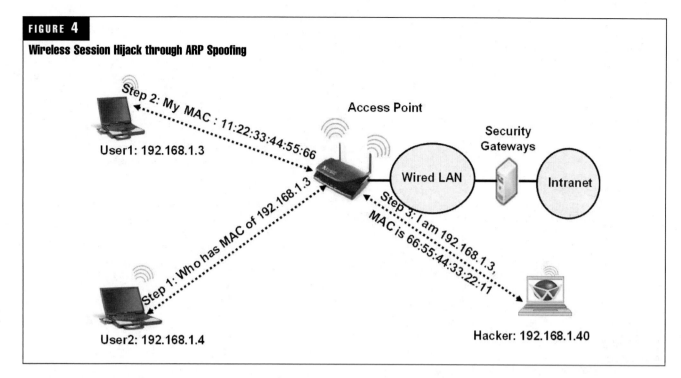

FIGURE 4

Wireless Session Hijack through ARP Spoofing

or her neighbor's WLAN to access the Internet. In the enterprise environment, however, it is considered a major security threat, and some malicious hacker might be able to read or change highly sensitive information. It is also possible for a hacker to use one network as a launching point for attacks on other networks.

Unauthorized access is one of the most challenging security threats to the enterprise network. An employee may put a rogue (unauthorized) AP on the network edge for ease of access and open the door for hackers to get in (*Figure 5b*). Similar to the rogue AP problem, Windows supports the bridge configuration between the wired and wireless LANs. With this bridge configuration, a laptop with wireless network interface card (NIC) can be hacked by external users to access the internal network (*Figure 5c*).

Replay
Packet replay attack on WLANs refers to the recording and re-transmission of message packets in the wireless network. A hacker may sniff and replay packets as illustrated in *Figure 6a*. WLAN packet replay is frequently undetectable but can be prevented by using packet time-stamping and packet sequence counting. There are two phases for replay attack on WLANs. The first phase is that the hacker monitors and captures transmitted packets between a target wireless client and the access point via a passive monitoring utility such as Airsnort. The second phase of the replay attack is to repeatedly (and selectively) transmit sniffed packets to and/or from the access point.

Tampering
In the tampering attack, a hacker sniffs and modifies the content of the intercepted packets from the WLAN and

FIGURE 5

Masquerading (a) and Unauthorized Access (b and c) Attacks

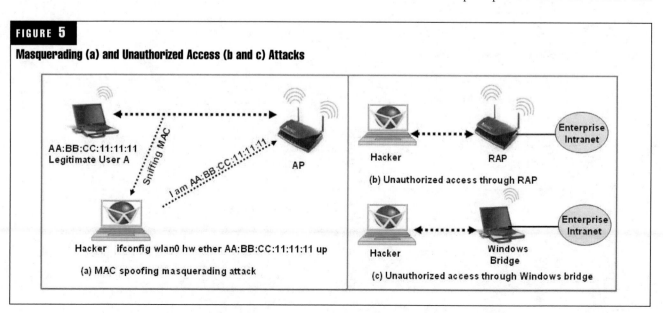

sends the modified packets to their destinations, which results in a loss of data integrity. For example, the hacker sniffs some packets, changes some part of the data, recalculates checksum and CRC to form a new packet and sends the modified packet to the victim (*Figure 6b*).

Denial of Service

WLANs are extremely vulnerable to DoS attack, which is to deny legitimate user access to the wireless network under one of the following situations:

- Wireless access point being out of service
- Wireless access point being unable to connect to a wireless station, and vice versa
- A wireless station being unable to communicate with another wireless station

There are four phases for WLAN communications, and DoS attacks could happen to any one of them, as illustrated in *Figure 7*. These phases are scanning for a wireless access point, authentication, association, and data transfer.

Physical Jamming DoS Attack

Physical jamming may prohibit victim nodes from accessing the wireless channel by using strong radio frequency (RF) signals to interfere with the communication between AP and wireless clients. In the case of 802.11b/g, a jamming device runs at 2.4 to 2.5 GHz, which can be a small RF transmitter, a powerful jammer built with microwave oven's magnetron, a flooding device built with a high-output wireless client card, or an access point. We conducted an experiment to demonstrate a jamming DoS attack. The experiment connects one wireless access point (jamming AP) to a high-capability traffic generator that sends frames with faked destination and source addresses. The jamming AP uses the same RF channel as the legitimate AP. When the jamming traffic increases, we observe significant performance degradation on the legitimate WLAN (*Figure 8a*). Homing/location tools such as Kismet are available to detect and identify the source of the jamming signal.

Authentication and Association Request Flooding DoS Attacks

Access point association and authentication buffers are used

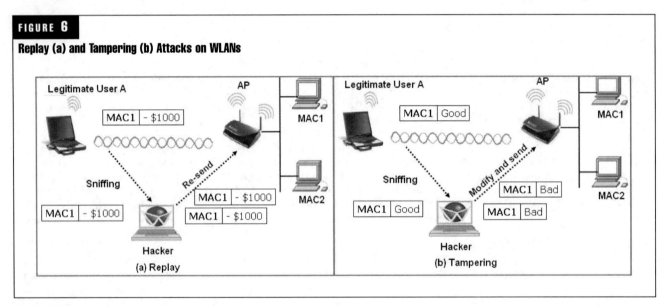

FIGURE 6

Replay (a) and Tampering (b) Attacks on WLANs

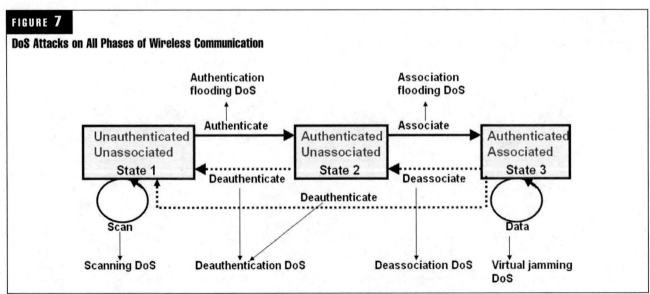

FIGURE 7

DoS Attacks on All Phases of Wireless Communication

to hold information during authentication and association processes. However, these buffers can be easily overflowed without implementation of appropriate protection. A hacker could continuously flood the access point with authentication or association request frames, making the access point incapable of accepting more client requests. We conducted another experiment to demonstrate authentication request flooding and association request flooding. The target WLAN involves one AP and one wireless station, and the wireless station pings the AP with a packet size of 64 bytes and speed of 1 packet per second. When these DoS attacks are launched, the AP will not accept new wireless connection, and the round-trip time (RTT) for the existing communications (the ping session) increases significantly (*Figure 8b*). Once the flooding is stopped, RTT of the wireless communication drops immediately to the normal level, and the AP starts to accept new wireless connections.

Deauthentication and Disassociation DoS Attacks

Deauthentication is a notification between AP and wireless clients, and it cannot be refused. A wireless client performs deauthentication by sending a deauthentication management frame (or group of frames to multiple stations) to advise the termination of authentication, which terminates an authenticated relationship. To terminate an existing association, stations may send disassociation requests to AP, which will remove the station from the wireless network. Hackers can launch sophisticated deauthentication and disassociation DoS attacks by configuring a station to operate as a rogue AP. The AP then floods the airwaves with persistent "deauthenticate or disassociate" messages that force all stations within range to disconnect from the legitimate AP.

Virtual Jamming DoS

The 802.11 standard has an optional procedure where senders and receivers exchange request to send (RTS) and clear to send (CTS) messages before data transmission. Because a node that receives the RTS/CTS frame must defer its transmission, a misbehaving node may randomly send out a large number of spurious RTS or CTS frames addressed to a possibly nonexisting node to block other well-behaved nodes in the neighborhood, thereby successfully carrying out virtual jamming DoS attack (*Figure 9a*) [18]. Virtual jamming results from the fact that a node does not check the validity or correctness of the network allocation vector (NAV) field in the RTS/CTS frames. A hacker can arbitrarily set the value of the duration (i.e., NAV) field, which reserves the channel for an extended period of time and blocks neighboring nodes from accessing the network.

Scanning DoS Attack

If there are multiple APs in the premise, wireless clients are usually connected to the AP with the strongest signal. When an internal user installs a rogue AP, wireless clients may disconnect from the legitimate AP and switch to the rogue AP. This may result in loss of network services because of security setting and/or virtual LAN (VLAN) configuration. For example, the mobile host A (MH-A) (*Figure 9b*) is connecting to the legitimate AP when the rogue AP appears close to MH-A. MH-A may disconnect from the current wireless network and switch to the route AP. As a result, it loses the intranet connections and service.

Solutions against Attacks on WLANs

In the last section, we discussed nine major attacks on WLANs, and these attacks can be classified into two categories: crypto attacks and non-crypto attacks. Crypto attacks include masquerading, unauthorized access, MITM, eavesdropping, replay, tampering, and session hijack. Non-crypto attacks include DoS and traffic analysis (*Table 2*).

Crypto attacks are related to the weaknesses of authentication and encryption, and they can be resolved with cryptographic solutions with strong authentications, encryptions, and data integrity. Non-crypto attacks are related to the features of the 802.11 protocol and the physical medium (i.e., RF), and these attacks cannot be resolved using cryptographic methods. There are no standard solutions to address non-crypto attacks, and it requires a combination of

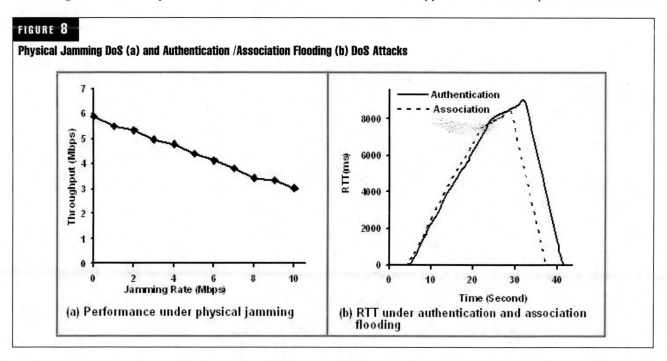

FIGURE 8

Physical Jamming DoS (a) and Authentication /Association Flooding (b) DoS Attacks

(a) Performance under physical jamming

(b) RTT under authentication and association flooding

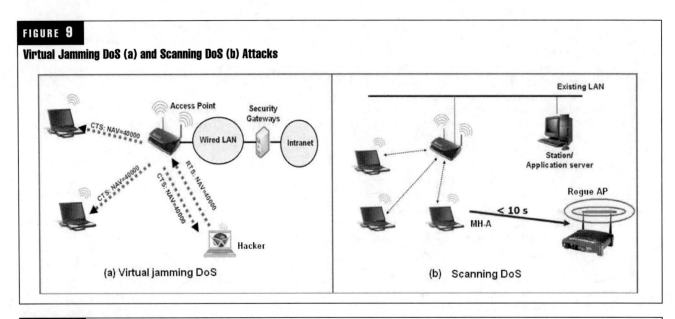

FIGURE 9

Virtual Jamming DoS (a) and Scanning DoS (b) Attacks

(a) Virtual jamming DoS

(b) Scanning DoS

TABLE 2

Attacks on WLAN and Security Solutions

Attack Type	Security solution Type
Crypto Attacks 1. Unauthorized access 2. Man-in-the-middle 3. Masquerading 4. Eavesdropping 5. Session hijack 6. Replay 7. Tampering	**Cryptographic Solutions** 1. 802.11 WEP 2. 802.11i (TKIP, CCMP) 3. VPN (PPTP, IPsec, L2TP-IPsec) 4. Integrated security approach of VPN over 802.11i
Non-crypto attacks 8. Traffic analysis 9. DoS	**Non-cryptographic solution** 1. Limit RF cover-area, reduce power, adjust antenna position 2. Physical detection and taking action 3. Modify 802.11 MAC protocol and change management frame processing mechanisms

policies, practices, and tools to address non-crypto attacks. Sections 4 through 7 discuss cryptographic solutions and non-cryptographic solutions in detail.

Wired Equivalent Privacy

WEP Authentication and Key Management
WEP uses MAC address filtering and a shared WEP key for device authentication. Since each client station attempting to connect to the access point has a unique MAC address, it can be used for authentication via limiting access to only those MAC addresses of the authorized devices, and reject other users who do not have the access permission. Another often used authentication method is to allow connections from client stations that have a valid WEP key. In the case of shared WEP key, the client station starts the authentication process by sending an association request to the access point, which then responds with a string of challenge text. The client then encrypts the challenge text using WEP key and returns the encrypted data. If the challenge text is encrypted correctly, the client is allowed to communicate

with the access point and move on to the next step of data transfer.

WEP has very simple key management system, which allows up to four manually distributed static keys. In many cases, one single WEP key is shared among all wireless clients and the access point. Since synchronizing the change to the WEP key is tedious and time-consuming, the single WEP key is seldom changed. If this WEP key is cracked, the WLAN security would be compromised.

WEP Data Encryption and Integrity
After associating with the access point, each data packet transferred between the access point and the authorized client station is encrypted, and the encryption involves the packet's data payload, 24-bits initialization vector (IV) and a secret 40- or 104-bit WEP key (*Figure 10a*). The resulting encrypted packet is then transmitted across the air. When the client station receives the packet, it uses the same WEP key to decrypt the encrypted data. Message integrity code is used to ensure the data transmitted and received are the

same. WEP uses a cyclic redundancy check (CRC), a type of hash function, for data integrity. The result of the CRC function is a small integer (checksum), which is calculated before and after transmission to ensure the data sent and received is the same. The checksum is also called the integrity check value (ICV). However, ICV is calculated based on the payload only and does not include the frame header.

Security Holes of WEP

The weaknesses and issues of WEP are widely publicized. We categorize WEP problems into the following five areas:

- *Weak authentication:* Using a wireless sniffer, a hacker can sniff the wireless traffic over the air (i.e., RF) and easily find the MAC addresses of valid users. After that, the hacker can spoof the MAC address to mimic one of the valid wireless users. For the shared key authentication, a hacker with the right tool can sniff both the plain text and the ciphertext to crack the key stream used for encryption (Figure 10b). The cracked key stream allows the hacker to crack the WEP key for launching further attacks on the compromised WLAN.

- *Initiation vector (IV) reuse attack:* RC4 stream cipher used to encrypt data has been proven insecure. The packet sent over the airwaves contains the 24-bit IV followed by the encrypted data. By passively listening to the encrypted traffic, the hacker can see the 24-bit IV out of the data stream. Because there are limited permutations of the 24-bit IV sequences for the RC4 encryption algorithm to use, WEP has to reuse the old IV to continue encrypting data. With a WEP sniffing tool, the hacker is able to sniff two ciphertexts that are encrypted with the same WEP key stream to decipher the encrypted data without even learning the encryption key (*Figure 11a*).

- *WEP key cracking:* Because of the vulnerability of the key scheduling algorithm (KSA) from the RC4 stream cipher, several weak IVs can reveal key bytes after statistical analysis [19, 20, 21], and WEP keys with a length of either 40 or 104 bits can be derived by passively collecting particular frames from a WLAN. Depending on the

actual key used, a hacker can take between 1,000,000 and 2,000,000 packets to crack a 104-bit key. On high-traffic wireless LANs, it can take less than an hour to collect up to 2,000,000 packets and less than a minute to crack the key. WEP sniffing tools such as WEPCrack, AirSnort, and Aircrack are available to crack WEP keys.

- *Forgery attack:* WEP does not protect the AP address, the source address, and the destination address used by WEP MPDU, and a hacker can modify the destination address of the sniffed packets and later retransmit the modified packet to a station on the wired network. In this case, the packet is decrypted by AP and then forwarded to the hacker station.

- *WEP data integrity attack:* ICV is based on CRC32 checksum function, and it is possible for a hacker to modify the higher-layer payload such that the CRC will still be valid for the messages [22]. Since the modified message is considered corrupted by the higher-layer protocol, the checksum would fail. As a result, the receiver discards the modified message and sends the well-known plain-text error message back to the sender. The hacker can intercept the error message and use it to crack the WEP key (*Figure 11b*).

In summary, WEP provides a primitive scheme for WLAN security. It is considered appropriate for home and SOHO networks; but it does not meet the stringent security requirements for enterprise or government networks. Because of the WEP problems, enterprises that have deployed WLAN are relying on VPN or other proprietary schemes from the vendors.

801.11i – 802.1X, TKIP, and CCMP

Spurred by data insecurities and a lack of key management with WEP, the IEEE Task Force of 802.11i developed a new standard for wireless security protocol to resolve WEP security issues. The 802.11i standard was ratified June 24, 2004 [23]. The 802.11i standard adopts 802.1X for user authentication. For data encryption, it specifies a short-term solution of TKIP and a long-term solution of common mode with

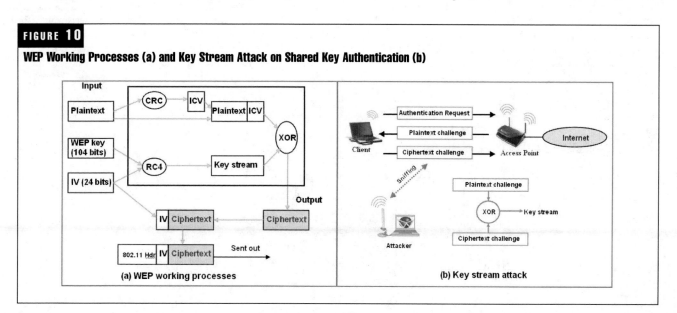

FIGURE 10

WEP Working Processes (a) and Key Stream Attack on Shared Key Authentication (b)

(a) WEP working processes

(b) Key stream attack

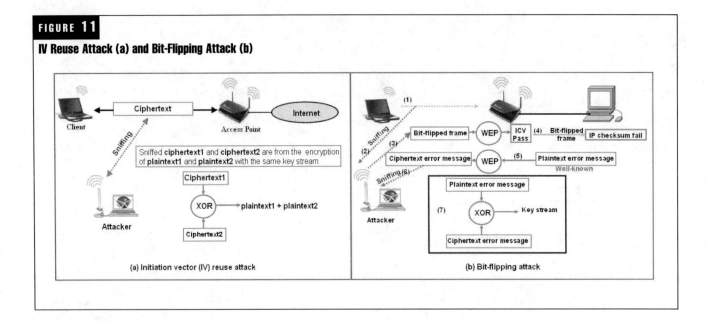

FIGURE 11

IV Reuse Attack (a) and Bit-Flipping Attack (b)

(a) Initiation vector (IV) reuse attack

(b) Bit-flipping attack

cipher block chaining message authentication code protocol (CCMP). The design of 802.11i is to address all known WEP deficiencies.

802.1X–User Authentication

WEP was designed to provide authentication and privacy, but did not accomplish either. To solve user-authentication problems, the 802.11i standard adopts the 802.1X standard, which provides "per-port user authentication," and requires user authentication before granting network access. The adoption of 802.1X for use in WLAN is an improvement in security over shared WEP key authentication. The 802.1X standard, which is based on IETF Extensible Authentication Protocol (EAP, RFC 3748), is a port-based access control protocol. It provides the MAC layer security at the authentication stage and provides an end-to-end transport for authentication between the supplicant (i.e., wireless client) and the authentication server (AS). AP proxies EAP messages between supplicants and the AS and supports the backend protocol to interwork with the AS. EAP is a general protocol that supports multiple authentication methods (*Figure 12*) such as message digest 5 (MD5, RFC 1321), transport layer security (TLS, RFC 2716), tunneled TLS (TTLS), lightweight extensible authentication protocol (LEAP), and protected extensible authentication protocol (PEAP).

The authentication process for 802.1X-EAP consists of device authentication and user authentications. An 802.1X-EAP authentication system has three entities (*Figure 13a*): The supplicant (wireless device), the authenticator (access point), and the authentication server (e.g., a backend remote authentication dial-in user service [RADIUS] server). Before associating with an AP, the station needs to accomplish device authentication, which only involves the supplicant and the authenticator. It uses the open-system authentication as illustrated in *Figure 13b*. IEEE 802.11i specifies a robust secure network information element (RSNIE) to carry RSN security information about authentication and cipher algorithms. RSN–capable stations shall include the RSNIE in beacons, probe response, association and reassociation request, and these RSNIE data will be exchanged between the supplicant and the authenticator during the device authentication.

IEEE 802.1X also defines an EAPoL (EAP over LANs) that encapsulates EAP messages between the supplicant and the authenticator. When a client requests access to the access point, it needs to provide a set of credentials to accomplish the user authentication. The client supplies some form of credentials, the access point then forwards the credentials to a RADIUS server for authentication and authorization. 802.1X–EAP also stipulates mutual authentication of client-to-

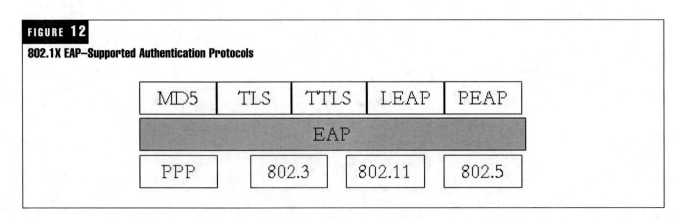

FIGURE 12

802.1X EAP–Supported Authentication Protocols

MD5	TLS	TTLS	LEAP	PEAP
EAP				
PPP	802.3	802.11	802.5	

FIGURE 13

802.1X–EAP Components (a) and Device Authentication Processes (b)

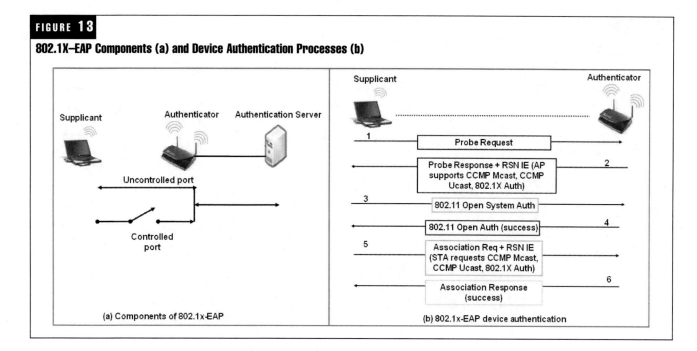

(a) Components of 802.1x-EAP

(b) 802.1x-EAP device authentication

AP and AP-to-client, which prevents MITM attack. There are two forms of 802.1X–EAP user authentication—password-based authentication and certificate-based authentication.

Cisco's lightweight EAP (LEAP) is an example of password-based user authentication, which can be used for mutual authentication and delivering keys used for WLAN encryption (*Figure 14a*). The authentication steps for this user-name/password-based scheme can be described as follows.

The wireless client sends an EAPOL–start message to an AP, which then sends an access request on behalf of the RADIUS server to the wireless client.

The client responds back to AP with its username, which is forwarded to the RADIUS server.

The RADIUS server sends back a challenge, which is forwarded to the wireless client by the AP.

- The wireless client processes the challenge and the user password with a LEAP algorithm, and sends back the processed value to the AP, which forwards it to the RADIUS server.

- The RADIUS server compares the received value with the one from its own calculation. If the two values are the same, the RADIUS server will send the wireless client a success message via the AP.

- After it is authenticated, the wireless client sends a challenge to the RADIUS server for the server authentication. If the RADIUS server is successfully authenticated, a communication channel will be opened between the wireless client and the AP.

- After finishing mutual authentications, unicast and broadcast keys will be generated and distributed to the wireless client and the AP for unicast and broadcast data encryptions.

- EAP–TLS is a certificate-based authentication protocol, which requires both the wireless client and the RADIUS server to provide digital certificates during the authentication processes (*Figure 14b*). The authentication process of EAP–TLS is discussed as follows.

- The wireless client sends an EAPOL–start message to an AP, which then sends an access request on behalf of the RADIUS server to the wireless client.

- The client responds back to AP with its username, which is forwarded to the RADIUS server.

- The RADIUS server sends its digital certificate to the wireless client for validation.

- If the RADIUS server is validated successfully, the wireless client will send its digital certificate to the RADIUS server.

- If the RADIUS server successfully validates the wireless client, it will send back the authentication success message. Otherwise, the failure message will be sent out.

- After finishing mutual authentications, unicast and broadcast keys will be generated and distributed among the wireless client and the AP for encryptions of the unicast and broadcast traffic.

There are many advantages of using 802.1X in 802.11 WLANs. The first advantage is that 802.1X allows a network to restrict access at the edge. In this way, the unauthenticated intruders were stopped from ever gaining access to the network. The second advantage is to provide a framework that allows a system to use dynamic session encryption keys. This scheme significantly enhances security by eliminating static keys and foiling attacks that require the collection of large amounts of data encrypted with a single key. The third advantage is that 802.1X is highly scalable. It

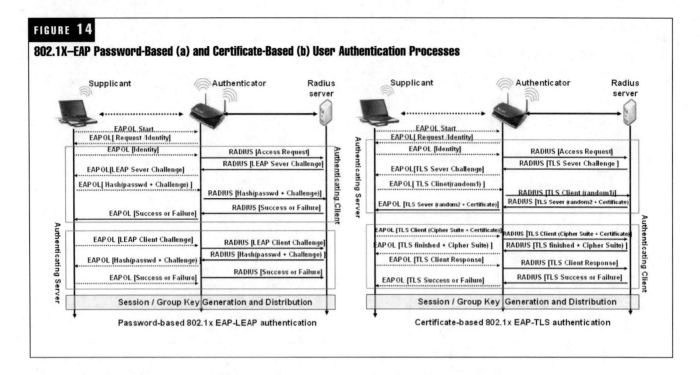

FIGURE 14

802.1X–EAP Password-Based (a) and Certificate-Based (b) User Authentication Processes

does not involve encapsulation and does not add per-packet overhead. 802.1X can be implemented on existing access points with little effect on performance. Another advantage is that it provides mutual authentication mechanism to protect against the MITM attack.

Key Distribution and Management
One issue with WEP is a lack of key management mechanism, and the encryption key must be manually configured on the wireless clients and AP. This issue is resolved in 802.1X. At the end of successful user authentication, the AS and the supplicant have established a session and possess a mutually authenticated master key, which is used to derive a pairwise master key (PMK) used during four-way handshake authenti-

cation and key distribution (*Figure 15a*). The four-way handshake process establishes a fresh pairwise key bound to the wireless station and the AP for this session. Then, the group key handshake provides a group key to all wireless stations for broadcasting communication. Meanwhile, the session key and group key could be dynamically updated for that session, which will prevent hackers from intercepting a large number of packets needed to crack the key. The key distribution structures of 802.11i are shown in *Figure 15b*.

TKIP Data Encryption and Integrity
TKIP is intended as an interim solution that allows deployed systems to be software or firmware upgradeable where the current WEP hardware implementation remains

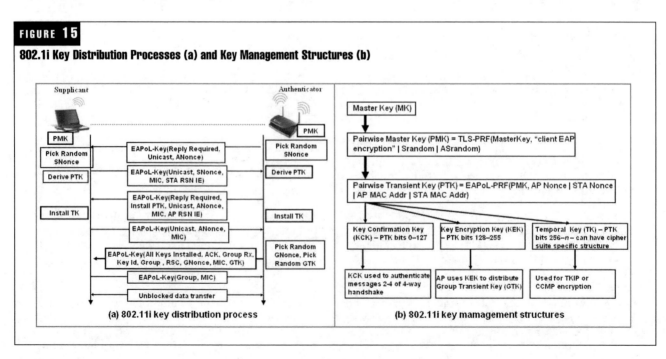

FIGURE 15

802.1i Key Distribution Processes (a) and Key Management Structures (b)

unchanged. The purpose is to minimize the performance degradation imposed by the TKIP upgrade. TKIP wraps WEP (*Figure 16a*) in the following new elements:

- A message integrity code (MIC), called Michael, to defeat forgeries
- A packet-sequencing discipline, to defeat replay attacks
- A per-packet key mixing function, to prevent encryption key cracking

The temporal key hash function takes as input the temporal key (TK), the transmitter address (TA) and the 48-bit IV, and outputs a 128-bit per-packet key stream where 24 bits are derived from the IV. TK is generated through key management system (see Section 5.2). TK values do not repeat, even across reboots. The 48-bits initialization vector (IV) value is related with each sent packet, which is implemented by a counter, starting with zero. TA, TK, and IV are hashed to generate the RC4 key stream, which is different for each station. In this way, the traffic sent by a wireless client to AP will use a different RC4 key stream than the traffic sent from the AP to the wireless client. As a result, it becomes a lot more difficult for hackers to crack a TKIP encryption key. One of the central ingredients in TKIP is a new MIC, also called Michael, which is a new keyed cryptographic hash function to produce a 64-bit output. MIC is computed from the frame header and payload as follows:

MIC = hash function (TK, source address, destination address, payload)

MIC is used to detect errors in the frame header as well as in the payload, either because of transmission errors or purposeful alterations (forgery attack). When the receiver gets a packet, it recalculates and checks the MIC. If the MIC check fails, the receiver drops the packet and initiates an alarm.

CCMP Data Encryption and Integrity

A major issue with TKIP is that it still inherits all the problems of RC4, so TKIP is viewed as a short term solution. CCMP is the preferred encryption protocol in the 802.11i standard. CCMP is based upon the CCM mode of the advanced encryption standard (AES) encryption algorithm, and used 128, 192 or 256 bits key to encrypt and decrypt data (*Figure 16b*), and its CBC–MAC mode is used to deliver data integrity. This encryption algorithm is proved to be more robust than TKIP, and has been certified for government use. Since CCMP is based on CCM mode of AES, CCMP can provide strong security against all known attacks. In addition, MIC is supported in CCMP to protect the forgery attacks against the frame header.

CCMP requires new hardware for 802.11 wireless devices, and the current WLAN devices cannot be upgraded to support CCMP. TKIP is backward-compatible with WEP and supports software or firmware upgrade of existing WEP devices. As a result, more TKIP–enabled devices are available than CCMP–enabled devices. 802.11i (TKIP or CCMP) adopts the 802.1X standard to provide mutual authentications between the AP and wireless clients, and it prevents forgery attacks such as MITM, masquerading, and unauthorized access. In addition, TKIP provides a per-packet key mixing function and changes temporal keys dynamically to prevent eavesdropping and session hijacking. Furthermore, a MIC and a packet sequencing mechanism are implemented to defeat the tampering attack and the replay attack respectively. However, 802.11i is a Layer-2 solution, which only covers the wireless area and cannot prevent attacks targeting the wireless-to-wired communications (*Figure 17a*).

Integrated Solution against Crypto Attacks

Since 802.11i does not prevent attacks on the wireless-wired communication, IP–based VPN is needed to protect the end-to-end security. VPN security technologies mainly include IPsec (RFC 2401, RFC2406-2409), IPsec–L2TP (RFC 3193)

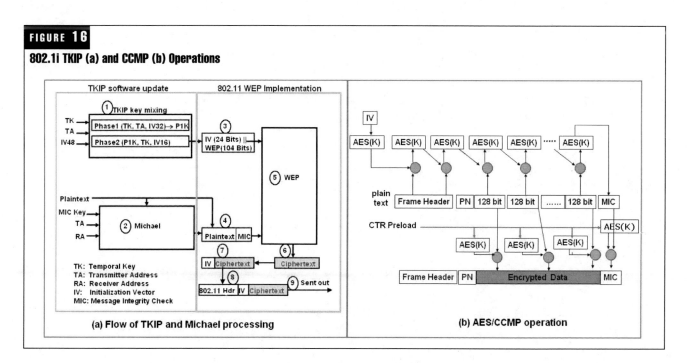

FIGURE 16

802.1i TKIP (a) and CCMP (b) Operations

(a) Flow of TKIP and Michael processing

(b) AES/CCMP operation

and PPTP (created in 1996 by 3Com, Microsoft, Ascend, and ECI Technologies). VPN security technologies can be used to fully protect the wireless-to-wired data communication through a secured channel, as illustrated in *Figure 17b*. Although VPN provides an effective way to protect wireless-to-wired communications from attacks such as eavesdropping, session hijacking, replay, and tampering, it cannot prevent Layer-2 attacks [24] such as unauthorized access, masquerading, and MITM.

We propose an integrated approach of VPN over TKIP/CCMP to defeat crypto attacks on WLANs. The network design involves virtual LAN (VLAN), VPN, and WLAN and is illustrated in *Figure 18a*. In this design, wireless users of building one are on VLAN1, and wireless users of building two are on VLAN2. All wired users (including buildings one and two) are on VLAN3. Because of the nature of wireless communication, VLAN1 and VLAN2 are at a higher risk of being attacked than VLAN3. For this reason, the wireless data communications involving VLAN1 and VLAN2 need to be fully protected using the integrated security approach of VPN over TKIP/CCMP (*Figure 18b*). This design does not change the existing infrastructure of wired

LAN. VPN over WEP was discussed in the literature [15]. Because of the inherited security weaknesses in WEP, VPN over WEP is not sufficient to protect enterprise WLANs from problems such as forgery attacks on the frame header.

The integrated security approach provides strong protection for WLANs that is inherently more effective than an individual Layer-2 or Layer-3 security approach. To understand the integrated approach's effect on performance, we conducted an experiment to measure VPN over TKIP, and compared the results with VPN and TKIP respectively. The lab results show that the performance overhead of using VPN on WLANs (with or without TKIP) is insignificant. The TCP throughput of TKIP only is 5.26M bps versus 4.66M bps for VPN (IPSec) over TKIP. This finding supports the use of VPN over TKIP–enabled WLAN to maximize the security protection. Although we did not have the environment to test VPN over CCMP, we expect the results to be comparable to VPN over TKIP. In summary, the integrated approach of VPN over TKIP/CCMP provides an efficient and effective mechanism of end-to-end communication on wired and wireless networks.

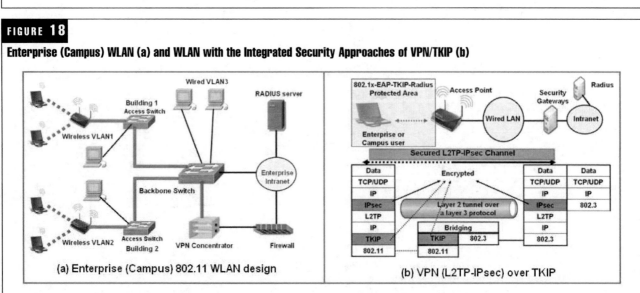

FIGURE 17

The 802.1x-EAP-TKIP (a) and VPN (b) Systems to Protect WLANs

(a) 802.1x-EAP-TKIP-Radius

(b) VPN---L2TP-IPsec

FIGURE 18

Enterprise (Campus) WLAN (a) and WLAN with the Integrated Security Approaches of VPN/TKIP (b)

(a) Enterprise (Campus) 802.11 WLAN design

(b) VPN (L2TP-IPsec) over TKIP

Solutions to Resolve Non-Crypto Attacks of DoS and Traffic Analysis

Major attacks on WLAN include crypto attacks and non-crypto attacks, and previous sections present the 802.11i standard to protect against the crypto attacks. For the non-crypto attacks of traffic analysis and DoS, there are no efficient and effective solutions available yet. This section provides sound practices to alleviate the effect of such attacks.

Practices to Protect Traffic Analysis

Traffic analysis is an attack where a hacker collects or guesses confidential information about the target WLAN even if the transferred messages are encrypted and cannot be decrypted. The collected information can be used for future attacks. The following non-cryptographic practices can be used to reduce the risk of traffic analysis attack:

Place APs near the center of buildings to reduce the signal strength outside exterior walls

Use sectional antennas (instead of omni-directional) for the AP and redirect the signal to avoid covering public areas such as parking lots, lobbies, and adjacent offices

Inject random garbage traffic to prevent hackers from finding patterns of useful traffic

Practices to Protect DoS Attacks

Although a hacker who implements a DoS attack does not steal enterprise information, the attack prevents target users from access network services. Currently, it is impossible to distinguish whether 802.11 management frames are sent by a hacker or by a valid AP. As a result, it is very difficult to prevent a DoS attack. Based on our studies, we recommend a few practices to reduce the effects of DoS and possible directions for future research and development.

Physical Jamming and Scanning DoS Attacks

Because of the nature of RF medium and the design of the core 802.11 protocols, we cannot protect against physical jamming DoS attacks and scanning DoS attacks on WLANs. However, these attacks can be detected and minimized. One common practice is to ensure continuous vigilance against rogue APs. RF monitors should be installed to continuously monitor 802.11 traffic (2.4G and 5.0G) within their range to identify the jammer and rogue AP. In addition, position detection device should be considered to pinpoint exactly where the rogue AP or the jammer device is, and security personnel can use the information to locate and remove the device.

Virtual Jamming DoS Attack

To eliminate the virtual jamming attack, one solution was proposed to verify the duration field of every overheard frame at the MAC layer. Obviously, an overheard DATA or ACK frame should be discarded if the reservation time is not equal to the fixed constant.

Authentication/Association Request Flooding DoS Attacks

For the authentication request flooding and the association request flooding attacks, AP needs to have the capability to distinguish valid requests from hacking requests. Unfortunately, the current 802.11 (or 802.11i) standard does not support authentication of such requests, and further research is needed.

Deauthentication/Deassociation DoS Attacks

WEP and 802.11i are not able to prevent deauthentication and disassociation DoS attacks. However, for a short-term solution, some modification of current 802.11 protocols may reduce the risk of these DoS attacks. Major approaches to prevent these attacks can be described as follows:

- One approach is to delay deauthentication or disassociation requests on the AP for 5 to 10 seconds. If a data packet arrives within the time interval, the original request is discarded. This scheme is based on the concept that a legitimate client would never generate packets in that order.

- Another approach to detect such DoS attacks is to compare the sequence number of the received deauthentication or disassociation frame with the previous frames sent by the access point.

These methods against deauthentication/deassociation flooding are not part of standard implementations. The long-term solution would be to authenticate deassociation and deauthentication frames without causing much performance overhead.

Conclusions

In this paper, we review, analyze, and categorize nine attacks on WLANs, and these attacks are classified as crypto attacks and non-crypto attacks. For crypto attacks, unauthorized access, MITM, and masquerading can be prevented with strong authentication technology such as 802.1X-EAP. For other crypto attacks of eavesdropping, session hijack, tampering, and replay, they can be prevented using dynamic key management, strong encryption, and MIC as specified in the new 802.11i standard (TKIP and CMCP). We conducted extensive experiments on WEP, VPN, and TKIP, and the results show little performance overhead of VPN on WLANs (with or without TKIP). This finding supports the development of an integrated approach of VPN over TKIP to fully protect WLANs against crypto attacks. For the non-crypto attacks of traffic analysis and DoS, there are no standard or effective approaches at this time. Technically, authentication- and association-related DoS attacks could be resolved with the authentication of wireless management and control frames; however, this mechanism of authentication is not specified in the 802.11 standard. For future research, we will focus on the modification of 802.11 management and control frames to ensure that the APs and wireless clients accept only authenticated management and control frames to prevent such DoS attacks.

References

1. LAN MAN Standards Committee of the IEEE Computer Society, "Part 11: Wireless LAN Medium Access Control (MAC) and Physical Layer (PHY) Specifications," standards.ieee.org/getieee802/download/802.11-1999.pdf.

2. B. Nikita, I. Goldberg, and D. Wagner, "Intercepting Mobile Communications: The Insecurity of 802.11," The proceedings of the Seventh Annual International Conference on Mobile Computing and Networking, July 16–21, 2001, www.isaac.cs.berkeley.edu/isaac/mobicom.pdf.

3. Intel Corporation, "VPN and WEP: Wireless 802.11b Security in a Corporate Environment," www.intel.com/cd/ids/developer/asmo-na/eng/20507.htm.

4. IEEE 802.11i Task Group, standards.ieee.org/getieee802/download/802.11i-2004.pdf.

5. Intel Corporation, "Wi-Fi Protected Access and Intel Centrino Mobile Technology Deliver a Robust Foundation for Wireless Security," www.intel.com/products/mobiletechnology/docs/wpa_cmt_security.pdf.

6. R. Fuller, "Building a Cisco wireless LAN," Syngress publishing, 2003 chapter 8, pages 375–446.

7. Symantec Enterprise Security, "Wireless LAN Security: Enabling and Protecting the Enterprise," securityresponse.symantec.com/avcenter/reference/symantec.wlan.security.pdf.

8. L. Phifer, "Understanding Wireless LAN Vulnerabilities," Business Communications Review, 2002, www.corecom.com/external/bcrmag/bcrmag-wlansec-sep02.pdf.

9. R. A. Stanley, "Wireless LAN Risks and Vulnerabilities," 2002, www.isaca.org/ContentManagement/ContentDisplay.cfm?ContentID=13592.

10. N. Cam-Winget, R. Housley, D. Wagner, J. Walker, "Security flaws in 802.11 data link protocols," Communications of the ACM, Vol. 46(5), 2003, pages 35–39.

11. S. Convery, D. Miller, S. Sundaralingam, "Cisco SAFE: Wireless LAN Security in Depth," www.cisco.com/en/US/products/hw/wireless/ps430/products_white_paper09186a008009c8b3.shtml

12. J. Walker, "802.11 Security Series Part II: The Temporal Key Integrity Protocol (TKIP)," or1grebe.cps.intel.com/cd/ids/developer/asmona/eng/19181.htm, M. Disabato, "Wi-Fi Protected Access: Locking Down the Link," June 2003, www.wi-fi.org/OpenSection/pdf/Wi-Fi_ProtectedAccessWebcast_2003.pdf.

13. S. Gayal, S. A. Vetha Manickam, "Wireless LAN Security Today and Tomorrow," www.fts.gsa.gov/2003_network_conference/3-8_wireless_lan.

14. R. Abdul Hamid, "Wireless LAN: Security Issues and Solutions," 2003, www.sans.org/rr/whitepapers/wireless/1009.php.

15. T. Karygiannis, L. Owens. "Wireless Network Security 802.11, Bluetooth and Handheld Devices," NIST publication, Nov. 2002, csrc.nist.gov/publications/nistpubs/800-48/NIST_SP_800-48.pdf.

16. C. Javier Castro Pena and J. Evans. "Performance Evaluation of Software Virtual Private Networks (VPN)," Annual IEEE Conference on Local Computer Networks, 2000, pp 522–523.

17. A. Godber, P. Dasgupta, "Countering Rogues in Wireless Networks," Proceedings of the 2003 International Conference on Parallel Processing Workshops (ICPPW'03), csdl.computer.org/dl/proceedings/icppw/2003/2018/00/20180425.pdf.

18. D. Chen, J. Deng, P. K. Varshney, "Protecting wireless networks against a denial of service attack based on virtual jamming," www.sigmobile.org/mobicom/2003/posters/11-Chen.pdf.

19. S. Fluhrer, I. Mantin, A. Shamir, "Weaknesses in the Key Scheduling Algorithm of RC4," www.drizzle.com/~aboba/IEEE/rc4_ksaproc.pdf.

20. A. Stubblefield, J. Ioannidis, A. D. Rubin, "A key recovery attack on the 802.11b wired equivalent privacy protocol (WEP)," ACM Transactions on Information and System Security (TISSEC), Vol 7(2), 2004, pages 319–332.

21. D. Hulton , "Practical Exploitation of RC4 Weaknesses in WEP Environments," www.dachb0den.com/projects/bsd-airtools/wepexp.txt.

22. Cisco Systems. "A Comprehensive Review of 802.11 Wireless LAN Security and the Cisco Wireless Security Suite," 2002, www.cisco.com/warp/public/cc/pd/witc/ao1200ap/prodlit/wswpf_wp.pdf.

23. C. He, J. C. Mitchell. "Analysis of the 802.11i 4-way handshake," Proceedings of the 2004 ACM workshop on Wireless security, 2004, pp 43–50, byte.csc.lsu.edu/~durresi/7502/reading/p43-he.pdf.

24. Hewlett-Packard Development Company, L.P., "Executive Briefing: Wireless Network Security," docs.hp.com/en/T1428-90017/T1428-90017.pdf.

Note

1. CCMP is one of the encryption methods of AES and 802.11i chose CCMP for data encryption.

An Architecture for Enhancing Internet Control-Plane Security

Ram Ramjee

Technical Manager
Bell Labs, Lucent Technologies

T. V. Lakshman

Director
Bell Labs, Lucent Technologies

Thomas Woo

Director
Bell Labs, Lucent Technologies

Krishan Sabnani

Senior Vice President
Bell Labs, Lucent Technologies

Introduction

Network operators worldwide are contemplating a move toward a converged Internet protocol (IP) network in which they expect to carry all types of traffic, including voice, video, and data. For example, British Telecom is launching a major initiative to move toward a converged 21st-century IP network. IP routers comprise the basic network element in these converged IP networks. Thus, a closer examination of the architecture and functions of IP routers and networks is critical. Re-examining the network architecture and distribution of router functions has been a topic of recent research interest (Clark et al., 2002; Clark et al., 2003; Feamster et al, 2004; Yang, Weatherall, and Anderson, 2005; Lakshman et al., 2004).

David Clark argues that, as the Internet moves from a research network to a critical component of everyday use, several new requirements have emerged that require a redesign of the Internet architecture (2002). In his view, the next-generation architecture should be positioned to accommodate the tussle between the many users of the Internet such as service providers, government, content providers, and end users. In the final summary of the Defense Advanced Research Projects Agency (DARPA)–funded New Arch project, the authors conclude that while the fundamental aspect of the current Internet, i.e., application-independent data transport service, needs to be preserved in the new architecture, many attributes of this architecture, including the pure datagram assumption, global addressing, and universal transparency, need to be re-examined (Clark et al., 2003). The focus of this paper is on a new-network architecture that improves the security of the control-plane infrastructure.

Attacks on the Internet infrastructure have been increasing steadily in the past few years. Thus, security has been a key area of focus in research and protocol development (Kent

and Atkinson, 1998; Dierks and Allen, 1998; Kent, Lynn, and Seo, 2000; Ateniese and Mangard, 2001; Yang et al., 2004). A number of secure protocols such as IP security (IPSec) (Kent and Atkinson, 1998), transparent local-area network (LAN) service (TLS) (Dierks and Allen, 1998), secure border gateway protocol (S–BGP) (Kent, Lynn, and Seo, 2000), and domain naming system security extensions (DNSSEC) (Yang et al., 2004) have been developed and form an important part of increasing the security of the Internet. However, one of the key drawbacks of the current Internet control-plane architecture is that control messages are sent in-band and share the same resources as the data plane, thus exposing the control plane to many possible attacks.

In this paper, we present the soft router architecture, which separates the router control plane from the forwarding plane, and argue that introducing this separation in the Internet architecture has significant benefits in terms of increasing the security of the Internet-control infrastructure. An analogy of a similar migration is readily found in the telephone network: In the 1980s, the signaling in the telephone network was in-band and suffered from security problems because of the "phone phreakers" who emulated the signaling tones to obtain free phone calls. The move toward a separate signaling network called the signaling system 7 (SS7) network helped significantly improve the security of the telephone network infrastructure.

The rest of the paper is organized as follows: In Section 2, we present related work; in Section 3, we present an overview of the soft router architecture; in Section 4, we discuss the security benefits of the soft router architecture; and finally, in Section 5, we present our conclusions.

Related Work

Security of Internet protocols has been a major research theme in recent years. IPSec provides mechanisms for authenticating and encrypting the end-to-end connection at the IP layer (Kent and Atkinson, 1998), while TLS provides end-to-end privacy and data integrity over a transport protocol such as transmission control protocol (TCP) (Dierks and Allen, 1999). Routing protocols such as open shortest path first (OSPF) and border gateway protocol (BGP) have their own security measures, though IPSec could be used as an underlying mechanism to authenticate the peering routers at the link level. OSPF relies on password authentication or a Message Digest 5 (MD5)–based cryptographic authentication of protocol messages. BGP uses the TCP MD5 signature option to authenticate a peer and thwart attacks based on simple TCP segment spoofing and resets. S–BGP presents a thorough approach to the problem of securing the entire BGP update message (Kent, Lynn, and Seo, 2000). The domain name server (DNS) infrastructure has also been a victim of attacks called "cache poisoning," which can result in hijacking of user connection requests. DNSSEC adds security to the DNS to protect against these attacks (Yang et al., 2004). While all these proposals play an important part in securing the Internet-control plane, this paper proposes a complementary approach of enhancing the security of the control plane through the introduction of a new architecture that separates the control plane from the forwarding plane.

Denial of service (DoS) attacks are a major source of infrastructure weakness today, and novel architectures such as using capability-based routing are being investigated (Yang, Weatherall, and Anderson, 2005). In this approach, a source initially obtains "capabilities to send packets" from its destination; the source then adds this capability information to each packet, which is then verified by the routers before delivering the packets to the destination. This approach stops the DoS attack at the source itself, since a router near the source can drop packets with incorrect capabilities. In Feamster et al. (2004), the authors make a case for separating BGP from the routers and centralizing it, thus helping avoid the numerous configuration problems that plague today's Internet service providers. While this separation provides some of the security benefits discussed in this paper, a full separation of all control protocols as proposed in our soft router architecture provides a comprehensive approach to control-plane security.

Soft Router Architecture

In this section, we present an overview of the soft router architecture that was originally introduced in Lakshman et al., 2004. We first describe the network entities that constitute this architecture. We then present two manifestations of this architecture—the logical and physical separation of the control plane and the forwarding plane. Finally, we present the protocols that enable this architecture.

Network Entities
There are two main types of network entities in the soft router architecture—the forwarding element (FE) and the control element (CE), which together constitute a network element (NE) (router).

The FE is a network element that performs the actual forwarding and switching of traffic. In construction, an FE is very similar to a traditional router; it may have multiple line cards, each in turn terminating multiple ports, and a switch fabric for shuttling data traffic from one line card to another. The key difference from a traditional router is the absence of any sophisticated control logic (e.g., a routing process such as OSPF or BGP) running locally. Instead, the control logic is hosted remotely. The exact nature of forwarding function can be packet forwarding, which includes Layer-2 (MAC–based switching) and Layer-3 (longest-prefix match) forwarding; label switching, an example of which is MPLS forwarding; and optical switching, in which the traffic can be time-switched, wavelength-switched, or space-switched among the links. In each of these cases, the switching function is driven by a simple local table, which is "computed" and "installed" by a CE in the network.

The CE is essentially a general-purpose computing element such as a server. It connects to the network like an end host, except that it is typically multi-homed to the network via multiple FEs so that it is not disconnected from the network when a single link fails. A CE runs the control logic on behalf of FEs, and hence "controls" them. In principle, any control logic typically found on a traditional router can be migrated to the CEs, including routing protocols such as OSPF and BGP as well as protocols such as RSVP, LDP, mobile IP, etc.

At a high level, the NE is a logical grouping of FEs and the respective CEs that control those FEs. Given this wide spectrum of possibilities of FE and CE combinations, we focus on a restricted but practical case where the FEs making up an NE are part of a contiguous "cloud." Physically, this represents the clustering of neighboring physical FEs into a single NE. A typical scenario is that of several routers being connected back to back in a central office. From a routing perspective, this clustering-based definition of the NE results in a natural hierarchy, thus reducing the inter–NE routing complexity.

Network Architecture
There are two possible ways of separating the CE (control plane) from the FE (data plane). In a logical separation (*Figure 1*), a soft router network is not significantly different from a traditional routed network, except for the addition of a few multi-homed servers (CEs). The control-plane protocol messages continue to traverse the data plane for communication between adjacent routing peers (they are tunneled between the CE and the FE so that neighboring routers/FEs are unaware of the separation of control from forwarding). This results in an architecture that very closely resembles the current architecture except for the decoupling of the control plane and the resulting benefits of improved scalability and reliability. Since it mimics the current network architecture, minimal routing protocol changes are needed for proper functioning.

In a physical separation (*Figure 2*), the control plane is physically separated from the data plane, similar to the way the SS7 signaling network is separate from the telephony network. Thus, all controllers in the routing server farm form their own private network topology that is independent of the underlying forwarding plane topology. This provides a very high-security environment for the network, in addition to the improved scalability and reliability advantages mentioned above. However, the downside of this architec-

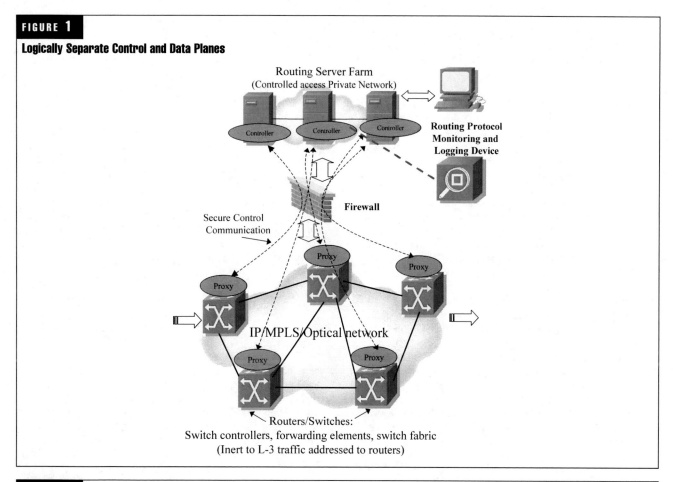

FIGURE 1

Logically Separate Control and Data Planes

Routing Server Farm
(Controlled access Private Network)

Routing Protocol
Monitoring and
Logging Device

Secure Control
Communication

Firewall

IP/MPLS/Optical network

Routers/Switches:
Switch controllers, forwarding elements, switch fabric
(Inert to L-3 traffic addressed to routers)

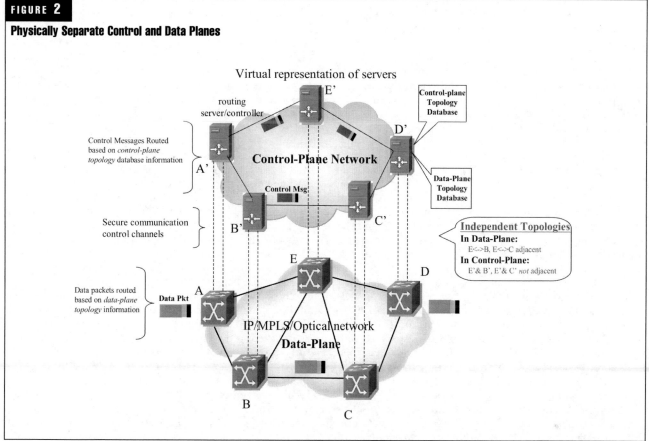

FIGURE 2

Physically Separate Control and Data Planes

Virtual representation of servers

routing
server/controller

Control-plane
Topology
Database

Control Messages Routed
based on *control-plane*
topology database information

Control-Plane Network

Control Msg.

Data-Plane
Topology
Database

Secure communication
control channels

Independent Topologies
In Data-Plane:
E<->B, E<->C adjacent
In Control-Plane:
E'& B', E'& C' *not* adjacent

Data packets routed
based on *data-plane*
topology information

Data Pkt

IP/MPLS/Optical network
Data-Plane

ture is that changes to existing routing protocols are needed (e.g., the protocols need to keep track of two network topologies—one for the data network and other for the control network).

Protocols

In this section, we present an overview of the protocols that are necessary for separating the control plane from the forwarding plane. For a detailed description of these protocols, please refer to Ramjee et al., 2006.

- *Dynamic binding protocol*: The binding between an FE and a CE is established via a protocol that discovers CEs and FEs, and also maintains these bindings in the face of network disruptions. In the most general case, a FE can bind to any available CE, and a CE can perform control functions for any FE, thus yielding maximal resiliency and minimal configuration overhead. This dynamic binding ability is a core feature of the soft router concept. While control separation from the forwarding plane is important, the real power of the soft router architecture comes from the ability of an FE to dynamically bind to any CE on the network at any time, and vice versa. This provides a significant degree of freedom in the design and deployment of the control plane, which in turn leads to improvement in resiliency, operational complexity, scalability, management, and security. In contrast, the static bindings in the traditional model make it less reliable and less flexible with respect to redesigning the Internet control plane.

- *FE/CE protocol*: There are two parts to the FE/CE protocol—data and control. For the data part, it supports tunneling of routing protocol packets between FEs and CEs so that a routing protocol packet received by a FE can be sent to CE for processing. For the control part, once a binding is established, the FEs and the CEs use this protocol to communicate state and perform control. On the uplink (FE to CE) direction, this control protocol provides link and forwarding state information (e.g., link up/down signal) to the CE. On the downlink direction, the protocol carries configuration and control information (e.g., enable/disable a link, forwarding information base [FIB], etc.). The Internet Engineering Task Force (IETF) is working on standardizing a protocol between the CE and the FE in the ForCES working group (Yang et al., 2004). Although the current focus of the working group is limited to a single-hop, direct connection between the CE and the FE, the set of protocols developed can be easily enhanced for our purposes.

- *CE/CE protocol*: In a pure soft router network, especially one with a physically separate signaling network, a CE/CE protocol is necessary for the CEs to discover each other and determine the routes between them. In this case, the CEs would also exchange control packets (e.g., OSPF database exchange packets) directly between themselves. On the other hand, in a heterogeneous network or a logically separate signaling network, a CE/CE protocol is not strictly necessary. In this case, interoperability with existing routers can be preserved by using the FE/CE protocol to tunnel any control packets received from or sent to any of

the ports in the FE (e.g., all OSPF protocol packets if the CE is running OSPF for the FE).

Security Benefits

In this section, we discuss the security benefits of the soft router architecture. First, we need to ensure that the new protocols introduced in the soft router architecture—to be able to flexibly separate the control plane from the forwarding plane, namely, the dynamic binding, FE/CE and CE/CE protocols—do not introduce any new vulnerabilities to control-plane attacks. We note that these protocols execute only on the FEs and CEs of a single autonomous system and they communicate over either a logically or physically separate links. Thus, they are not amenable to man-in-the-middle or DoS attacks. Further, these protocols can be protected by standard security mechanisms such as password-based authorization or hash-based authentication of protocol messages.

We now highlight the security benefits of the soft router architecture. The soft router architecture adopts a multilayered approach to protecting the Internet control-plane infrastructure. We discuss each of these layers below.

Separate Control Plane

As discussed in Section 3, the soft router architecture provides for either a logical or physical separation of the control plane from the forwarding plane. In the current Internet architecture, control protocol messages (such as BGP or OSPF) are sent on the same links as the data packets and share the bandwidth of the links in a best-effort manner. Thus, a DoS attack that attempts to flood the data links connecting two routers has an undesirable side effect of causing congestion and packet drops of control messages.

A logically separate control plane differs from today's architecture in that a certain amount of bandwidth resources on each link are reserved for control messages. This protects important control messages such as the OSPF Hello protocol message that maintains router adjacency and live-ness of links in the OSPF link state database. When bandwidth is not reserved for control messages, a DoS flood attack that saturates the link can result in loss of multiple OSPF Hello messages; this would result in OSPF declaring that the connecting link is down, causing packets (including attack packets) to be routed around this link. In turn, this reroute would cause subsequent OSPF Hello packets to be successfully exchanged over the original link, resulting in the link being declared as up and the attack traffic redirected back over this link. This oscillation of link up/down events in OSPF can cause other cascading effects throughout the network, resulting in an unstable network. Similar events can also be constructed in the case of the I–BGP protocol, where protocol messages between peers may be exchanged over multiple links instead of just neighboring router links as in the case of OSPF. A simple bandwidth reservation approach, as depicted in the logically separate control-plane architecture, can easily avoid these kinds of attacks.

A physically separate control plane adds another layer of security as compared to the logically separate control-plane architecture. In this case, data packets cannot traverse the control network; thus, the control network is completely shielded from DoS attacks on the data plane. The addresses of the control servers in the network can be private (in fact, the control servers need not even be IP–addressable) and thus unreachable from outside the network. Finally, since the control-plane topology is independent of the data-plane topology, the control servers can be hosted in more physically secure locations than the data-forwarding elements, further increasing the security of the control-plane infrastructure.

Firewall Protecting the Control Servers
In today's Internet architecture, the border routers of an ISP–autonomous system terminate external border gateway protocol (E–BGP) sessions with border routers of other autonomous systems. A common approach to secure these E–BGP sessions is to use the transmission control protocol (TCP) MD5 signature option, which authenticates each TCP packet sent from the peer router. However, since the verification of the MD5 signature is a processor intensive operation, a simple DoS attack can be launched at the BGP processing router by sending a flood of BGP packets. Since these packets will not have the correct MD5 signatures, they will eventually be dropped, but only after expending significant processing resources; this could result in buffer overflow of legitimate protocol packets, resulting in a DoS attack. Moreover, if these packets were sent using the spoofed address of the peer router as the source IP address, simple IP–based filters will be ineffective in preventing such an attack.

In the soft router architecture, a hardware-based firewall that performs MD5 signature verification before forwarding valid packets to the control server can be deployed. This can thwart the E-BGP DoS attack described above and is cost-effective, since there are only a few control elements in an autonomous system in this architecture. Contrast this with the case of today's architecture, where hundreds of border routers may terminate E–BGP sessions with other autonomous systems, thus rendering the use of such an approach very expensive.

Higher-Capacity Control Servers
Control processor blades in current routers have limited processing capability because of numerous constraints on router design such as power, slot, and cooling. In contrast, a blade-server-based control server does not suffer from these constraints. The availability of higher processing capacity in the control servers can be very useful from a security point of view. For a straightforward example of this observation, note that the deployment of S-BGP has partly suffered because of the compute-intensive cryptographic operations of the protocol that cannot be easily supported by the processing capabilities in today's routers (Kent, Lynn, and Seo, 2000). The use of control servers in soft router architecture can easily address this issue. Similarly, the processing requirements of control-plane protocols can surge during unexpected or malicious events such as during the initial stages of worm propagation. For example, the number of BGP updates and withdrawals surged almost tenfold during the Slammer worm attack in January 2003 (Lad et al., 2003). The soft router architecture with high-capacity control servers can better manage such sudden surges in processing.

Higher processing capacity in control servers can be useful in deploying other security applications as well. For example, the deployment of sophisticated statistical-analysis-based intrusion-detection software may require the availability of ample processing and storage resources in the network.

Open-Source-Based Router Operating System
In the soft router architecture, the control servers run an open-source-based operating system. This is in contrast to current router vendors that provide proprietary operating systems to manage their routers. There has been a lot of debate in the security community with respect to the impact of open source on security. Open-source software allows everyone, including attackers and defenders, to explore the vulnerability of the system. If we make the assumption that there will be fewer attackers than defenders, then the open-source system will benefit from the increased scrutiny. Furthermore, once a problem is detected, it can be fixed immediately, unlike a closed-source system. While the impact of open source on security is still being debated, the consensus among researchers is that open-source systems have important properties that enable it to be more secure than closed-source systems (Wheeler, 2003). Deploying control servers on these open-source-based operating systems provides yet another layer of defense against attacks on the control infrastructure.

Conclusions

We first presented the soft router architecture, which separates the router control plane from the forwarding plane. We then argued that this separation has significant benefits in terms of increasing the security of the Internet control infrastructure, mainly because of the following four reasons: a separate control plane shields the control plane from DoS attacks on the data plane; fewer control servers than routers allows the cost-effective deployment of hardware-based special-purpose firewalls; higher processing capacity of these control servers help deploy more secure protocols and handle unexpected overload; and use of open-source operating systems in these control servers has the potential to improve the security of the Internet control platform. As security increasingly becomes a major concern, new architectures that address this vulnerability will play a key role in the migration toward a next-generation Internet.

Bibliography

G. Ateniese and S. Mangard. 2001. A new approach to DNS security (DNSSEC). Proceedings of eighth ACM Conference on Computer and Communication Security.

D. Clark, J. Wroclawski, K. Sollins, and R. Braden. 2002. Tussle in cyberspace: Defining tomorrow's Internet. Proceedings of ACM SIGCOMM Conference, August 2002.

D. Clark et al. 2003. NewArch: "Future generation Internet architecture." Final report; available at www.isi.edu/newarch/iDOCS/final.finalreport.pdf.

T. Dierks and C. Allen. 1999. The TLS Protocol Version 1.0. RFC 2246, Internet Engineering Task Force.

N. Feamster, H. Balakrishnan, J. Rexford, A. Shaikh, and J. van der Merwe. 2004. The case for separating routing from routers. SIGCOMM FDNA workshop.

S. Kent and R. Atkinson. 1998. Security Architecture for the Internet Protocol. RFC 2401, IETF.

S. Kent, C. Lynn, and K. Seo. Secure border gateway protocol (S-BGP). Institute of Electrical and Electronics Engineers Journal on Selected Areas in Communications, 18 (4):582–592, April 2000.

M. Lad, X. Zhao, B. Zhang, D. Massey, L. Zhang. 2003. Analysis of BGP update surge during the Slammer worm attack. IWDC.

T.V. Lakshman, T. Nandagopal, R. Ramjee, K. Sabnani, and T. Woo. 2004. The SoftRouter archictecture. HotNets.

R. Ramjee et al. 2006. Separating the Control Software from Routers. COM-SWARE 2006.

L. Yang et al. 2004. Forwarding and Control Element Separation (ForCES) Framework. RFC 3746.

X. Yang, D. Weatherall, and T. Anderson. 2005. A DoS-limiting Network Architecture. Proceedings of ACM SIGCOMM Conference.

D. Wheeler. 2003. *Secure Programming for Linux and Unix HOWTO*. Available at www.dwheeler.com.

Telecommunications

The Open Telecommunications Services Model

Peter Briscoe

Founder, President, and Chief Executive Officer
Convedia Corporation

Voice, data, and video services are merging in the Internet protocol (IP) network. This evolution will take many years, but it is inevitable. The move to an Internet model will change the way many businesses function because the Internet model empowers end users and cuts out middlemen. This move hollows out businesses by shortening the path between the creation of value and the consumer of that value. In this environment, telecommunications companies will find that they will lose the cash cow that voice services have been for them until now. However, if they make the right moves, telecom companies will more than make up for this loss with a greater breadth of services and a higher volume of users. This paper outlines an open telecommunications services model (OTSM) that telecom companies could use to stimulate the market and drive many new services.

The Old Way

Historically, telecommunications services such as voice conferencing, voice mail, and auto-attendants have been delivered by telecommunications companies, which developed powerful, expensive servers that sat in the network. This tied the creation of new services and features to these companies' development cycles and to their limited knowledge of the businesses where the services were actually being used.

The interface to the user has followed a predefined service model that could not be altered. This model has usually been designed to be generic enough to capture a large cross section of the end user's requirements. The services themselves have followed strict state machine-based routines that eliminate creativity and force the end user into a very limited service. The net result of this has been a paucity of services (especially services that can stand alone) and services that do not integrate well with desktop applications. While service platforms have gotten better with service creation environments (SCEs) and telecom developer protocols such as Jain and Parlay, the basic premise is still that a group of telecom developers creates the services, which are then offered as predefined packages to end users. This is in direct contrast to the way the Internet works.

How the Internet Changes the Model

The Internet model is inherently different from the historical telecom services model. In the Internet model, users are empowered as control and information moves out from the core to the edge of the network. The market gets distributed across the Internet, rather than as a large centralized company owning the market. The music and book industries have learned this; the auto industry is under siege; the movie industry is beginning to see this; and the telecommunication industry will be no exception.

Since this change is inevitable in telecommunications, it is important to embrace the new model and to try to find the best way to capitalize on it. Telecom service providers that are not successful in this will become pipe providers for peer-to-peer traffic or be bypassed altogether as shown in *Figure 2*.

The key to success in this new order is getting on the side of and being part of the new model. The telecom industry must embrace the Internet culture to create an attractive framework for the millions of creative people who use the Internet and take advantage of telecommunications capabilities in an open way. This would empower the end-user community to apply telecommunications services to their specific business requirements while allowing the network providers to make money.

The Open Telecommunications Services Model

The OTSM is built upon the following fundamental concepts:

- There are capabilities that are naturally suited to the core network or competitive within the core network. These are called telecom service building blocks.
- In the OTSM model, telecom service building blocks are developed and offered within a standardized framework and offered to the end users as open source. End-user applications that access the telecom service building blocks are then charged for usage.

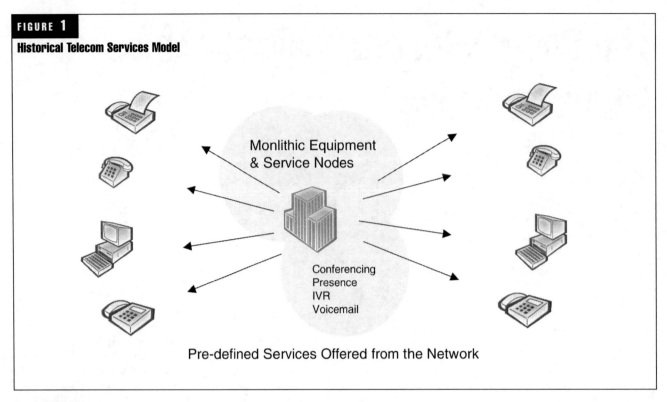

FIGURE 1

Historical Telecom Services Model

Monlithic Equipment & Service Nodes

Conferencing
Presence
IVR
Voicemail

Pre-defined Services Offered from the Network

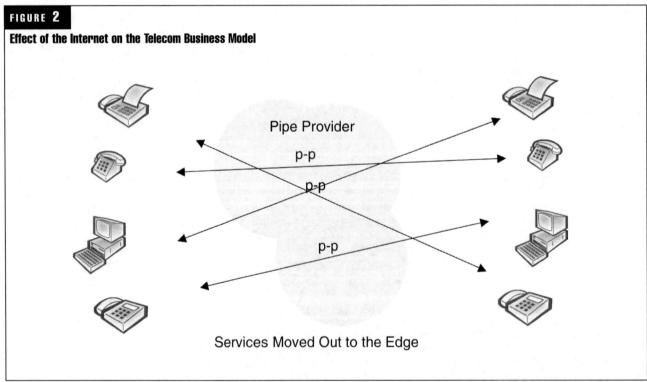

FIGURE 2

Effect of the Internet on the Telecom Business Model

Pipe Provider

p-p

p-p

p-p

Services Moved Out to the Edge

- Access to telecom service building blocks is through a simple but robust protocol that is published with open source for all end-user application developers to take advantage of.
- The telecom network provider must be able to protect the telecom equipment and make money on the resulting capability.

Telecom Service Building Blocks
In the OTSM model, the telecom service building blocks sit in the network. They are accessed on the fly by end-user applications at a desktop, by the enterprise computer system, or by the IP private branch exchange (PBX) in the end user's organization. The services are invoked as required, and then released for other users.

FIGURE 3

End Users Accessing Network-Based Service Building Blocks

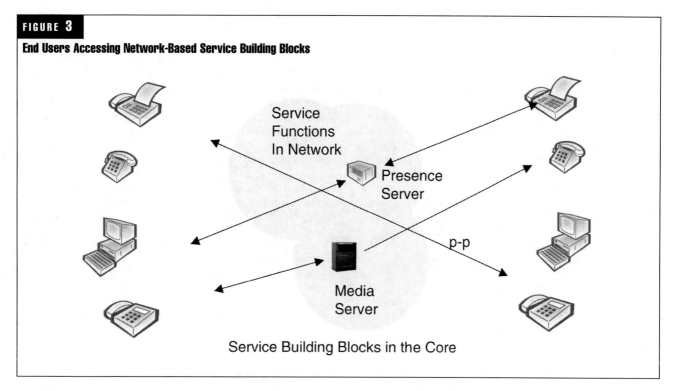

The following are some sample building blocks that could provide capabilities that companies could offer and charge for. These could all be offered in the network and accessed using simple commands.

- Name server information
- Presence server information
- Conference servers, including audio, wideband, and video
- Store-and-retrieve servers
- Speech recognition and speech synthesis servers
- Security servers, including image recognition, finger-print, retina scan, and voice print

Building Block Access Protocol

The building block access protocols must fit into the typical application development environment for the Internet. They must also be simple enough to have low latency, yet sophisticated enough to provide adequate security and allow for instant authentication.

Protocol details are beyond the scope of this paper. However, this framework should be built on a combination of industry-proven open-standard mechanisms such as SIP, XML, and DES, as shown in *Figure 4*.

Implementation Challenges

Implementation challenges to be resolved in this model include the following:

- There must be strong access security mechanisms such as encryption, denial-of-service prevention, and anti-hacking mechanisms.
- Everything possible must be done to reduce latency in executing the building block function, or else the whole model will be unworkable.
- The service provider must be able to authenticate each access in order to charge for the event, and the authentication process must be extremely fast to keep latency low.
- The service provider must devise a charging model that will entice use of the building blocks but still allow charging. Monthly charges, a per-usage charge, or the use of built-in advertising could be used to build a business.

FIGURE 4

Protocol

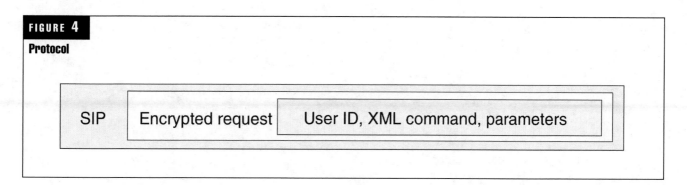

Benefits

Succeeding with such a model would have enormous benefits, including the following:

- Telecommunications capabilities will be designed directly into applications where they are required.
- End-user application developers such as automotive, textiles, and entertainment groups can tailor their telecom services to their varied and specific requirements.
- Creative developers can try new ideas continually and harvest the successful ones into moneymaking ventures.

The ability to create telecommunications services will no longer be limited to a few telecom developers. The service provider will no longer be compelled to invest large amounts of money just to try out telecom services or forced to make them generic and universal. The telecom industry will be able to focus on developing new base building blocks that make sense to be offered out of the network. It will be able to offer them within a framework that makes it simpler and less expensive to develop, deploy, and scale up as demand warrants.

Conclusions

The telecommunications industry and market are going through a commoditization period brought on by increased competition and the advent of the Internet and IP technology. The Internet is having the effect of hollowing out businesses as the end user is empowered and middlemen disappear. This will cause immense business pressure on telecom companies to find new revenue streams. Telecom service provider business models must change to match the new reality and the business model of the Internet. There are some telecommunications capabilities that are competitive to offer in the network. These telecom service building blocks should be accessed by a simple protocol that is embedded in end-user application programs. This will have the benefit of moving the creativity out to the end-user application program developers, which increases the potential to find many next hot services that will drive demand, traffic, and revenue for telecommunication service providers.

How Telecom Providers Are Using Advanced Technologies Today

Jack B. Grubman

Managing Partner
Magee Group, LLC

While the Market Was Licking Its Wounds, Telecom Engineers Were Busy

While the stock market was licking its collective wounds over the meltdown in tech and telecom stocks over the past five years, network engineers were still busy working. In essence, we may be going from a period of "negative disarray" to "positive disarray." But it is disarray nonetheless, which will present both opportunities and challenges to all players in the broad telecom/media/tech ecosystem. The result has been great advancements in standards and traffic engineering (i.e., multiprotocol label switching [MPLS]) to make Internet protocol (IP) networks closer to carrier-class quality, all of which occurred at a time of rapid deployment of broadband access over a variety of media (i.e. broadband over power line [BPL], Wi-Fi/WiMAX, fiber, digital subscriber line [DSL], cable modems). Furthermore, Ethernet has evolved from a premise-based protocol to become the universal Layer-2 data link with carrier-class attributes. This makes Ethernet, not asynchronous transfer mode (ATM), the enabler of converged services and applications.

IP and broadband will have a powerful symbiotic relationship that will drive the growth of one another, much as has occurred with operating systems and microprocessors in the computing world. The consequence will be service features and applications that go well beyond existing offerings, providing a great deal of flexibility and value thanks to the virtues of IP and broadband.

Whether it is videostreaming applications like DisneyTV or downloads of specialized content to cellphones or enterprise data-heavy multicasting, the pipes are finally getting filled, but application/content developers not "bit haulers" will be where value accrues. Also, the proliferation of IP and broadband will be extremely disruptive across the entire ecosystem, whether it is IP television (IPTV) devaluing ad-driven media or voice over IP (VoIP) rendering wireline phone services obsolete. Furthermore, IP networks will retain the reliability and quality attributes of more expensive and less robust legacy networks. Attractive content and useful applications will always "find" users; this becomes even truer in a converged IP/MPLS/Ethernet network world with multiple broadband pipes into customer locations.

In essence, the application and content are key to value creation, because without it, transport pipes have little value. In regards to the so-called "net neutrality" issue, the telcos and MSOs are so strident because water pipelines do not have the value of products that are largely water but with added coloring or packaging. They realize the value is in the content contained in a packet and not that one is providing a fast, fat pipe. My view is that if a website can be downloaded faster due to a DSL or cable modem link, then the customer of the carrier is paying for that already and the carriers should be smart enough to charge appropriate monthly rates to garner a positive return on investment. To charge an additional website fee is like charging GM or Ford every time a driver of one of their cars pays a toll on a turnpike. However, if a content developer wants extra "network engineering," such as prioritization of packets or higher levels of QoS or SLAs in order to deliver specialized services, then we would argue that a carrier deserves additional economic compensation. This is the "nexus" between carriers and content developers that will be addressed later in the paper.

The result will be a converged, multiservice network versus the historic construct of separate voice, data, and video networks. Instead of networks built around an application (i.e., voice or data), all services and applications will be packets riding on an IP/MPLS core, enabled by a universal Ethernet layer and carried to end users over myriad broadband access pipes. Furthermore, IP addressing will allow for seamless migration of content and applications across a panoply of devices resulting in the true intersection of IP and mobility. This means things are far more complex in terms of relationships between networks and services and where new applications reside versus the neatly compartmentalized world of the past. Applications will not be tethered to physical network infrastructure and will be the greatest source of value creation versus either the platform (i.e. IP backbone network) or the "power" (i.e. broadband access). IP will allow for various kinds of wireline and wireless networks to carry applications that may be hosted on a network but is not part of one, and widespread deployment of broadband will allow such applications to proliferate but not necessarily be the domain of the owner of the physical broadband access pipe. Therefore, creativity at the application layer will derive more value than having a gold-plated transport network.

If true, we will see a proliferation of IP–based, Web-centric services for businesses and consumers. The network pipes may be the necessary condition, but applications are the sufficient condition for creating value. Businesses are likely to be the earlier adopters of packet-based broadband services, given their need to enhance productivity and/or strategic position (i.e., Ethernet replacing frame relay and ATM for virtual local-area network [LAN] or virtual private LAN service [VPLS] Layer-2 virtual private network [VPN] services). Consumer applications will likely lag because of later buildout of networks (ubiquitous fiber-to-the-home [FTTH] will come later than widespread, packet-aware, metro Ethernet networks in central business districts) and the fact that consumers are cognizant of overall household budgets and do not have as much "strategic" reason to be a leading-edge user of IP–based services as businesses do.

The ubiquity of IP, with billions (soon to be hundreds of billions, thanks to IP version 6 [IPv6]) of addresses in cyberspace that can be reached anytime, anywhere—versus legacy-based/location-connected devices—will portend an array of new feature-rich services and applications that, over time, will diminish the importance of owning infrastructure. Furthermore, the notion of a "universal jack" (i.e. an Ethernet plug) means services and applications are even more portable than electricity (unlike electricity, there are not differing "voltage" standards for Ethernet around the world). Therefore, Ethernet is truly a plug-and-play, highly scalable enabler of and source of wide-ranging applications. These applications will be able to be stored and retrieved anywhere in cyberspace as long as one has the appropriate

IP address and can access the site where the application is hosted or where the information resides.

For example, corporate executives can access company-wide applications by connecting to an Internet link via a secured password, which allows for greater flexibility and presumably enhanced productivity. Managed Ethernet services such as VPLS brings together the virtues of Ethernet and MPLS. Unlike Layer-3 VPNs, VPLS is a Layer-2 VPN that allows an enterprise to control routing of packets while retaining attractive attributes such as network management, applications management, and dynamic service provision. Consumers could click on a Web browser from anywhere and have content delivered to a home recording device that has an IP address (an example of multicasting). Moreover, the "industrial-strength" engineering of IP networks will make these services usable for both business and residential customers.

Not to be overlooked, security takes on an entirely new and more critical meaning. In the old Bell system days, we had (and still do in the time division multiplex [TDM] world) monolithic networks and standards with managed devices (i.e., phones) that were physically connected to the network, and features were embedded inside of physical elements of the network. In the new world order, we have a variety of IP networks (wired, wireless, etc.) with evolving standards with a myriad of "unmanaged" devices (i.e., PDAs, PCs, TVs, cell phones, iPods, etc.) accessing applications that may reside or be hosted on these networks or simply are accessed via one of these networks. The ability to hermetically seal a network is not possible like it used to be, so advances in network security as we have seen in enterprises will be essential.

My view is that we are evolving from a TDM/synchronous optical network (SONET)–based physical layer with an

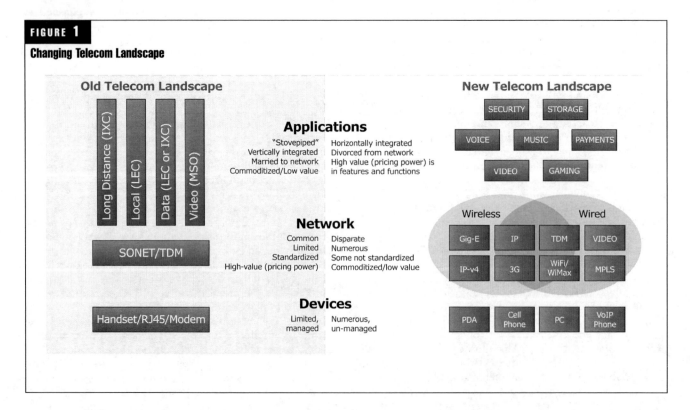

FIGURE 1

Changing Telecom Landscape

ATM data link and "connected" networks, where services are very much tethered to the physical infrastructure, to IP as the ubiquitous network layer with MPLS core providing connection-like carrier-class specs, and where Ethernet is the universal Layer-2 data link, given its position as a true global standard in enterprises and its growing importance in carrier networks. Most important, IP–based applications will no longer be tethered to the network infrastructure; they will exist as hosted offerings on a multitude of networks that will change the value calculus away from pipes and onto application developers and further empower end users. In some respects, this will mean a resurgence of the application service provider (ASP) model.

I believe we will see IP over optical (IPO) transport of packet-based services, with an MPLS core providing carrier-class capabilities and Ethernet allowing for seamless transmission of services between locations across carrier networks without the need to break down and reconstruct packet streams. All of this will reach customers over a menu of broadband access choices such as BPL, WiMAX, high-frequency fixed wireless, DSL, and cable modems. Furthermore, these new services will, increasingly, be created at a different layer of the open systems interconnection (OSI) stack than before, when most services evolved out of Layer 1 or Layer 2 physical infrastructure. If true, this will forever alter the definition of "telecom" and will result in a kluge of networks and applications versus the clearly demarcated stovepipes of the past.

I see the IP/broadband relationship resulting in the same kind of symbiotic relationship that the microprocessor and operating system enjoyed in the computing world for two decades—advances in broadband will enable or power increasing IP–based applications in much the same way that advances in chips powered more features and functions on operating systems. The reverse was also true—as operating systems became more feature-rich; the onus was on the microprocessors to keep up. The same is true in telephony, where new Web-based applications such as multicasting, storage and retrieval of files, and IPTV, are demanding higher bandwidth—not just in core networks, but also through the metro network and into the access fields. IP and broadband will drive each other's growth much the same way as operating systems and microprocessors have for one another.

The punch line is that IP, enabled by broadband delivery, will be the great disruptor to traditional telecom and media operators in much the same way that distributed computing and power on the desktop was to the mainframe. The analog in telecom is that the old stovepipe structure will be dissolved, with IP acting as the dissolvent, and applications will be "up the OSI stack" and not necessarily tethered to the physical infrastructure. Value drivers will, increasingly, be from applications versus transport and those who are "bit haulers" will need to be smart about forming a nexus with new applications developers, lest the owners of physical assets be relegated to being dumb pipes.

Rhetorical Question

Before adding some detail to the thoughts above, I think it is a good idea to ask a rhetorical question, namely "What or

who is a telecom provider?" A short history lesson is instructive in knowing how not to answer this question. One has to go back more than 70 years ago to the Telecom Act of 1934, which really set the definitions that, for the most part, still exist today. The piece of legislation envisioned a very clear separation of "church and state," so to speak, by the way the authors laid out compartments of types of companies via what was known as "titles." For example, Title I was for ancillary services, Title II was for telegraph, Title III was for wireline and Title VI was for broadcast services (I am not sure what happened to Titles IV and V). Well, after all these years, wireless operators are regulated under Title II, landline telcos are Title III, and cable is Title VI with so-called information services operating under Title I regulation, which is the most lax. In other words, the companies today are nothing more than direct descendants of the regulatory stovepipes put in place more than 70 years ago.

The so-called landmark Telecom Act of 1996 only tweaked what was then a 60-year-old piece of legislation, but essentially adhered to the traditional definitions of what constituted a carrier. This fatal flaw neglected to anticipate the arrival of an industrial-strength IP network layer, the proliferation of devices that are unmanaged endpoints not connected to a monolithic network, and the transcending of applications across myriad protocols and networks. Hence, the structural change being brought about in this industry went right by the members in Congress, and what was created by the act only ended up causing havoc among new and old carriers who were trying to play with somewhat new rules but thinking the overall structural framework of the industry had not materially changed. We will not even go into the changing regulatory rules, which only served to confuse and frustrate new entrants, or the economic framework of the act being a disincentive to resale. This last point caused a huge buildout of network facilities, which did not allow new players the luxury of building market share via resale—a la MCI circa 1973 to 1985—but instead caused them to raise capital (largely by going into debt) to fund capital expenditures that, among other things, led to the problems we all witnessed.

However, the real point of this discourse on the 1996 act is to suggest that it should have either been created six years earlier or four years later. If the former had occurred, we may have actually had an orderly transition to a competitive local market, with competition among players operating in a firmly established TDM world. More interestingly, if the act had been passed four years later, we would have at least had glimpses of developing feature and functionality of IP, which could have drastically changed definitions of markets and, subsequently, could have resulted in differently defined but perhaps better-equipped new players.

So the answer is, a telecom player is not simply "good old Ma Bell," despite efforts by some to resurrect her, but now includes all sorts of other entities. While Verizon or Vodafone, obviously, still qualify, clearly the likes of a Comcast or a Com Edison qualify and, increasingly, companies such as IBM Solutions, Accenture, Cisco, Google, and scores of application/content developers are all taking their place in the vast telecom ecosystem. This is all coming about thanks to IP–based, Web-enabled services delivered over an increasingly divergent variety of broadband-access media.

Hence, my view is that IP is the great disruptor, the solvent that crumbles the 70-year-old stovepipes. For this to occur, the underlying network technology must continue to evolve, and these smart guys must develop applications and services that people will actually pay for using.

IP and Broadband Have a Symbiotic Relationship Similar to That of Operating Systems and Microprocessors

I view IP and broadband as akin to having the same type of symbiotic relationship that has existed between operating systems and microprocessors. In this analogy, I view IP as the operating system or software partner and broadband as the supplier of the power or "memory" that allows for more sophisticated applications. Just as we have evolved from simple spreadsheets and word processing on operating systems, the continued evolution of the IP network from a "best-efforts" to a carrier-class network will allow for a multitude of connectionless, Web-based services and hosted applications (i.e., multicasting), which can be stored and delivered to or retrieved by users with IP addresses regardless of physical location. Similarly, the evolution from 286 chips to Pentium-class processors has given PCs far more memory and storage to allow for the enhancements to operating systems to be seamlessly delivered to computers.

Broadband access plays much the same role vis-à-vis IP as microprocessors have done with respect to operating systems. The so-called information superhighway had dirt roads for on- and off-ramps until we began to see accelerating deployment of broadband access. The continued rapid deployment of broadband access will enable the proliferation of IP–based, Web-centric applications via more bandwidth (the analog to memory in PCs) directly to end users. Thus, broadband access enables IP–based services much in the same way as faster processors enable more sophisticated operating systems, and advances in IP will create more need for bandwidth, much like advances in operating systems drove the development of faster processors with more memory.

This type of technology cross-pollination should result in a flood of new services with the result being a blurring, if not a complete obliteration, of the lines of demarcation between traditional telecom, media, and content players. However, if the only thing that comes out of all this is simply cheap transport, then it would have been a huge waste of time, money, and brainpower. For example, VoIP is still largely marketed as "cheap minutes," which reminds me of MCI circa 1980—been there, done that. Rather, VoIP should be positioned as a platform for packet-based services such as hosted private branch exchanges (PBXs), unified messaging, simultaneous ring, conferencing services, and a full array of multicast capabilities. More important, a fully engineered, ubiquitous IP network layer powered by widespread broadband access has the potential to take the concept of on demand to an unforeseen level for a wide range of services that are content- or application-oriented. In the enterprise space, the advent of an IP/multiprotocol label switching (MPLS) core will allow for scalable VPNs using the same IP–network backbone, as opposed to the current case where each customer VPN rides a separate IP network.

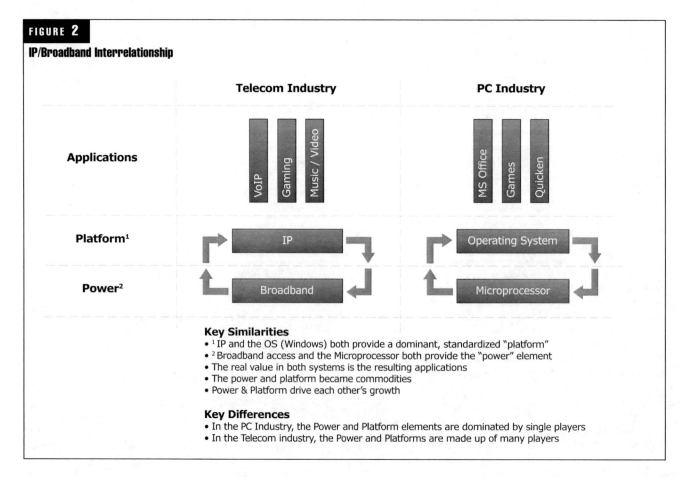

FIGURE 2

IP/Broadband Interrelationship

Key Similarities
- [1] IP and the OS (Windows) both provide a dominant, standardized "platform"
- [2] Broadband access and the Microprocessor both provide the "power" element
- The real value in both systems is the resulting applications
- The power and platform became commodities
- Power & Platform drive each other's growth

Key Differences
- In the PC Industry, the Power and Platform elements are dominated by single players
- In the Telecom industry, the Power and Platforms are made up of many players

Clearly, the emergence of a converged, multiservice IP/MPLS network platform with Ethernet in metro networks and a variety of broadband access alternatives has great potential, but it also will be the source of the most significant structural change in the telecom industry. Unlike in the computing world, where Microsoft and Intel were the two pillars of dominance for operating systems and microprocessors, respectively, in telecom, there are no such players, so the proliferation of IP and broadband will enable a multitude of players to develop value-added applications. Ironically, it is the big "bit haulers," namely, the few remaining large carriers, who must figure out how to be in the critical path of creating value other than simply having low-cost, high-bandwidth transport.

Ramifications of a Converged Multiservice Network Driving IP–based, Web-Centric Services and Applications versus Separate Transport Networks for Distinct Services

In no particular order, the following points represent what, I believe, to be the major ramifications of the dawning of an IP/broadband world. The combination of an IP network layer that is well beyond "best efforts" and wider-scale deployment of broadband in access networks will result in the following:

- Collapsing of the OSI stack, especially Layers 3 through 6. In an ironic twist, voice goes from the bottom to the top of the stack. Specifically, in a TCP/IP world, IP is at the network layer (3) with TCP at Layer 4 (the transport layer). In the legacy seven layer, OSI stack Layers 4 to 7 are responsible for interoperability and "logical" connections between two streams of bits over physical connections (i.e., altering an e-mail from a PC to be understood by a mainframe). I envision the presentation and session layers (5 and 6) being subsumed within Layers 7 and 3 or 4, respectively. As is the case today with TDM services, IP network services will depend on the data link layer (2) to deliver, in this case, packets (versus bits) from one hop to the next on the network. The difference between yesterday and tomorrow is that Layer 2 will, increasingly, be Ethernet as opposed to ATM, and while Layer-2 services will continue to exist, services will increasingly emanate from the IP network layer (hence my comment on voice going from the bottom—i.e., Layer-1 circuit switched—to the top—VoIP at Layer 7).

- Applications move up the OSI stack and are separate from the physical layer. A Web-centric, on-demand game application will reside in Layer 7 versus legacy services that evolved from the physical layers (i.e., private lines), and access to this hypothetical gaming application will be via point-and-click by the end user and, unlike a competitive local-exchange carrier (CLEC) needing a loop from a Bell, the Layer-7 application will ride the customer's broadband access without ever needing to engage the underlying carrier of said access. Thus, the current regulatory ruling limiting the resale of carrier broadband access is irrelevant to a Layer-7 application provider whose contact with the end user is via the Web. This is in contrast to a competitive Internet access carrier who needs to rent the DSL line from the Bell.

- Horizontal versus vertical competition—the competition for share of incremental value will be driven by competing layers versus competition between two stovepipes. As time passes, the question will be, "At what layer of the OSI stack is most value created (i.e., applications versus network infrastructure)?" as opposed to "Which stovepipe wins (i.e., cable versus telco)?" The multiservice IP core, combined with a growing choice for broadband access, means it will be difficult for any one entity to control end-to-end solutions, thus, "Where on the value chain (i.e., OSI stack) will the most value accrue?" is the open question.

- Bit carriers will become dumb pipes if all they provide is cheap transport. Owners of Layer-1 and Layer-2 network infrastructure will need to form a nexus with application and content developers perhaps in the form of guaranteed quality of service (QoS) and service-level agreements (SLAs) on packets going to a given application. Development of content is not generally found within the DNA of a carrier, so the aforementioned type of nexus could allow a carrier to participate in creating value driven by new IP–based services.

- Layer 2 will diminish in importance. ATM has been the legacy Layer-2 data link for more than a decade, and services such as frame relay and other legacy packet services actually originate on Layer 2. Going forward, Layer 2 will deliver IP services but not be involved in service origination.

- Ethernet will be the universal Layer-2 protocol. IP, without question, will be the de facto network layer, and MPLS will enable QoS and SLAs for multiservice platforms to allow packet over optical solutions with SONET being ultimately replaced by packet-aware networks.

Necessary and Sufficient Conditions for IP to Fully Displace Legacy Services

Despite the collective efforts of a lot of smart engineers over the past five years, IP–based services have yet to really make much headway in replacing legacy network offerings. The Internet was not designed to be a carrier-class network and is a "connectionless" network. This means there are no pre-ordained paths versus circuit switch nets with dedicated paths or an ATM type of packet-switched network that creates virtual circuits that are erected and broken down. However, ATM carries with it a big overhead. IP, as a protocol that is becoming a network layer, is evolving from being a "best-efforts" network to having carrier-class capabilities. This will allow for IP–based offerings to handle new applications while enabling the migration of existing legacy services to IP networks without sacrificing QoS or reliability.

Today, services such as voice, data, and video—not to mention music and games—still largely transverse separate networks, despite the fact that we are seeing some convergence at the device level. That said, the ultimate goal is to have multiservice network platforms based on IP with things such as MPLS ensuring legacy-type network reliability.

For IP–based services to flourish, there are both necessary and sufficient conditions that must be met. My view is that,

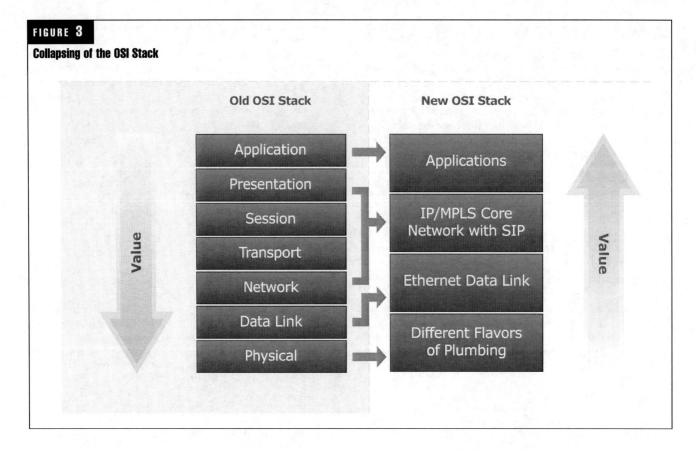

FIGURE 3

Collapsing of the OSI Stack

Old OSI Stack

New OSI Stack

Value

Value

Application
Presentation
Session
Transport
Network
Data Link
Physical

Applications
IP/MPLS Core Network with SIP
Ethernet Data Link
Different Flavors of Plumbing

in the ultimate irony, the necessary conditions that must be satisfied revolve around the ability of new converged, multiservice network platforms to, in fact, have many of the "reliability" underpinnings of legacy TDM/SONET/ATM networks so that IP–based, bandwidth-consuming services can provide the necessary QoS/SLAs on a service-by-service level. Of course, having a network that is at once robust and capable of providing flexible, high-bandwidth services all with legacy-type guarantees is only half the battle. Thus, the sufficient condition for IP to flourish is that the new services are, in fact, commercially viable. IP–based, Web-centric offerings must address the business needs of enterprises and/or the lifestyle needs of consumers for IP services to become pervasive.

Can SIP/MPLS/Ethernet Do for IP What SS7/SONET/ATM Does for TDM?

The overriding question for engineers is, "Can we achieve in an IP world the same QOS and SLAs and overall notion of network reliability as we have come to expect in the legacy TDM world?" This entails perfecting carrier-class standards in things such as session initiation protocol (SIP), MPLS, Ethernet, and packet-aware networks. To do so would go a long way in establishing IP as the clear network platform of the future, with packet-based services riding networks that offer dynamic, real-time reliability dependent on the requirements of a particular service. For example, instead of the 100 percent 1-to-1 protection of SONET, which is expensive and unnecessary, IP/MPLS networks could label and tag packets by services and offer varying levels of protection depending on the service being carried by a given packet stream.

If IP networks can deliver the flexibility and robustness they promise while offering service guarantees and security where required at the same time, this will go a long way toward quickening the migration from expensive and separate legacy networks to a true multiservice network platform. Of course, the market ramification of this will be revenue generation coming from creative applications versus simply reliable transport.

Let's Examine Some of the Developments in the Different Areas to Achieve This Goal

- SIP clearly is a session control mechanism needed for flexibility in application development, supports real-time packet services, and allows setup of sessions (i.e., services or applications) on IP. This is roughly an analog to signaling system 7 (SS7), otherwise known as "common channeling signaling" or "out-of-band signaling" in the TDM world. SS7 is what allows an 800 number to be translated into a plain old telephone service (POTS) number, for example. SS7 also is vital in the dynamic routing of calls. Similarly, SIP will set up sessions for packet-based services on IP networks.

- Ethernet is, arguably, the only true global protocol, being the instruction set (which is what a protocol is) that runs every LAN on the planet. I would argue that Ethernet is evolving from what I would call "enterprise class" to carrier class as Ethernet works its way out of the four walls of office buildings into metro networks and potentially long-haul networks (thanks to MPLS, which will allow Ethernet to travel longer distances via so-called tunneling, which forwards

Ethernet packets through routed networks). Since, up until now, Ethernet was really enterprise-focused, the protocol was not designed for carrier-class network quality, not paying attention to bit-error rates, restoration of lost packets, round-trip delays, etc., and Ethernet speeds dovetailed with ports on equipment. Now, thanks to things such as resilient packet rings (RPR), Ethernet is being equipped with SLAs and the ability to modulate bandwidth in increments not solely tied to port speed of enterprise equipment. Given that corporate enterprises love Ethernet and would love to have multi-location services such as virtual local-area networks (VLANs), the proliferation of Ethernet into carrier networks is simply a matter of how quickly it occurs. This will eliminate the requirement to multiplex/de-multiplex between Ethernet packets coming out of an office building and TDM circuits in a carrier network and then back to Ethernet at the destination site. Pervasive Ethernet in carrier networks will enable the deployment of cost-effective, feature-rich enterprise services on a multi-location basis. Thus, I believe Ethernet will replace ATM as the Layer-2 data link since Ethernet is far less costly and is backward-integrated into enterprise networks. Ethernet will serve as a delivery mechanism for packet-based, IP services with much less costly overhead than ATM.

- The sacred cow of SONET will slowly but surely fade away as traffic engineering at the MPLS layer allows for SONET-like capabilities with the added advantage of the ability to handle bandwidth-consuming services. Just like Ethernet was not designed to be carrier-class, SONET was not designed to handle the transport of hundreds of terabits of packets—which will eventually occur if FTTH really is widespread—not to mention IP–based services such as multicasting and a proliferation of "on-demand" services. The idea of a stack of SONET rings the height of the Empire State Building, along with all the cross-connections that would be required, is unimaginable as a solution going forward for the transport of IP–based bandwidth-consuming packet services.

- MPLS, at the core of IP networks, is developing the traffic engineering that will allow legacy network-like quality and reliability but with the ability to more efficiently handle increasing requirements for bandwidth driven by IP–based packet applications. For example, MPLS can now offer SLA features such as bandwidth guarantees and management. MPLS will label packets that will give IP a "connection-like" quality, in essence, mimicking a private virtual circuit (PVC) with far less cost than ATM; packets will also be tagged (i.e., classified) over MPLS, which will allow for SONET–type protection but on a dynamic basis—namely, a voice packet will get 1:1 protection, but a less time-sensitive packet will get 1:N protection, with N dependent on the nature of the traffic within the packet. This reduces the requirement for costly overhead found in SONET for 1:1 protection for all traffic when, in reality, well over half the traffic on carrier networks do not need that level of protection. Advances in the QoS/SLA capabilities of MPLS will allow for packet over optical

(versus packet over SONET [POS] or ATM) riding IP/MPLS core multiservice network platforms.

- Packet-aware networks will become pervasive in carrier metro networks, eventually replacing most of the elements of old TDM infrastructure. Packet-aware networks will easily interconnect with Ethernet interfaces, which will be important because, as we mentioned earlier, Ethernet will become pervasive within metro networks. Thus, the ability to seamlessly interconnect carrier metro networks with enterprise networks at the packet level will allow for metro-wide offerings of such services as virtual LANs. In addition, packet-aware metro networks will be equally important as FTTH and so-called triple-play services are rolled out to households. I believe switched Ethernet versus passive optical networks (PONs) will be the architecture of choice in residential rollouts (more dedicated bandwidth per home and far cheaper than ATM delivery), and the consumer services being envisioned will be packetized; thus, having packet-aware metro networks will be important in carrying packets between homes and carrier installations. The end result will be a multiservice IP/MPLS core backbone network supporting packet over optical services, with packet-aware metro networks and Ethernet being the universal data link and SIP controlling session flows. This will give the IP/broadband world legacy-like network quality and reliability while efficiently supporting myriad IP–based, Web-centric, bandwidth-consuming packet services.

If the Above Occurs, Will IP–Based Services Gain Commercial Acceptance?

We discussed the idea that for IP to become a pervasive network from which an entirely new array of services is offered, the "necessary" condition is getting the network elements ready for prime time. However, someone has to buy these services. Thus, the "sufficient" condition for IP to proliferate is the creation of applications and services that are doable because of the virtues of an IP/broadband network and the satisfaction of either the commercial needs of enterprises or the lifestyle needs of consumers. The following are a few illustrative (but very real) examples of the types of services made possible in an IP world that will likely resonate with a large segment of either businesses or consumers.

One of the most widespread uses of MPLS will be the ability of carriers with MPLS core networks to build scalable VPNs using the same IP backbone. This is in contrast to today, where each customer's VPN rides a separate IP backbone. MPLS will be able to ensure the isolation of each customer's traffic, which will allow multiple VPNs to share one IP backbone. Multicasting services will be a huge set of IP–based applications and is truly a class of services that could not exist without IP. An example of such a service would be what is known as a multimedia conference, where audio, video, files, whiteboards, etc., are shared to pre-approved IP addresses. The application will reside on a hosted platform on an IP/MPLS core network and is invariant to the type of end device or access network as long as the device has the capability to handle the storage that is being streamed and the broadband access pipe has enough bandwidth.

Both of the above, along with things such as virtual LANs, which will be enabled by pervasive Ethernet at Layer-2 and packet-aware metro networks, are easily believable as types of service offerings that corporate enterprises would find useful in increasing their productivity. For small-to-medium enterprises (SMEs), IP–based services off VoIP platforms such as hosted PBXs, simultaneous ring, desktop-to-desktop conferencing, and Web-based user control of handset features will be very attractive, especially for those SMEs with tight budgets and lack of internal expertise.

On the consumer side, I believe, there is less clarity. Businesses will pay for services that help increase productivity or enhance sales, etc. However, most households have a limit on what they will spend broadly on entertainment/media/communications products. Thus, while IP–based services will certainly cannibalize legacy services in business (think frame relay), there could still be an overall expansion of the market if the take-up on the aforementioned services among business users is significant. In contrast, the consumer segment is more likely to be a zero-sum game with pricing being a much more important part of the value proposition, more so than it is for enterprises, which do not have non-price attributes to consider.

For consumers, the big buzz today is the so-called triple play of voice, video, and data over either a telco or cable company network. However, these are just bundling of legacy services at a big discount. The virtue of IP coupled with broadband access for consumers will likely be found in applications such as IPTV, which could allow a consumer to channel-surf on the Web, making on-demand video something that does not have to be part of a larger bundle that includes ad-driven content. In general, IP will allow a slew of on-demand services such as on-demand ads. Clearly, streaming audio and video for music downloads or gaming will also be an example of IP–driven consumer applications. Also, the ability to seamlessly transfer content between devices (i.e., PCs, TVs, cameras, etc.) will require the ability to recognize IP addresses resident on different devices. Last but not least is VoIP.

Ironically, it may indeed be the case that the coming of age of IP combined with the accelerating deployment of broadband may turn out to be the undoing of the so-called consumer triple play, which was first envisioned more than five years ago in what was still a very solidly stovepipe-defined world. Now that IP can enable applications that transcend transport networks, the ability of the two big stovepipes serving the home (cable and telco) to dictate to the consumer what services they are offered may be undermined. IP will allow the applications developers to be the brand along with or perhaps instead of the network infrastructure provider. In fact, IP–based services are likely to be billed directly to consumers, bypassing either the Bell or the MSO, and such services may prove to be more appealing on an as-needed basis than having to write a sizable monthly check for services one does not largely use. Of course, it is not surprising that if IP is indeed a disruptive force, it will be as ubiquitous in its disruption as it is in its virtue.

Open Questions

There remains a slew of open questions that will be answered over the course of time. The following are a few to ponder:

- Broadband access is what powers IP. Will broadband penetration double from current levels over the foreseeable future? Currently, only 5 percent of broadband homes have VoIP. Can that grow, and can other IP–based services achieve better take rates with consumers who have broadband connections?

- The virtue of IP is the ability for applications and services to be hosted or reside anywhere given the ubiquity of IP addresses and the pervasiveness of IP networks. The commercial question will be, where will new, value-added applications reside? Will they be on carrier networks, within enterprise networks, in a router in the basement of an office building, or as part of a systems integration solution? This will go a long way in determining the relative share of value creation amongst the various players in the telecom ecosystem.

- As we mentioned at the outset, we are evolving from a monolithic network with managed endpoints to an environment with unmanaged endpoints with packets carrying services transversing different networks with varying protocols. Will intrusion protection be able to scale for wide-area networks (WANs) as it is beginning to do for enterprise software and operating systems?

- We have only barely mentioned wireless in this discussion because wireless networks are really access networks—be it mobile or fixed—with the laws of physics limiting the amount of bandwidth that can be carried on radio frequency (RF) spectrum. Having said this, a real question is, can WiMAX—with issues surrounding unlicensed spectrum, lack of standards, no mobility, and still high cost points—become a true broadband alternative? As we speak, Wi-Fi is a huge success from a user perspective (I am not sure any money has been made), and not just in places such as Starbucks. Enterprises love having internal Wi-Fi access points as a perfect way to take advantage of the virtue of IP and Ethernet so that a worker can travel with their laptop and still be able to be contacted anywhere in a corporate environment (SIP phones would do the same thing for a phone number), but Wi-Fi is range-limited, so other fixed wireless solutions need to succeed from an economic and capacity standpoint. As far as mobile wireless is concerned, the question is, what are the tradeoffs between more bells and whistles and the ergonomics of the phone and the power consumption?

- Which market segments will most benefit from the rollout of new IP–based applications? It seems, at this point, an almost barbell-shaped market is forming with either Fortune 2000 enterprises or residential homes being targeted by the various large suppliers or carriers for rollout of new service offerings. The vast middle—namely small and medium-sized businesses—has so far been largely neglected. Clearly, the two big recent telecom mergers were largely driven by the desire of the acquirers to gain traction in the large enterprise space. Meanwhile, the FTTH rollouts and overall triple-play offerings by cable companies and telcos are targeting the residential user. Thus, a question remains: To what degree will the IP–based, Web-centric solutions be actively marketed to the SME segment? The irony is that SMEs, who do not have the in-house budgets of large enterprises, are prime candidates for hosted IP applications. Perhaps this time around there will, in fact, be a place for niche carriers serving this market segment.

- What will the public policy be? Our view is, the less the better—let the marketplace decide.

- Finally, at the end of the day, what does all this mean to the industry revenues, profits, and structure? Over the next five to ten years, what will the net impact of the rollout of new IP–based services be? Will the cannibalization of some legacy services be offset by growth in new services? Will the industry be able to respond appropriately to a revenue model that is moving away from transport of bits to usage of applications? Will the industry continue to have just a few behemoth carriers, or will IP truly be the solvent that results in a much more fragmented industry structure, with players and brands in segments that one would not necessarily envision in these segments today?

The Greening of Telecom

Environmental Issues and Policies Related to the Elimination of Telecom E-Waste and the Encouragement of E-Sustainability

John Gudgel

Director, Industry Alliances
McGraw-Hill Construction

Abstract

This paper examines the ecological impact of telecom activities on the global environment. The magnitude of the telecom e-waste problem will be outlined along the major areas of the "green telecom" debate. Public policy issues will be discussed along with current government and corporate initiatives to encourage e-sustainability in future telecom activities.

Introduction

Technological development, like any other human activity, produces waste as a by-product. According to the United Nations Environment Program (UNEP), between 20 and 50 million tons of technological waste (e-waste) is generated worldwide each year, and another estimate states that electronic waste is growing three times faster than solid waste.[2]

This paper discusses the problem of e-waste from a telecom perspective. Telecom e-waste will be defined along with the magnitude of the problem both in the United States and the rest of the world. "Green telecom" will then be defined in the context of some of the major areas of debate, including telecom design, supply chain management, labeling, energy consumption, CO2 emissions, recycling, and total life-cycle cost. Public policy issues will be identified, and the paper will then conclude with a discussion of both government and corporate initiatives being implemented to encourage long term global e-sustainability.

Telecom E-Waste

Electronic waste, or e-waste, is defined by the eWaste Guide as "the term used to describe old, end-of-life electronic appliances. It includes computers, entertainment electronics, mobile phones, etc., that have been disposed of by their original users. While there is no generally accepted definition of e-waste, in most cases, e-waste is comprised of relatively expensive and essentially durable products used for data processing, telecommunications or entertainment products for private households and businesses."

It is estimated that the United States' population alone is producing more than 2 million tons of e-waste every year from obsolete and discarded televisions, packaging, commercial and household electronics, personal computers (PCs), and monitors (*Figure 1*). This e-waste will include 500 million computers and 130 million mobile phones by 2007. Currently e-waste comprises 2 to 5 percent of U.S. municipal solid waste stream, and the problem is growing as the production of electronics grows and computers, through technological advances, become obsolete sooner. The average life span of a computer in 2005 is two years, versus four to six years in 1997.

E-waste is also a global ecological problem affecting every continent (*Figure 2*). More than 4 million computers are discarded every year in China, and by 2010, it is estimated that 14 million computers, 39 million telephones and 18 million televisions will reach the end of their life span in India.

Much of this e-waste is telecommunications products, including mobile phones, cable modems, private branch exchange (PBX) equipment, and network switches. This telecom waste also includes all of the hazardous materials and by-products produced in the manufacturing of these products and also necessities such as lithium batteries and accessories such as headsets.

Green Telecom

People are becoming aware of the affect telecom products are having on the environment. Several areas of research activity are focusing on "tecology" or "green telecom" issues.

One area of study involves the design and manufacture of telecom products. Many hazardous products are used in the manufacture of telecom materials, including lead, cadmium, mercury, hexavalent chromium, polybrominated biphenyl (PBB) and polybrominated diphenyl ether (PBDE) flame retardants. One of the goals of "green telecom" is to reduce the amount of these hazardous materials used and to make

FIGURE 1

E-waste in the United States[6]

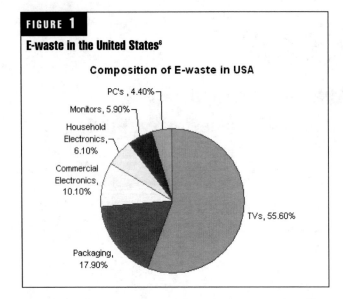

Composition of E-waste in USA

PC's , 4.40%
Monitors, 5.90%
Household Electronics, 6.10%
Commercial Electronics, 10.10%
TVs, 55.60%
Packaging, 17.90%

FIGURE 2

E-waste Around the World

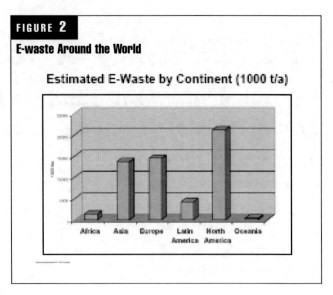

Estimated E-Waste by Continent (1000 t/a)

Africa   Asia   Europe   Latin America   North America   Oceania

the manufacturing process more energy-efficient, consequently reducing the release of carbon dioxide and other greenhouse gases.

Another "green telecom" area of study is recycling and disassembly of telecom products. Research is being conducted to find a way to reduce the environmental impact of waste telecom equipment by encouraging recycling and reuse. One issue of particular concern is the development of home-grown computer-recycling systems in China and India. Recycling can be a very profitable business, but if it is done improperly, the process can result in far more ecological damage than simply placing the equipment in landfills.

Public Policy Issues

There are several public policy topics that are particularly related to e-waste and the green telecom movement.

First, there is the issue of the ecology and the growing impact that telecommunications is having on the global environment. This is part of the overall struggle between the need to preserve the earth for future generations and the desire to use technology for economic development. This issue also includes evaluation of some of the positive ways telecommunications can impact global climate change and natural resource consumption through the encouragement of telecommuting, etc.

A second public policy issue is the impact that e-waste is having on the public health. E-waste not only has the potential for damaging the environment, but also can pose a public health risk. For example, UNEP statistics indicate that as much as 40 percent of the lead in landfills is related to the disposal of e-waste.

Risk management is also a major concern of many governments and corporations. The green telecom movement has brought the recognition that governments and corporations have a social responsibility to make sure that telecom products do not cause harm to people or the environment. Thus there is the recognition that there could be political repercussions or litigation resulting from improper handling of e-

waste and the need for government and corporate entities to manage their risk.

Finally, there is the issue of the costs for making telecom products green. One of the benefits of technological development has been the dramatic reduction in the price of electronic products. As pointed out earlier, this price reduction has directly contributed to the growth in e-waste as customers are encouraged to buy new technology and dispose of the old. Likewise, if green telecom products are to become accepted by business and individual consumers, they must be sold at an affordable price or they must have associated tax incentives that make the price acceptable.

Government and Corporate Initiatives

As previously noted, governments and corporations are recognizing the risks associated with e-waste and the need to develop policies on sustainable growth.

One of the best examples of a government initiative is the pending passage of two e-waste directives by the European Union. The first directive, on the restriction of certain hazardous substances (RoHS), sets maximum concentration levels of six hazardous substances utilized to manufacture all new electronic products sold in the EU as of July 2006. This directive could have a major impact on U.S. suppliers of computer and telecommunications equipment since all products imported into the EU must comply with this directive.

The second EU directive, on waste electrical and electronic equipment (WEEE), is designed to reduce the environmental impact of WEEE by encouraging recycling and reuse. This directive requires that producers cover the cost of recycling their products. Business consumers can request that producers collect old equipment, and producers can pass the cost of recycling onto consumers.

Many telecommunications corporations are also implementing "green" initiatives. For example, Cisco Systems recently launched its Surplus Product Utilization and Reclamation (SPUR) program that has the ultimate goal of eliminating e-waste. There are also companies such as Better World Telecom that utilize their mission to "reduce ecological impact" as a strategic marketing tool.

Conclusions

Telecom e-Waste is a major global issue with many environmental, political, and economic ramifications. There is a "green telecom" movement that is conducting research in a number of areas, and there are numerous public policy issues that that are being debated. Governments and corporations are responding to this movement by implementing green policies and initiatives. However there is a long way to go before the dream of eliminating e-waste will become a reality.

Bibliography

Better World Telecom Homepage, found at www.betterworldtelecom.com.

E-Waste Guide Homepage, found at www.ewaste.ch.

Griffith University Environmental Engineering Web site, United Nations Environmental Program (UNEP) and Other Facts and Figures, found at www.griffith.edu.au/school/eve/ewaste.

Integrated Waste Management Board, April 2004, "Best Management Practices For Electronic Waste," found at www.ciwmb.ca.gov/Publications/Electronics/63004005.pdf.

Telcology, September 2005, "E-cycling puts vendors in a spin," found at www.telcology.com.

Williams, E., 2005, "International activities on e-waste and guidelines for future work," United Nations University, Proceedings of the Third Workshop on Material Cycles and Waste Management in Asia, National Institute of Environmental Sciences: Tsukuba, Japan. Found at www.it-environment.org/publications/international%20ewaste.pdf.

Notes

1. Griffith University Environmental Engineering website, United Nations Environmental Program (UNEP) and Other Facts and Figures, found on the Internet at http://www.griffith.edu.au/school/eve/ewaste/

2. Definition found at www.ewaste.ch/welcome/ewaste_definition.

3. Integrated Waste Management Board, April 2004, "Best Management Practices for Electronic Waste," found at www.ciwmb.ca.gov/Publications/Electronics/63004005.pdf.

4. Ibid.

5. Griffith University Environmental Engineering website, United Nations Environmental Program (UNEP) and Other Facts and Figures, found at www.griffith.edu.au/school/eve/ewaste.

6. Griffith University Environmental Engineering Web site, United Nations Environmental Program (UNEP) and Other Facts and Figures, found at www.griffith.edu.au/school/eve/ewaste.

7. Williams, E., 2005, "International activities on e-waste and guidelines for future work," United Nations University, Proceedings of the Third Workshop on Material Cycles and Waste Management in Asia, National Institute of Environmental Sciences: Tsukuba, Japan. Found at www.it-environment.org/publications/international%20ewaste.pdf.

8. Griffith University Environmental Engineering Web site, United Nations Environmental Program (UNEP) and Other Facts and Figures, found at www.griffith.edu.au/school/eve/ewaste.

9. Telcology, September 2005, "E-cycling puts vendors in a spin," found at www.telcology.com.

10. Found at newsroom.cisco.com/dlls/hd_013003.html.

Development Trends of Telecom Bearer Networks

Chunguang Lin

Marketing Director and Specialist
Huawei Technologies Co., Ltd.

Summary

Operators are facing the challenges of network transformation. To reduce capital expenditures (CAPEX) and operational expenditures (OPEX), the advanced operators seek a unified bearer network for telecom services. IP is the next generation of telecom bearer network technology. There are many problems to be solved, such as quality of service (QoS), reliability, security, operability, and manageability. The trends of telecom bearer networks are discussed in this paper. Research has proved that with careful planning and considerate deployment, IP networks can support the transition of telecom and carry telecom services. IP telecom network (IPTN) resource and admission control subsystem (RACS) is one of the solutions.

From TMD to IP

The telegraph was the first telecom service in history. More than a century since the Bell's invention of the telephone, the telephone is still one of the anchor telecom services. Each major technological invention pushed the telecom industry to a new stage. The invention of the stored program control (SPC) switch made communication an affordable daily presence in people's lives.

Parallel to the development of the SPC switch, synchronous digital hierarchy (SDH) led telecom bearer networks into the era of time division multiplex (TDM).With the development of computer technologies in the 1980s, it became urgent to implement the interconnection of heterogeneous computers. At that time, the transmission control protocol (TCP)/IP came into the picture. The fledgling IP was carried over TDM in most cases at that time.

People soon realized that the future of communication is multimedia communication—that is, voice, data, and image need a unified carrying platform. By integrating the strengths of public switched telephone networks (PSTNs) and packet networks (especially IP networks), asynchronous transfer mode (ATM) as a new generation of carrying technology was proposed. As the Internet developed in an explosive manner, the ATM market was squeezed. Because of the lack of ATM applications, complications of the technology, and other factors, ATM failed to replace TDM as a telecom bearer network.

The healthy interaction between the Internet and computer unified the communication protocol of wide-area network (WAN), local-area network (LAN), and desktop systems. People developed a large number of services over IP networks, including Web, e-mail, e-commerce, voice over IP (VoIP), and IP television (IPTV). The IP traffic amounts to half of the entire network transmission today. People may wonder if IP is our next generation of telecom bearer network technology.

Opportunities and Challenges of IP Bearer Networks

Problems of Network Operators

At the end of the 20th century, technological development and regulation change brought about the unprecedented prosperity of the telecom industry, and at the same time, bubbles in the high-tech field. The overinvestment of network operators resulted in the recession of the telecom industry. Now, fixed network operators are faced with the problem of increasing business with fewer revenues. They are all looking elsewhere and seeking better opportunities. The major problems of network operators are as follows:

- Open telecom industry leads to more competition. With the end of the monopoly of the telecom industry, the number of network operators increased sharply, which caused unprecedented competition. The increase in subscribers cannot cover up the loss caused by price drop. The revenues of the conventional voice service in a global context decreases by 3.9 percent annually, as shown in *Figure 1*.

- Business is being channeled from fixed-network communication to mobile communication. Mobile communication satisfies customers' desire to communicate with each other anytime, anywhere. In terms of voice service, mobile communication has surpassed the fixed network communication.

- VoIP poses a threat to long-distance voice call. The development of VoIP toppled the long-distance business model. In 2005, SBC Communications Inc. purchased AT&T, a company that lasted for more than 100 years, for $16 billion, which marked the end of the distance-based operation model of voice service.

- The development of the Internet deprived network operators of the control over services. In the past, telecom services could not go without telecom networks, and for this reason, network operators controlled the services. Nowadays, the Internet-based e-commerce, gaming, IPTV, and other services have shaken off the harness of network operators, who have been reduced to mere channel providers. As a result, the network operators slid to the low end of the value chain.

The mainstream network operators in the industry have realized the seriousness of the situation and are making an effort to explore transition of management, service, marketing, business, and network. IP–based construction of the next-generation telecom bearer network becomes a consensus in network transition reached among network operators.

Mainstream Network Operators' Thoughts on Network Evolution

Telecom operators deploy new generations of services over IP by cooperating with strong IP enterprises. For example, SBC cooperated with Yahoo—and Verizon with MSN—to provide broadband subscribers with portal, e-mail, and other services. DT achieved the same end through internal consolidation.

The 21CN plan of BT was to construct an IP/multiprotocol label switching (MPLS)–based unified bearer network, thereby cutting operations, administration, and maintenance (OA&M) expenses by half. Another aim of the plan was to urge its transition into an "integrated information provider," providing communication and information technology (IT) services for the government and enterprises.

In the RENA plan of NTT, a carrier-class IP bearer network of high security and reliability will be constructed, with session-based service control and video/streaming network capabilities. It aims to provide subscribers with network and content services, and enterprise users with communication, IT services, and outsourcing services.

In sum, in the matter of network transition direction, mainstream network operators seemed to have an unsaid agreement by all steering toward IP–based new-generation bearer network. They hope to implement cost cuts, more flexible service-provisioning capability, and value chain expansion capability through a unified network infrastructure.

Major Problems of IP Bearer Networks

IP networks as next-generation bearer networks have a number of key technologies that need to be reformed. According to the statistics of the next-generation network (NGN) softswitch experiment, 60 to 70 percent of the problems are related to the bearer network in the form of QoS, security, and operable and manageable faults (See *Figure 2*).

IP networks have two areas of concern with respect to QoS problems. The first is related to the IP network itself, as shown in *Figure 2*. The average delay of a tested connection is only 9 ms, but the transient delay can reach up to 1 s. The average packet loss rate is only 0.14 percent, but the transient packet loss rate can reach 50 percent. The IP network has relatively good average performance and a bad tran-

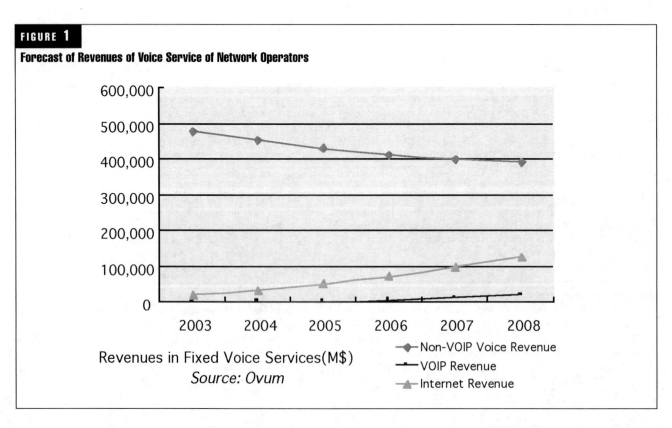

FIGURE 1

Forecast of Revenues of Voice Service of Network Operators

Revenues in Fixed Voice Services(M$)
Source: Ovum

- Non-VOIP Voice Revenue
- VOIP Revenue
- Internet Revenue

FIGURE 2

Delay and Packet Loss Tests of a Certain Metropolitan-Area Network

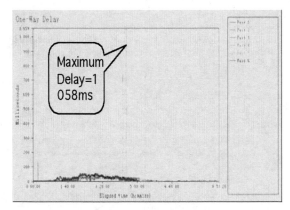

Average Delay=9ms

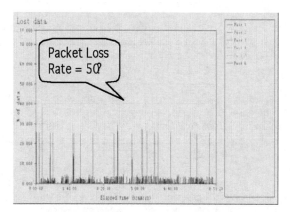

Average Packet Loss Rate = 0.14%

sient feature. The IP network features traffic in bursts and unexpected traffic. Even in the case of a light bearer network, the IP network cannot guarantee QoS. The second is that the IP bearer network cannot sense services. It is the IP network that can only identify packets and cannot identify the services borne. Currently the IP bearer network does not have a unified network resource control system and cannot gather proper resources according to the telecom services to satisfy the quality requirements of services.

Reliability of IP Networks
The reliability of the conventional IP network is an important factor that affects QoS. Although the IP network has dynamic protocol, redundancy connection, and other reliability technologies, its level is far from the requirement of carrier class. A common IP network fault will result in service interruption for a period from a few seconds to minutes. The features can satisfy the requirements of conventional Internet service carrying, but they cannot satisfy the QoS requirements of real-time voice and video services.

Security of IP Network
The network security includes the guarantee of the availability, privacy, and completeness of data. Proven by practice, the Internet has serious security problems, plagued with viruses and suffering from hacker attacks. Important applications may require the application of legal, complicated encryption technologies, but the amount of the overhead of encryption technologies is sizable. For that reason, it cannot satisfy the needs of massive telecom services, and a secure network structure is urgently called for.

Operability and Manageability
The conventional IP network, which is carried as the next-generation multiservice bearer network, lacks an effective operation and management model. If the IP bearer network is seen as a flat network, it cannot perform effective traffic calculation and planning. In actual applications, QoS and necessary bandwidth guarantee the need for reasonable network resource configuration. The configuration includes IP network node equipment processing capability, port resource allocation, and trunk bandwidth allocation. It needs to conduct forecast of the streaming bandwidth, and also needs to learn from the successful experience of the conventional telecom network in creating call model and traffic planning. The IP network is generally constructed for the Internet and data services and cannot satisfy the requirements of telecom services. It requires the new construction or reform of the existing IP network based on the requirements of multiservices.

Development Trends of Telecom Bearer Networks

Application of IP Technologies and Equipment to Construction Unified Bearer Network
Constantly Developing, Highly Reliable Technologies
Unlike the Internet, the IP network as a telecom bearer network must be highly stable and reliable. Currently the mature equipment has redundancy backup capability for key ports and important boards and components. For example, the switching of the active main control board to the standby one will not result in service interruption. Commercial routers will have the label-switched path (LSP)–, virtual private network (VPN)–, and IP–based fast rerouting capability. They utilize multiprotocol label switching (MPLS) OA&M, bidirectional forwarding detection (BFD) and other end-to-end, section-to-section detection technologies to implement the backup of faulty nodes and links within 50 ms and end-to-end protection within 200 ms. In this way, telecom services are free from service interruption during the operation.

Bettering QoS Technologies
Each service carried can be provided with QoS on demand. VoIP and other real-time services can be provided with QoS similar to that of the conventional PSTN network. This standard of QoS can justify the IP network as IP backbone network for carrier-class services.

The International Telecommunication Union Telecommunication Standardization Sector (ITU–T) G.114 recommendation suggests that the full-process delay not exceed 300 ms and the unidirectional delay not exceed 150 ms. The delay jitter of streaming in network transmission is required to be less than 20 ms. In the streaming transmission, packet loss rate should be less than 3 percent. Currently, the gateway equipment is equipped with self-adapting buffers and can work normally with a packet loss rate of 10 percent. There have been recent breakthroughs in the structure of the bearer network QoS. The industry universally recognizes the RACF/RACS structure. The bearer network can sense services, and can reserve network resources and designate routes for telecom services, with the call admission control (CAC) capability of services.

All-Round, All-Dimensional Network Security Capability

To implement all-round security requirements, it is necessary to first divide a number of security sections according to the secure networking principles of network planning and network design. Huawei has accumulated a lot of experience in this aspect. Second, the interconnection of security sections adopts proper security equipment, for example, high-performance firewall and session boarder controller (SBC). The SBC will develop toward high-performance, large-capacity, and multiprotocol processing and will be eventually integrated into the RACF/RACS structure. Last, it is the security of the equipment against hackers and viruses.

Networking Flexibility and Service Expandability

According to the increase and changes of future services, the network can be smoothly expanded and upgraded, maximally reducing the adjustment to the network structure and existing equipment. First, multiple access means should be supported, including xDSL, Ethernet, hybrid fiber coax (HFC), Wi-Fi, WiMAX, and fiber-to-the-x (FTTx). Second, network planning should be supported. For example, segmental label-switched path (LSP) and traffic engineering are applied to solve the N2 issue in traffic engineering. Third, multiple topologies should be supported—for example, full interconnection, tree, and ring.

Manageability

The network is monitored in a centralized manner, and managed with different levels of authorities. Advanced network management platform is selected, with such functions as the management of equipment and ports, the statistical analysis of service traffic, the provisioning of end-to-end network performance monitoring, and the provisioning of fault automatic alarm.

Implementation of Multiservice Transfer over IP Bearer Networks

The future 3G, IMS/NGN, IPTV, and other services are all carried over IP in a unified manner. The bearer network needs to simultaneously support voice, video, data, enterprise interconnection, and other services, and implement the transfer of the three media streams, audio, video, and data. The operation indices of the three media streams are required as follows:

- *Voice service*: Small delay and delay jitter, small packet loss rate
- *Video service*: High bandwidth, and small transmission delay and delay jitter

- *Data service*: Non-real-time burst service, with low requirements on delay and jitter, but a low error rate required

For signaling and network management system, the bandwidth required is small compared with the service traffic. The delay requirement is lower than the voice requirement, but the error rate should be low. They are included into the data service that poses a high requirement.

Lowered CAPEX and OPEX

To accept the new challenge, the general need of the operators is to reduce the CAPEX and OPEX by doing the following:

- Restructuring the supply chain and seeking suppliers with a high price/performance ratio. Now that the telecom bubble has popped, the advanced European operators are faced with equipment upgrade issues, and the original suppliers cannot fully satisfy the needs of the requirement to lower the CAPEX. In that regard, they look to the east for suppliers such as Huawei, who can provide advanced equipment.

- Ensuring operators and equipment vendors form strategic partnerships. To improve the diversified competitiveness, operators require that the equipment vendors provide tailor-made equipment according to their needs as a way to reduce network operation expense. For example, BT and Huawei formed a strategic partnership, and Huawei entered BT's shortlist.

- Constructing a unified telecom bearer network. The construction of a unified telecom bearer network will greatly reduce the maintenance and operation expense. The goal of BT is to cut the maintenance and operation expense by half.

Quickening Process of Open Standardization

Currently the structure and the QoS issues of the IP bearer network are hot research topics in the industry. Mainstream equipment vendors invest enormously into product and technology R&D, and the ITU–T, the Telecommunications and Internet Protocol Harmonization over Networks and Services and Protocols for Advanced Networks (TISPAN), the third-generation partnership project (3GPP), the Multiservice Switch Forum (MSF), and the Internet Engineering Task Force (IETF) are picking up the pace in their research of bearer network structure and QoS.

In 2004, the ITU–T and the European Telecommunications Standards Institute (ETSI) TISPAN all proposed basically similar NGN structures. In particular, concerning bearer network resources and QoS management, the RACS was defined. This subsystem can be shared by the four subsystems on the service control layer.

An agreement has been reached concerning structure. The ITU–T has published Y.1291 (An architectural framework for support of QoS in packet networks) and H.360 (An architecture for end-to-end QoS control and signaling). In May 2005, in the plenary meeting held by the ITU-T SG11, the Q.rcp.cci and the Q.rcp.nci were passed, based on Diameter and RCIP protocols.

Postscript

Telecom services are integrated with IP, whereas IP networks are getting more involved with telecom. Experiment has proved that with careful planning and considerate deployment, the IP network can support the transition of telecom and can carry telecom services. In the long run, to ultimately solve the IP bearer network problems, the industry proposed multiple structures. The IPTN RACS solution proposed by Huawei Technologies gained the recognition of the industry and is highly consistent with the structures proposed by the ITU–T and the TISPAN. Currently the standard organizations have focused their research on specific protocols. Though bearer network technologies and standards are being developed, it is believed that with the shared efforts of the industry, the IP bearer network can eventually satisfy the service requirements of NGN and 3G networks, offering solutions to QoS, security, and operation and management issues.

A Policy-Driven Methodology for Managing Telecommunication Networks

Sarandis Mitropoulos

Visiting Lecturer, Department of Informatics
University of Piraeus, Greece

Christos Douligeris

Associate Professor, Department of Informatics
University of Piraeus, Greece

Abstract

In the age of high-speed development and continual changes in technological infrastructures, policy-driven management is necessary to address the complex problems that arise in the management of corporate and telecommunication networks. Policy-driven management is an effective approach for structuring the management task in large-scale networked environments with numerous resources. Management of change is an additional benefit to be gained by the use of policy-driven management. In this paper, we provide a state of the art on the subject, present policy support middleware platform services as well as recent implementations, and, at the end, propose a methodological approach for developing coherent policy sets on systems and networks. A case study is used to exemplify the methodology and point to a number of open issues.

Introduction

Telecommunication networks and systems management involves numerous activities, services, actors, and resources due to its inherent distributed nature. The main goal of management systems is to ensure availability, reliability, and performance over all the dimensions of networks and systems. Two of the main problems that usually arise on large-scale networked systems are heterogeneity and complexity. These two problems drive the system and network management developments to support mechanisms for activity consistency, conflict resolution, and relationships handling.

Focusing on telecommunication networks that are in modern times a tied part of various business activities, including response time, reliability, and quality of service (QoS), is usually a critical parameter in corporate or service provider operations. Networks are not just a means of exchanging information, but are involved in more sophisticated activities, such as collaborative work, tele-working, and video-conferencing. Network management is the function that faces all the above issues, transforming the network into a managed resource as a whole, as well as a manageable collection of network elements.

Among network management techniques, the distributed policy-driven network management is a powerful methodological approach to monitor and control telecommunication networks, ensuring security, availability, reliability, and efficiency. For example, rapid development in networked multimedia applications introduces demanding requirements for the available network bandwidth. The latter is a business-critical network resource for almost all telecommunication networks. Thus, adaptive policy actions according to service bandwidth pattern usages and bandwidth optimizations must be applied for performance enhancements and cost reduction.

Policy-driven network management can be defined as the organization of network elements, networks, and telecommunication application services into appropriate management domains for the purpose of applying interrelated policies on them for satisfying important business goals.

Organizing large-scale networks and network services into domains offers modularity and provides managers with the capability to enforce a variety of policies in different management domains. Organizing systems and network resources into domains can be done according to various criteria, while domains can be members of another domain, structuring in this way domain hierarchies based on the domain/sub-domain relationship and on the possible domain overlapping. Having organized network elements, networks, and the respective services into domains and policies can be defined for enforcement on them. Thus, policies are handled as separate entities, or managed objects, that can be also organized into domains and configured appropriately, so that policies can be enforced on policy domains.

Any policy must be referred to a managed domain, which includes the respective managed objects, which, in practice, is the sphere of influence of a specific manager or management application. The latter has the responsibility for the enforcement of its policies on that domain. It is clear that the above approach adds flexibility and structuring capability to the management activities. *Figure 1* depicts an example of organizing a distributed system in terms of management domains and management policies [17, 18, 21, 22].

Another issue that arises from that approach is that, at the end, a significant number of relationships between the policies is developed. These are high-level policies that usually derive from business goals, but there are also low-level policies that are directly enforced onto network resources. This implies the need to translate policies from high-level to low-level ones, and management policy hierarchies are formed this way. This translation is not an easy task and it becomes even more complicated in large-scale distributed systems and in telecommunication networks, where numerous policies are usually enforced. Many different factors must be taken into consideration to effectively complete such a task, like existing management systems and technologies, management structures, management domain construction criteria, roles of management authorities, and agents, etc. [13, 17, 19].

This paper deals with the requirements that arise when one develops a policy-driven network management framework. The paper gives both research and practical views of policy-driven management. It presents a middleware architecture for providing policy management and enforcement services, and also a methodological approach for developing hierarchically interrelated policies for managing systems and net-works, as well as it presents case study. Finally, it discusses open issues on the subject and future work.

The Problem Space in Network Management

Before we provide a state of the art on policy-driven management, we give some parameters of its problem space in the general area of network management.

These parameters are the requirements for a policy-driven network management, which must support the deployment of a wide range of management services for the satisfaction of these requirements. In other words, the support framework for a policy-driven network management must be integrated and provide modular services for adaptive use of network resources, something that is especially useful for ad-hoc networks, autonomous computing, wireless networks, and context-aware networks.

From a practical point of view, an effective policy-driven network management must support the management and policy enforcement for the following:

- A strategic monitoring over the manageable network resources as well as the network surveillance rules
- Classification of network traffic
- Congestion control
- Dynamic bandwidth (de-)allocation, reserved or on demand for diverse traffic patterns
- Customizable QoS to end users according to their legitimate requirements
- Admission control and its rules with respect to bandwidth usage
- Identification and discovery rules for user and services

FIGURE 1

The Use of Domain and Policy Concepts in System and Network Management

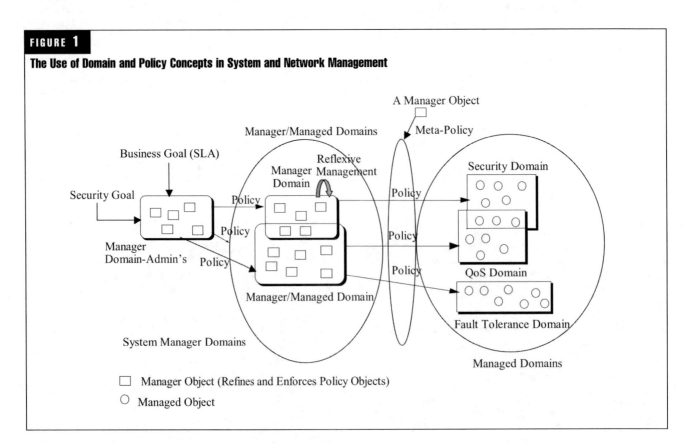

- Access control for security and optimization reasons over network resources, defining user specific network privileges and point of control, e.g., wide-area network (WAN) access point or point of origin
- Control over a pool of servers that offer processing power
- Number of communicating entities, e.g., users, applications
- Various imposed restrictions, such as temporal, of content, of traffic classes
- Priority rules, e.g., voice acceleration
- Event triggering and notification dissemination
- Accounting and billing based on traffic usage
- Performance benchmarking and tuning
- Control of configuration changes
- Reaction management, e.g., due to security violations, etc.
- Propagation of control down through the network implementation layers
- The complexity handling induced by many constraints and exceptions that usually arise and are almost impossible to be handled manually
- The network administration task, which presupposes a large variety of skills, i.e., in operating systems, network technologies, security systems such as firewalls and internal data services (IDS), application protocols, etc. [24, 25]

State-of-the-Art of Policy-Driven Network Management

Policy-driven network management is an event-triggered constrained action provisioning mechanism for an automated response on the network according to pre-defined policies.

Policies are rules of the general form: *ON <event> IF <condition> THEN <do actions>*. But, who is going to enforce policy on what object must be also defined. So, a policy, or better a policy object, is defined as the following set of attributes [7, 17, 21]:

PO = {Type, Subject, Target, Event, Actions, Constraints, Priority}, where:

- PO   = Policy Object
- Type   = Policy Type
- Subject   = Manager Objects
- Target   = Managed Objects
- Event   = On a specific Event Triggered *do* the Actions
- Actions   = Task-specific Actions
- Constraints = Restrictions on Policy Object Enforcement
- Priority   = An integer expressing a Policy Priority Type

Policy type (including modality) concerns negative or positive authorization and negative or positive obligation.

Some policy examples in the format shown above are given in section 7, along with a short case study. Some policy examples expressed in plain English are the following:

- The Access to the Medical Database Files is permitted only to Doctors during the working hours (*permission*).
- The Correction of Data Records is not permitted to non-Advanced Users at Any Time (*prohibition*).
- The Bandwidth must be allocated from the Network Managers to the Users according to the Service Type on User Registration (*obligation*).
- System Administrators do not have to perform hacking tests on Any System at Any Time (*refrain policy*).

Figure 2 gives a general idea of the use of a policy-driven management support system, which otherwise is called policy middleware [17, 18, 25]. The goal of the policy support system management is to automate the reactions when retrieving predefined policies that point out the actions that should be taken when an event happens. This automated function reduces the likelihood of human error, provides a more flexible management procedure, and accelerates reaction time, configuring a large number of resources with a single policy. The translation of high-level policies into low-level resource-specific configurations allows changes in policies without the need to change the corresponding management application code. In other words, policies parameterize the management applications [17, 21, 25].

Policy-based management leads to flexibility, adaptability, and scalability. These benefits arise when policies are correct, complete, valid, and consistent. These requirements force us to analyze policies to detect inconsistencies and to derive policies from high-level goals. The latter is not an easy task. For this reason, there are several approaches that try to automate the process, but we are still at the very beginning. Business goals are usually general statements expressed in plain text. Appropriate data policy models must be used to translate plain-text statements to the policy definition format provided above. The highest-level policies produced from the initial business goals must be also translated into lower level ones. In most cases, the produced new policy is strongly depended on the context of the application.

Of course, for such task specific policies (e.g., ad hoc network fault tolerance policies) best practices must be followed in order to the minimization of potential enforcement problems. For this reason, there is a need for an automated solution to audit policy results on the network. In fact, a policy model implies a context or domain within which the policy applies, e.g., network configuration or access control. Furthermore, it specifies when a policy is to be applied. The condition can be specified as a Boolean expression, which is the "if clauses" within a policy. Decision is policy guidance, which is the "then clause" within a policy.

A policy hierarchy is the means to cope with the enterprise and system management roles and management structures. A management policy specifies the authorizations and obligations for a group of managers, namely, the behavior expected from managers assigned to a particular management position. Equivalently, the role assigned to a manager is defined as a set of policies applying to a domain of managers, called position domain. Managers can be assigned to or removed from a role without re-specifying the respective

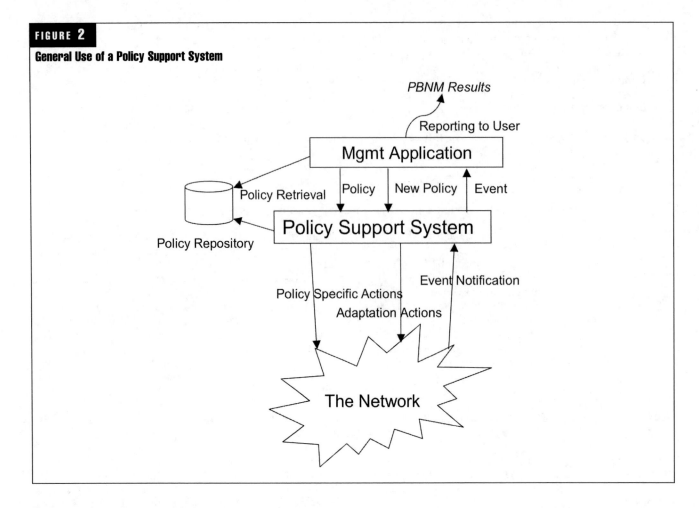

FIGURE 2

General Use of a Policy Support System

policies and manager roles can interact with each other [13]. Thus, we can easily understand that a role can be de-structured to a set of single policies and that manager hierarchies can be mapped to policy hierarchies. In other words, the management problem is almost the same with the management policy hierarchy creation, analysis, and optimization [15, 19].

Focusing on the policy hierarchy definition, we note the parent (meta-) or child (sub-) relationships between policies on the hierarchy. Dynamic environments require policy hierarchies to expand or contract in a flexible way. A policy hierarchy is determined by the following arrangement [17]:

PolicyHierarchy = (Names, Descriptions, Relationships) = (N, D, R), where:
- N = the names or IDs of the Policy Objects (PO), e.g., N = {PO1, PO2}
- D = the total descriptions of all the hierarchy policies, e.g., D={PO1=[PT1, S1, T1, E1, A1, C1, P1], PO2= [PT2, S2, T2, E2, A2, C2, P2]}, and

R = the set that contains for each policy the set of its meta-policies/parent policies, e.g., for PO1⇒PO2, it is: R={PO1= [ ],PO2=[PO1]}

Architecture of Policy-Related Management Services

For the deployment of a policy-driven network management, there is a need to develop appropriate distributed management platforms or, equivalently, policy-related management services. Administrator or system management applications use domain- and policy-related tools to structure their management task. These tools support capabilities for translating policy specifications of the highest level, i.e., business goals or service-level agreements (SLAs), to lower-level ones, continuing this process up to the network element control or monitoring actions that are implementable. Of course, network administrators will be provided with appropriate management consoles for editing, compiling, creating, and deploying policies and monitoring the results. The administrator must also be supported with management domain browsers (for creating and navigating domain hierarchies) and a role definition tool. Storage, searching, and retrieval tools for policy objects must also be provided. A policy service or control and decision making on policy evaluation and final policy selection and/or configuration must be also provided. Finally, a task-specific policy enforcement service must be implemented. Interoperability between all of these components must be ensured via various protocols such as common open policy services (COPS), lightweight directory access protocol (LDAP), single network management protocol (SNMP), etc. [1, 4, 8, 9, 17, 25]. *Figure 3* presents an architectural view of the previously mentioned policy-related services and components.

From the implementation architecture point of view, the services mentioned in *Figure 3* may be centralized or distributed. In large-scale distributed management services it is obvious that they will be distributed across a network of

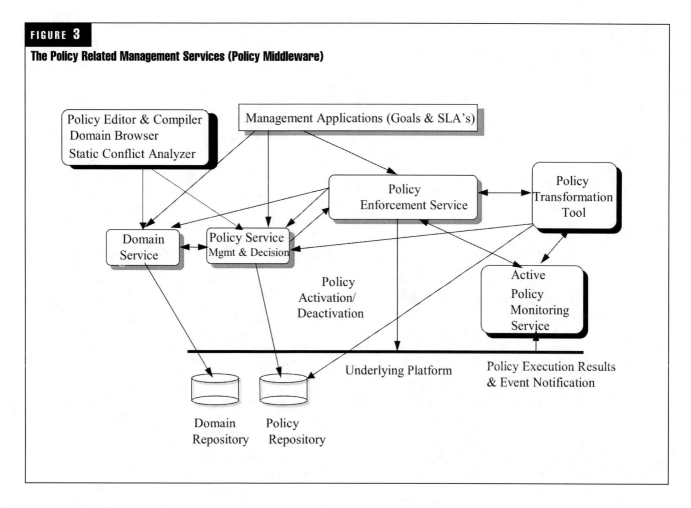

FIGURE 3

The Policy Related Management Services (Policy Middleware)

servers. For instance, a policy service may consist of cooperative distributed sets of policy services and policy repository servers to manage policies at endpoints of the network.

For the completeness of our presentation, in *Figure 4*, we show where the above common policy related management layer is buckled up (the underlying mechanisms).

Regarding the policy life cycle, we start from the abstract and high level policies. A policy classification which categorizes policies according to various criteria such as the open distributed processing viewpoints [11] or the system management functional areas (SMFAs) (*Figure 5*) is necessary in the beginning. Based on the policy template, which usually depends on the policy specification language or the policy model followed, e.g., rule-based or formal logic-based, the policy definition takes places.

FIGURE 4

Underlying Platform Layers

Underlying Distributed System Platform

Distributed Processing Services (File, Security, Time, Transaction, etc.)

Distributed Object Services (Corba/DME, SNMP, CMIP)

Communication Services (RPC, OSI, TCP/IP, Others)

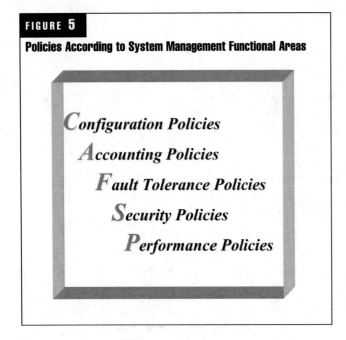

FIGURE 5

Policies According to System Management Functional Areas

*C*onfiguration Policies

*A*ccounting Policies

*F*ault Tolerance Policies

*S*ecurity Policies

*P*erformance Policies

ment usually concerns?" The answer can use a combination of previously presented viewpoints in open distributed processing (ODP), telecommunications management networking (TMN), and SMFA:

- The business viewpoint
- The computational viewpoint
- The informational viewpoint
- The engineering viewpoint
- The technological viewpoint
- The functional-area viewpoint
- The functional-level viewpoint
- The geographical viewpoint

It is out of the scope of this paper to describe in detail a task-specific policy enforcement mechanism through the underlying platform interfaces. Instead, we roughly provide a generic enforcement process. After a policy editing, static analysis and compilation, this policy is stored in a policy object repository for persistent storage. After the policy selection and analysis phase via the policy service, an enforcement service takes action for policy activation and enforcement of policy actions on the underlying system platform via the application programming interface (API) this platform offers. The enforcement service lists all policy members of the subject domain, as well as of the target domain using the domain service. Then, it defines the actions (object methods) to be invoked on the managed system, while in parallel examines potential conflicts with already activated policies, listing the invoked methods with their input and output parameters on the respective managed system. After this stage, it repetitively invokes the methods (policy actions). By the end of invocations, the pol-

These policies before being propagated up to the lowest level of managed resources, must go through an appropriate policy analysis (e.g., for conflict detection and resolution) and refinement [1, 2, 3, 4, 14, 18]. *Figure 6* depicts the policy life cycle.

The question which arises is, "According to the complicated views of a corporate networked system or a telecommunication network which are the viewpoints that policy enforce-

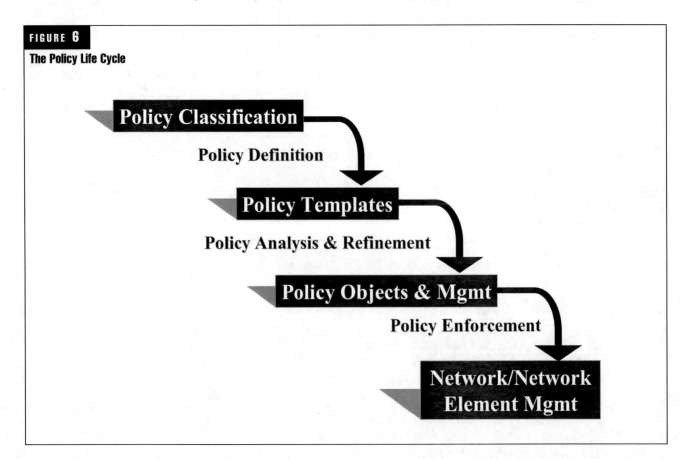

FIGURE 6

The Policy Life Cycle

Policy Classification

Policy Definition

Policy Templates

Policy Analysis & Refinement

Policy Objects & Mgmt

Policy Enforcement

Network/Network Element Mgmt

icy is enforced and its operational state becomes active or enabled. Finally, the enforcement service notifies the policy service for the policy activation. In [16, 17] the reader can find interesting enforcement architecture examples for network performance and QoS management.

A 10-Step Methodological Approach for Building up the Policy-Driven Network Management

In this section, we present a methodological approach for building up the policy-driven network management task. The question, which usually arises, is how we can construct an initial set of policies that governs the network. We may have setup tools, but this does not mean that we have setup the right policies. Thus, there is a need for developing a methodological way for setting up the initial policies which could be transformed to adapt to a business goal or SLA change (if we accept a policy from the top level) or a network configuration change (if we accept an appropriate event from the bottom level). Of course, this is a complicated task in networked systems or telecommunication networks due to the multiple relationships between the components and the relative policies.

Any policy can be analyzed in a triplet of the following type (subject, policy actions, target, etc.) [17]. Deploying the initial business goals down through the system in fact we create an important number of such triplets. Thus, the key steps and/or factors for such a deployment are the following:

1. Localize the existing management structures of the system and the network.
2. Build-up management domain hierarchies. For large-scale systems, this is quite complicated, and both top-down and bottom-up approaches must be followed. Overlapping between domains is permitted. Coherent membership rules for each managed object must be developed. Managed objects concern network elements, networks, services and applications.
3. For each managed object on the domain hierarchy setup initial authorizations associated with it, as well as initial configurations. This must be well stored from the beginning for effective adaptive event-based management of change on a later stage.
4. Localize and assign managers onto the previously constructed domains. Usually, there is no single source of management authority over the managed resources. Heterogeneity of management system technologies may play an important role here.
5. Localize the relationships between the domains developed for the purpose of policy applicability (management task).
6. Elaborate initial business goals and disjointing goals, and develop multiple strategies.
7. Define and deploy policies (authorizations and obligations) and delegate responsibilities, deploying in this way roles and behaviors. Authorizations for such delegations must be first defined.
8. Interpret/refine the high-level policies to sub-policies up to the lowest level, statically or at run-time. Policy conflict detection and resolution must take place at this step as well.
9. Optimize policy sets.
10. Manage the change set by new goals or driven by triggered events and make new policy decisions.

Current Implementation Approaches

The work done in policy-based management has a diverse of approaches. One of them concerns the formal policy specification. Ponder [7], developed by Imperial College, is a declarative language with three policy categories, authorization, obligation, and delegation with two modalities: positive and negative. A negative obligation is translated as "refrain". Furthermore, Ponder supports domain hierarchies for handling role relationships.

Event Calculus [3] is an important tool for policy implementation approaches because it has well understood formalism, models event-driven systems, uses deduction for simple property checks and uses reasoning to derive explanations for property violations. Furthermore, it is suitable for interaction with formal representations.

Logic-based languages are suitable for security policies extending modal logic operators for permission, obligation, and prohibition. In this context, role-based access control (RBAC) handles permissions, inheritance, and role reusability in a flexible way. The security policy language (SPL) is an event-driven access control policy language, while the OASIS standard extensible access control markup language (XACML) is an XML specification for defining access control policies over the Internet. XML–based specifications have the advantage of ease of parsing, compatibility with Web services, and seamless extensibility for adding new functions and data types. The Internet Engineering Task Force (IETF)/Distributed Management Task Force (DMTF) policy core information model (PCIM) has been created for representing policy information. DMTF supports instances and groups of PolicyRules that aggregate PolicyConditions and PolicyActions. Conditions can be combined in conjunctive or disjunctive sets, while actions can be sequenced. Currently there is work in support of generic Event-Condition-Action policy rules. The policy definition language (PDL) from Bell Labs uses an event-condition-action rule paradigm of active databases to specify a policy, while the enterprise language of the enterprise viewpoint of reference model for open distributed processing (RM–ODP) further adopted and elaborated policies and roles within the community concept [7].

Regarding the existing management platforms there is an important number of implementations, the presentation of which is out of the scope of this paper. We only mention an indicative example, the open software foundation (OSF)/distributed management environment (DME) which is an object-oriented infrastructure which can be added to fit on it a policy middleware layer. DME is based on common object request broker architecture (CORBA) and open software foundation (OSF)/data communications equipment (DCE) remote procedure call (RPC) and supports simple network management protocol (SNMP) and common management information protocol (CMIP) access through the x/open management protocol (XMP) API, management user interfaces based on OSF/Motif, distributed notification services, as well as other management services [20]. Using appropriate gateways, distributed management platforms can be bridged. For instance, an RPC/interface definition language (IDL) to CMIP/guidelines for the definition of management objects (GDMO) gateway provides the appropriate interface specification mappings from the first language to the second one [10], allowing applications running on a different platform to use functionality provided by another one.

Regarding commercial platforms and tools, we indicatively mention Bull ISM, Siemens Transview, and IBM Tivoli.

A Case Study in QoS Management

In this section, we present an indicative case study for the purpose of clarifying some of the concepts of policy-driven network management.

The example concerns differentiated services (DiffServ) based on transport characteristics (any varying demand from the user side must be reflected on the QoS management of network). QoS management implies techniques such as DiffServ per hop for different service types.

Policy-related services include, among others, decision making on the selection of the policy which best fits to the QoS change requirements. In other words, a network-wide service adaptation for implementing the new service requirements is required. The initial high-level policy selection process fires another policy selection process at a lower level, while underlying service mechanisms manage sessions and admission control. The final goal is to configure appropriately the QoS mechanisms of the DiffServ network. Initially, a new request from a management application will invoke the Policy Service to select from the Policy repository the high-level policy based on given parameter values. The high-level Policy after its activation among other actions will force the activation of a lower-level policy, which will come from a translation (refinement) process. The latter policy will be activated and triggered to configure DiffServ parameters of the network, usually associated with a path of routers within a DiffServ domain [3, 4, 16, 22].

Suppose that the following goal needs be satisfied and described in a plain text statement: Web Services (WS) Multimedia (MM) Applications from Service Domain X must get "Top QoS."

The high-level (service-oriented) policy in plain text is: "If the _service user_ is "WS MM Application" from _ServiceDomain X_, then provide _Top_ level service"

And, in formal specification:

PO	= Diffserv_Policy_High
Type	= Obligation
Subject	= Manager of "Service Grade"
Target	= Managed ServiceDomain_X
Event	= On Triggered Request for "Top Service"
Actions	= Activate_Policy(Diffserv_Policy_Low)
Constraints	= When Network_Service(Available) ^
	Time==Working_Hours
Priority	= High

With Policy Translation Rule from 1 to 2:
1. _WS MM Application_ is on the _ServiceDomain X_
2. _Top_ service is to provide a _Max bandwidth, Min_Delay and Min_Loss_ in _network_path={router1, router2, router3}_

After the translation of the high-level policy to a lower one, we have the following policy:

The low-level (network-oriented) policy in plain text is: "If the _network user_ is from _AdminDomain X_, then reserve a _Max bandwidth, Min_Delay_ and _Min_Loss_ in _network path = {router1, router2, router3}_"

And, in more formal specification:

PO	= Diffserv_Policy_Low_1
Type	= Obligation
Subject	= Manager of DiffServ
Target	= Managed Resources on AdminDomain_X {Router1, Router2, Router3}
Event	= On Triggered Request for "Top Service"
Actions	= Set_Bandwidth(Max); Set_Delay(Min); Set_Loss(Min);
Constraints	= When Resources(Available) ^
Bandwidth<	=Upper_Limit^Time==Working_Hours
Priority	= High

Of course, there must be a mapping between the _ServiceDomain_ and the _AdminDomain_, as well as between the service user and the network user.

A networked multimedia application may request different QoS guarantees at run time by upgrading or downgrading its own service parameter values. This leads to a new policy enforcement to adapt the network to the change. Managing this change we have the following Policy Objects at the high-level:

PO	= Diffserv_Policy_High_Change
Type	= Obligation
Subject	= Manager of "Service Grade"
Target	= Managed ServiceDomain_X
Event	= On Triggered Request for "New Service"
Actions	= Get_Parameters("NewService", Par_1, Par_2, Par_3); Activate_Policy (Diffserv_Policy_Low_Change)
Constraints	= When Network_Service(Available) ^
Time=	=Working_Hours
Priority	= High

With Policy Translation Rule from 1 to 2:
WS Application asks for Change on _ServiceDomain X._
Gold service is to provide a _Max bandwidth, Min_Delay_ and _Min_Loss_ in _network_path={router1, router2, router 3}_

PO	= Diffserv_Policy_Low_Change
Type	= Obligation
Subject	= Manager of Diffserv
Target	= Managed{Router1,Router2,Router3} on AdminDomain_X
Event	= On Triggered Request for "New Service"
Actions	= Set_Bandwidth("Par_1"); Set_Delay("Par_2"); Set_Loss("Par_3");

```
Constraints   = When Bandwidth<=Upper_Limit ^
Time=         =Working_Hours
Priority      = High
```

We note that for the purpose of the management of change, the policy objects must be parameterized. The explanation of the above policy examples is quite obvious and left to the reader.

Discussion and Future Work

Policy-driven management platforms, even if they have been extensively examined and researched by industry, are still behind regarding their wide deployment. In addition, even if policy frameworks have been formally defined, they are not widely used. The reason is that they are quite complicated to use and if we want to widely deploy them, additional tools such as artificial intelligence techniques are needed [12]. Domain hierarchy handling tools have been exploited due to the fact that they are easier to understand and they are really helpful in understanding better the policy applicability bounders. Appropriate visualizations, sophisticated user interfaces, and wizards might further help set up complicated policies and analyze their relationships.

A mature policy middleware supporting common policy-related management services could help developers and administrators achieve high-quality policy implementations, including effective policy analysis and conflict resolution, policy translation, validation, and policy set optimization. From the above, we understand that there is still a number of open issues in the policy-driven management research and development arena.

Future work could drive a standardized policy specification language that will be widely adopted by industry, to cope with heterogeneity and interoperability problems. But still many other problems on the subject remain unsolved. Resolution on policy conflicts and consistency between policies are two main topics for research. Validation, transformation, and optimization in policy objects sets are also at an immature stage and need further research. Role-based access control (RBAC) and dynamic assignment of roles to manage objects can be further examined to enhance security management. Separation of duties and delegation of authorizations and obligations can also be benefited by the flexible and effective use of policies.

We believe that further deployment of policy-driven management can be boosted if there is first a mature policy middleware development regarding the interoperability of the common policy-related management services and second adaptation of AI techniques, both for policies of generic type as well as for specific management application areas. For instance, appropriate AI algorithms and mechanisms could support automated derivation of policy actions, the distribution of actions between agents, the development of sophisticated modeling and simulation tools, the optimization process, strategies and complexity handling, etc.

Conclusions

In this paper, we presented a policy-driven methodology for managing corporate and telecommunication networks.

This method has been proved to be very effective in structuring the management task, as well as in making it more feasible in large-scale distributed systems where there are numerous resources and different kinds of management activity. Policy-driven management is most suitable for the management of change in networks and systems. Our focus was on presenting the added value brought by policy-driven management, examining a policy support middleware platform, discussing recent development efforts, and proposing in a methodological approach for building policies on systems. We also presented a case study in QoS management depicting the high applicability of the policy-driven management framework. Open issues and future work was also provided.

In conclusion, we emphasize that even if there is a considerable amount of work done on deploying policy-driven management services in the past two decades, many challenges still remain that could be approached via interdisciplinary scientific research.

References

1. Agrawal, D., Giles, J., Lee, K., Voruganti, K., Filali-Adib, K. (2004). Policy-Based Validation of SAN Configuration. *Proceedings of 5th IEEE Workshop on Policies for Distributed Systems and Networks (Policy 2004)*, IBM Watson Research Centre, New York, June 2004.

2. Bandara, A., Lupu, E., and Russo A. (2003). Using Event Calculus to Formalize Policy Specification and Analysis. *Proceedings of 4th IEEE Workshop on Policies for Distributed Systems and Networks (Policy 2003)*, Lake Como, Italy, June 2003.

3. Bandara, A., Lupu, E., Russo, A., Moffett J. (2004). A Goal-based Approach to Policy Refinement.*Proceedings of 5th IEEE Workshop on Policies for Distributed Systems and Networks (Policy 2004)*, IBM Watson Research Centre, New York, June 2004.

4. Beigi, M., Calo, S., Verma, D. (2004). Policy Transformation Techniques in Policy-based System *Proceedings of 5th IEEE Workshop on Policies for Distributed Systems and Networks (Policy 2004)*, IBM Watson Research Centre, New York, June 2004.

5. Cactano, A., Zacarias, M., Silva, A.R., and Tribolet, J.(2003). A Role-Based Framework for Business Process Modeling. *Proceedings of 38th Hawaii International Conference on System Sciences*, Island of Hawaii, June 2003.

6. Damianou, N., Bandara, A., Sloman, M., and Lupu, E. (2002A). A Survey of Policy Specification Approaches. *Dept of Computing, Imperial College of Science Technology and Medicine*, www.doc.ic.ac.uk/~mss/Papers/PolicySurvey.pdf, London.

7. Damianou, N., Dulay, N., Lupu, E., and Sloman, M. (2001). The Ponder Policy Specification Language. *Proceedings of Workshop on Policies for Distributed Systems and Networks (Policy 2001)*, Bristol, UK, January 2001, Springer-Verlag LNCS 1995, 18–39.

8. Damianou, N., Dulay, N., Lupu, E., Sloman, M., and Tounouchi, T. (2002B). Tools for Domain-based Policy Management of Distributed Systems. *Proceedings of IEEE/IFIP Network Operations and Management Symposium (NOMS2002)*, Florence, Italy, 15–19 April 2002, 213–218.

9. Dulay, N., Lupu, E., Sloman, M. and Damianou, N. (2001). A Policy Deployment Model for the Ponder Language. *Proceedings of IEEE/IFIP International Symposium on Integrated Network Management (IM'2001)*, Seattle, May 2001.

10. Forbici, E., and Penna, M. (1997). Implementation of a RPC/IDL to CMIP/GDMO Gateway. *Tina*, vol.00, 216.

11. ISO/IEC JTC1/SC21 (1995). Basic reference model of open distributed processing, part 2: Descriptive model. *ITU-T X.903-ISO/IEC 10746-3*.

12. Kephart, J., Walsk, W. (2004). An AI Perspecitve on Autonomic Computing Policies.

13. Proceedings of 5th IEEE Workshop on Policies for Distributed Systems and Networks (Policy 2004), IBM Watson Research Centre, New York, June 2004.

14. Lupu, E., and Sloman, M. (1997). Towards a Role Based Framework for Distributed Systems Management. *Journal of Network and Systems Management* vol.5(1), 5–30.

15. Lupu, E., and Sloman, M. (1999). Conflicts in Policy-based Distributed Systems Management. *IEEE Transactions on Software Engineering* vol.25(6), 852–869.

16. Lupu, E., Milosevic, Z., and Sloman, M. (1999). Use of Roles and Policies for Specifying and Managing a Virtual Enterprise. *Proceedings of 9th IEEE International Workshop on Research Issues on Data Engineering: Information Technology for Virtual Enterprise (RIDE-VE'99)*, Syndey, Australia, March 1999.

17. Lymberopoulos, L., Lupu, E., and Sloman, M. (2003). An Adaptive Policy based Framework for Network Services Management. *Journal of Network and Systems Management, Special Issue on Policy Based Management of Networks and Services* vol.11(3).

18. Mitropoulos, S. (2000). Integrated Enterprise Networking Management: Case Study in Intelligent Multimedia Message Handling Systems. *Journal of Network and Systems Management* vol.8(2), 267–297.

19. Mitropoulos, S., and Veldkamp, W. (1994). Integrated Distributed Management in Interconnected LANs. *Proceedings of IEEE/IFIP Network Operations and Management Symposium (NOMS' 94)*, Florida, USA, February 1994, 898–908.

20. Moffett, J., and Sloman, M. (1993). Policy Hierarchies for Distributed Systems Management. *IEEE JSAC Special Issue on Management* vol.11, 1404–1414.

21. OSF-DME (1992). *Open Software Foundation. OSF Distributed Management Environment (DME) Architecture*, May 1992.

22. Sloman, M. (1994). Policy Driven Management for Distributed Systems. *Journal of Network and Systems Management* vol.2(4), 333–360.

23. Sloman, M., Magee, J., Twidle, K., Kramer, J. (1993). An Architecture for Managing Distributed Systems. *Proceedings of 4th IEEE Workshop on Future Trends of Distributed Computing Systems*, Lisbon, September 1993.

24. Steven Van den Berghe, FilipDe Turck, PietDemeester (2004). Integrating Policy-based Management and Adaptive Traffic Engineering for QoSDeployment. *Proceedings of 5th IEEE Workshop on Policies for Distributed Systems and Networks (Policy 2004)*, IBM Watson Research Centre, New York, June 2004.

25. NetVeda, "Policy.Net: Policy Based Network Traffic Management," www.netveda.com.

26. Waller, A. (2004). Policy Based Network Management, INET2004, May, 2004.

Application of the Systems Approach to Telecommunications Strategic Management

Steven R. Powell

Professor, Computer Information Systems (CIS) Department
California State Polytechnic University

Systems thinking has been incorporated into many disciplines, and its development has produced a variety of approaches to solve problems. This paper addresses systems thinking in strategic management and investigates how the systems approach can be applied to telecommunications strategic management. Although some systems thinking can already be found in telecommunications strategic management, this paper suggests additional ways it can be applied.

Introduction

We are now leaving the machine age and entering the systems age (Ackoff 1999a). The machine age viewed the world as a machine based on cause-and-effect relationships. Machine-age thinking was analytical, focusing on structure and reducing problems to indivisible elements. The systems age views the world in terms of purposeful systems and a belief in the incompleteness of cause-and-effect relationships. Systems-age thinking is synthetic, focusing on function and expanding the objects under investigation. Analytical thinking increases knowledge; synthetic thinking promotes understanding. Systems thinking focuses on holism, in which a system is considered to be more than the sum of its parts. The whole gives purpose to the study (Jackson 2003).

Systems thinking and the systems approach have been incorporated into many disciplines. Philosophy, biology, ecology, political science, social science, cognitive science, control engineering, the physical sciences, telecommunications, and organizational management are some of the disciplines in which systems thinking has found a place (Bertalanffy 1968; Klir 2001). In the field of telecommunications, telecommunications operations, the telecommunications network, and even the telecommunications industry can be approached from a systems perspective (Smol, Hamer, and Hills 1976; Goldman 1995). In organizational management, organizations can be viewed as systems that interact with their external environments (economic, political, social, competitive, etc.) and consist of interrelated internal functional and management subsystems (Cleland and King 1972, Beer 1995). Systems thinking is important to organizational management for a number of reasons, including the following: organizations can be complex, requiring a holistic approach rather than an approach based on reductionism; systems thinking emphasizes process as well as structure, leading to innovative behavior; systems thinking is multidisciplinary, drawing on the strengths of each discipline; and systems thinking has an established track record for solving real-world management problems (Jackson 2003). Systems and the systems approach will be discussed in greater detail in the second section of this paper.

Systems thinking has been applied to strategic management, an important organizational management discipline. Strategic management plays a key role in organizational performance and is the set of managerial decisions and actions that enables an organization to achieve its long-term objectives. Strategic decisions are of a long-term nature, set precedents, and are rare and consequential (Hickson et al. 1986). Strategic management consists of environmental scanning, strategy formulation, strategy implementation, and evaluation and control (Wheelen and Hunger 2002). The systems approach can make the strategic management process more effective. How this can be achieved will be described in the third section of this paper, which will examine, among other approaches to strategic management, the strategic planning and implementation approach of Cleland and King (1983) and the interactive planning approach of Ackoff (1999b). The fourth section will investigate how the systems approach can be incorporated into telecommunications strategic management. Although some systems thinking can already be found in telecommunications strategic management, much more is possible. The paper concludes with a summary and some directions for future study.

Systems and the Systems Approach

While there is no unique definition for a system, it is often considered to be a complex whole, the functioning of which depends on its parts and the interactions among those parts (Jackson 2003). According to Churchman (1968), the following five considerations should be kept in mind when thinking about the meaning of a system: the performance measures of the whole system; the system's environment and the fixed constraints; the resources of the system; the components of the system, including their activities, goals, and measures of performance; and the management of the system. Elements common to systems include inputs, outputs, process, feedback, control, environment, and goals. Systems are sometimes affected by changes in their environment. Activities occurring within the system are endogenous; activities affecting the system that occur in its environment are exogenous. Systems that have exogenous activity are open; systems with no exogenous activity are closed. Feedback occurs when there is coupling between the input and output of a system (Gordon 1978). Models can aid in analyses of systems. A model is a representation or an abstraction of an object or representation of a particular real-world phenomenon. Models can be used to investigate cause-and-effect relationships and the interaction among key variables. Models are used extensively in operations research and management science to analyze the future consequences of decisions in complex, uncertain situations (Wagner 1975).

The systems approach allows the individual components to be viewed in relation to the whole. Just as there is no unique definition for a system, there is no unique approach to systems. Ackoff (1999a) views the systems approach as identifying a containing whole (system) of which the thing to be explained is a part, explaining the behavior or properties of the containing whole, and then explaining the behavior or properties of the thing to be explained in terms of its role(s) or function(s) within its containing whole. However, Churchman (1968) writes of a continuing debate between four kinds of approaches to systems: science (there is an objective way to view a system and to build a model of it), efficiency (identification of trouble spots), humanist (looking first at the human values of freedom, dignity, and privacy), and anti-planning (systems must be lived in and experienced, and not changed by grandiose schemes or mathematical models). Similarly, in his advocacy for creative holism, Jackson (1993) describes the following types of systems approaches for management: systems approaches for goal-seeking and viability by increasing the efficiency and efficacy of organizational processes and structures (e.g., hard systems thinking, system dynamics, organizational cybernetics, and complexity theory); systems approaches to improve organizational performance by exploring purposes and ensuring sufficient agreement is obtained among an organization's stakeholders about purposes (e.g., strategic assumption surfacing and testing, interactive planning, and soft systems methodology); systems approaches for ensuring fairness in organizations by eliminating discrimination and encouraging participation (e.g., critical systems heuristics); and systems approaches for promoting a diversity of ideas, thought, and emotions (e.g., postmodern systems thinking).

The Systems Approach to Strategic Management

The systems approach is well known in strategic planning and management. Planners feel that the way to look at a whole system is in terms of a plan (Churchman 1968). Elements of planning in an industrial organization from a systems perspective can be found in some early work of Beer (1959), and more fully in Ackoff (1970) and Cleland and King (1972). Later, these authors considered the subject in greater depth, which will be discussed in this section. This section also will investigate the work of Haines, a proponent of Ackoff's planning approach. For the most part, the differences in the strategic planning and management approaches these authors propose relate to the significance placed on the planning process relative to the plans generated, the comprehensiveness of the approach, the key participants identified in the process, the direction and degree of continuity of the process, the learning that takes place during the process, and the degree to which the process shapes the environment.

Beer (1979) applies cybernetic laws to organizations to formulate a model describing an organization's interactions with the environment. The model consists of five essential subsystems (implementation, coordination, operational control, development, and policy) with feedback loops and information flows. In Beer's view, planning is the organization's cohesive glue. It is a continual learning process that should be undertaken by managers rather than by professional planners. However, since the environment changes so rapidly, plans are immediately out-of-date, with the result that what is most important is the planning process itself, not the plans. Plans do not have to be implemented, since they will be aborted constantly by the manager's own decisions to commit resources to a different future. Ultimately, therefore, it is the manager's actions that constitute the plan.

Cleland and King (1983) also recognize the importance of the planning process to the organization, but they do not dismiss the plans or their implementation. Instead, they propose a comprehensive strategic planning and implementation system, based on the systems approach, that possesses a conceptual basis and defines an operational process and system for ensuring that the process is performed on an ongoing basis. The approach to strategic planning taken by Cleland and King is based on a model describing how planning unfolds in a cybernetic way through learning in feedback cycles (Jantsch 1973). Cleland and King feel that applying the systems approach to strategic management allows planners to see the big picture, enabling them to go beyond problem solving to exploit opportunities in the environment. However, understanding social systems is difficult, since they are extremely complex, with many interactions and interdependencies resulting in second-order effects, which can grow over time. The operational process for applying systems thinking to planning that Cleland and King propose consists of defining the organization in system terms by describing every relevant clientele group and organizational claimant in meaningful and (hopefully) measurable terms, defining the goals for each organizational element and claimant described, deciding on alternative courses of action, creating management systems to institutionalize the decision-making process, and integrating all relevant management systems. The strategic planning system Cleland and King propose to ensure that planning is carried out systematically consists of an interrelated system of the entirety of strategic, development, operational, and project plans generated within the organization; a planning process (e.g., establishing general goals, information collecting and forecasting, making assumptions, establishing specific objectives, and developing plans); a decision subsystem employing systems analysis; a management information subsystem to support planning; and an organizational culture and management system that facilitates planning and ensures that it is performed effectively.

For plan implementation using the systems approach, Cleland and King propose a conceptual framework based on a series of strategic choice elements that the organization possesses or, in other words, choices that the organization makes. These relate to its choice of mission ("business" the organization is in), objectives (desired future positions or roles for the organization), strategy (general direction for pursuing the objectives), goals (specific targets sought at specified points in time), programs and/or projects (activities through which strategies are implemented and goals pursued), and resource allocations (allocation of funds, personnel, etc., to various units, objectives, strategies, and programs). In addition to defining the strategic choice elements, it is important to analyze the interrelationship of these elements, which generally can be represented as a hierarchy with strategies, goals, and programs supporting the objectives, which in turn support the mission. While analyzing the interrelationship of the strategic choice elements is important, the key to effective implementation comes from a process of strategic program evaluation, which filters potential projects and programs based on their relationship to certain strategic criteria, including fit with mission, fit with objectives, consistency with strategy, contribution to goals, corporate strength base, corporate weakness avoidance, comparative advantage level, internal consistency level, risk level acceptability, and policy guideline consistency. The output of the evaluation process is a prioritized set of programs and projects. The system to achieve successful implementation on a day-to-day basis is a project management system that ensures that projects are evaluated, selected, funded, and executed appropriately. This system consists of a facilitative organizational subsystem that superimposes project teams on a functional structure in a matrix arrangement; a project control subsystem for the selection of project performance standards; a project management information subsystem; a set of operations research and management science techniques and methodologies to support project management; a cultural ambience subsystem describing how project management is practiced in the organization; a planning subsystem for proper project planning; and a human subsystem, which concerns the human aspects of project management.

Ackoff (1970; 1981; 1999a; 1999b), like Beer, views planning as a continual process with no natural conclusion or endpoint. Planning approaches a solution but never achieves it because an infinite amount of analysis is possible and both the system being planned for and its environment change during the planning process. A plan is not the final product of the planning process; it is an interim report. Ackoff advocates interactive planning, i.e., planning based on designing a desirable future and inventing ways to bring it about. The most important product in interactive planning is engagement in the planning process, not the formulation of plans. During the planning process, knowledge, understanding, and wisdom are generated. In interactive planning, the professional planners' role—one of encouraging and facilitating planning by others—is much more educational than advisory.

Operating characteristics of interactive planning include planning backward from where you want to be to where you are; a continuous process of planning and implementation with a continuous need to alter the plan based on erroneous assumptions or expectations; participative planning increasing knowledge, understanding, and wisdom in the organization and giving participants a vested interest in the plan's implementation; and a belief in the holistic principle, i.e., the more parts of a system and levels of it that plan simultaneously and interdependently, the better. Interactive planning consists of problem formulating (determining what problems and opportunities face the organization, how they interact, and any obstructions or constraints; this requires a systems analysis, an obstruction analysis, and preparation of reference projections); ends planning (determining the ideal or desirable future, objectives, and goals to be pursued by the organization; the idealized design involves selecting a mission, specifying desired properties of the design, and designing the system); means planning (determining the means by which the specified ends are to be pursued in terms of appropriate courses of action, practices, programs, and policies; this may require the construction and manipulation of models); resource planning (determining what resources will be required, how much, when, and how they will be acquired or generated; resources include inputs, facilities and equipment, personnel, and money); and implementing and controlling (determining who is to do what, when, and where, and how the implementation is to be carried out as expected and produces the desired result). A flow chart of the interactive planning process is presented in *Figure 1*.

Like Ackoff, Haines (2000) suggests the use of an interactive, participative, continual improvement process for strategic management, beginning with an ideal future and working backward, and, like Cleland and King, proposes a comprehensive strategy formulation and implementation system to achieve it. Features of the system include a yearly cycle and review, the support of line leadership, quantifiable outcome measures, integration into the business units' annual and daily decision making, team building, the involvement of many stakeholders, linking strategic planning with business unit plans and functional/department plans, individual goal setting and rewards, day-to-day decision making, an emphasis on implementation, and a facilitator role for the planning staff. The strategic management system can be described in terms of its environment, input, output, feedback, and process, which are determined in the following sequence: environment (environmental scan), output (an ideal future vision), feedback (key success measures and goals), input (current assessment and strategy development), and process (actions, implementation, and change).

To accomplish this, the strategic management process consists of the following steps: plan-to-plan (educate organiza-

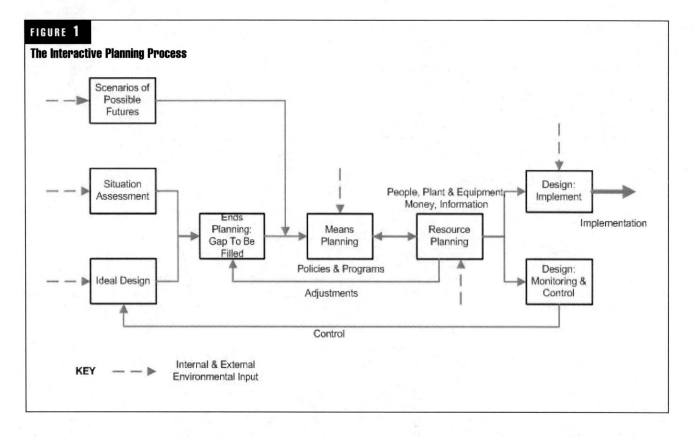

FIGURE 1

The Interactive Planning Process

tion about planning and organize planning team); environmental scan; ideal future vision (define a mission and core values; formulate an ideal vision); key success measures or goals (develop outcome measures); current state assessment (examine strengths, weaknesses, opportunities, and threats analysis and comparison versus vision); strategy development (develop core strategies to achieve ideal vision); business units and three-year business planning (identify and prioritize strategic business units [SBUs] in the portfolio and develop three-year business plans and pro forma financial statements); annual plans and strategic budgets (develop business unit and department annual plans with prioritized tasks and allocated resources); plan-to-implement (educate and organize senior management for plan implementation); strategy implementation and change (formulate individual plans, programs, and efforts, and tie a reward system to it); and annual strategic review and update (conduct yearly audit of the results and process).

The Systems Approach to Telecommunications Strategic Management

The previous section examined several systems approaches to strategic management. This section will investigate how the systems approach can be applied to telecommunications strategic management. Discussions of how to perform telecommunications strategic management can be found in the telecommunications management literature. In the opinion of Carr and Snyder (2003), telecommunications should be treated like a full-fledged business in support of the parent organization, and, consequently, the contents of the telecommunications strategic plan they propose are the same as that for the strategic plans of many businesses, namely, mission (explains the scope of the business and why the business exists), operating philosophy (explains how the

values, beliefs, guiding principles, and mission will be followed), vision (provides strategic direction in five to 10 years), objectives (describes what end results are to be achieved), strategy (explains how the vision will be achieved), long-range plans (describes three to five years ahead); annual plan (executes the strategy for the year); and financial plan (displays the budget to make it happen). In Schoening's (2005) view a sound enterprise telecommunications planning process should address all telecommunications technology areas, translate investment requirements into business English and numbers, and identify the financial benefits of the telecommunications investment. The plan should consist of a vision statement (statement elements are timeless and always appropriate), strategic plan (beyond three years), tactical plan (one to three years), and operational plan (present). According to Gable (1999), strategic planning involves assessing the technological capability and capacity of all products and services that serve the company, making an effort to monitor technological trends for the future, and assessing the future needs of the company.

Green (2001) views the telecommunications strategic plan as a business plan in which telecommunications resources play a key role. The telecommunications strategic plan should be subsidiary to the strategic business plan, supportive of the company's mission, and compatible with the information service (IS) plan. Strategic telecommunications planning concerns the organization and its environment, emphasizing that strategy formulation is achieved as a fit between internal business requirements and organization and external threats and opportunities in the technological, regulatory, vendor, and competitive environments. The strategic planning process is formal and usually embedded in the company's long-range and annual plan cycles. Periodic reviews are required, but ad hoc reviews may be conducted as necessary.

Datapro (1986) breaks down the strategic planning process into the following phases:

- *Review of business and management information system (MIS) strategies and their effects on telecommunications*: The key objective in this phase of the process is to ascertain the business and organizational trends of the company and translate these trends into demand for services and products.

- *Projection of environmental trends and effect on service levels and costs*: The goal of this phase is to analyze the key technology, regulatory, standards, and vendor trends and forecast the effect these trends will have on the cost of telecommunications services and products and on the levels of service offered by the providers.

- *Evaluation of alternative methods of satisfying the communications requirements in terms of cost and service and recommendation of a set of alternatives*: Some of the issues in the evaluation and selection of network alternatives are architecture, topology, protocols, services, vendors, and network management.

An investigation of Ackoff's three-step approach to systems (identify a containing whole, explain the properties of the whole, and explain the properties of the thing to be explained in terms of its role in the whole) demonstrates that the systems approach's influence can already be felt in telecommunications strategic planning. Viewed as a corporate entity, the telecommunications unit is a subsystem of the parent organization, receiving planning input in the form of strategic plans and policy directives from it and delivering planning output to it, and is a peer to the other functional units in the parent organization, including IS, receiving planning input from it. When viewed as a telecommunications entity, the telecommunications unit is a subsystem of the telecommunications industry, and the telecommunications unit receives planning input from telecommunications industry peers, such as service providers, equipment manufacturers, regulators and standards-setting bodies. Telecommunications strategic planning attempts to describe the telecommunications unit's behavior and its role within the parent organization and telecommunications industry by means of its mission, vision, objectives, goals, programs, etc.

Nevertheless, the systems approach has a great deal more that it can contribute to telecommunications strategic management. The "predict and prepare" philosophy underlying the way telecommunications strategic management is performed (i.e., formulating the best strategic alternatives based upon planned business requirements, current unit strengths and weaknesses, and forecasted developments in technology, standards, product and/or service offerings, and other environmental factors) is not proactive in nature. Given the increasingly rapid rate at which developments occur in the telecommunications environment, a proactive approach such as interactive planning, which attempts to shape the environment rather than solely forecast it, seems more appropriate. Funding R&D projects and collaborating with researchers on them, becoming active members of standards and user groups, testing new products and services,

and lobbying policy makers are some of the ways to make the future better fit the ideal vision. It also can be achieved organizationally. Ackoff (1999b) relates how the telecommunications unit serving corporate headquarters at Kodak initiated an idealized design that led to significant improvements and subsequent preparation of an unconstrained design, which combined several telecommunications departments into one. Corporate management accepted the proposal. Later, the centralized computing and telecommunications centers jointly prepared an idealized design that combined the two functions within one organizational unit. Corporate management also accepted this design. Finally, the integrated computing-and-telecommunications unit conducted studies to determine whether the resulting performance was as good as could be obtained externally. Kodak was able to achieve further improvements in service at reduced costs by entering into joint ventures with IBM and DEC to provide the company with computer and telecommunications services.

Besides advocating planning backward, Ackoff and the other proponents of interactive planning emphasize the importance of the process and argue that it be performed continuously, flexibly, and in a participative manner. Currently, telecommunications strategic management generally follows rigid time schedules, and the activity is performed by a small group of planners. The methodology seems to emphasize the plan rather than the process, de-emphasizes the role of feedback and learning, and emphasizes planning rather than implementation. Haines' suggestions for planning and implementation reviews, audits, comprehensive measurement criteria, enlisting senior and line management in the process, increasing stakeholder involvement, promoting team building, and offering financial rewards and other incentives can make strategic management more of a continual and participative activity. Finally, telecommunications strategic management traditionally has been very analytical and scientific in its approach, employing hard systems thinking but neglecting much of the softer systems thinking approaches developed in management, sociology, political science, and philosophy. Applying Jackson's creative holism—which advocates the use of a variety of systems approaches, both hard and soft, to solve problems—can increase the effectiveness of telecommunications strategic management.

Conclusions

As the environment becomes more complex and changes faster, organizations are increasingly incorporating systems thinking into the way they perform strategic management to compete more effectively. This paper has examined systems, the systems approach, and ways the systems approach has been incorporated into strategic management. The paper also has examined telecommunications strategic management and found some systems thinking in the way it is being performed. However, more can be done in telecommunications strategic management from a systems perspective, especially with regard to making it more interactive, continuous, participatory, and holistic. Future study should be directed toward learning about how specific telecommunications units use systems thinking to perform strategic management and, as new strategic management methods of a systems nature are developed, how these methods can be applied to telecommunications strategic management.

References

Ackoff, Russell L., A Concept of Corporate Planning, John Wiley & Sons, New York NY, 1970.

Ackoff, Russell L., Creating the Corporate Future, John Wiley & Sons, New York, NY, 1981.

Ackoff, Russell L., Ackoff's Best: His Classic Writings on Management, John Wiley & Sons, New York, NY, 1999a.

Ackoff, Russell L., Re-Creating the Corporation, John Wiley & Sons, New York, NY, 1999b.

Beer, Stafford, Cybernetics and Management, John Wiley & Sons, Inc., New York, NY, 1959.

Beer, Stafford, The Heart of Enterprise, John Wiley & Sons, Chichester, UK, 1995.

Bertalanffy, Ludwig von, General Systems Theory, George Braziller, New York, NY, 1968.

Carr, Houston H., and Snyder, Charles A., Management of Telecommunications, 2nd ed., McGraw-Hill Irwin, New York, NY 2003.

Churchman, C. West, The Systems Approach, Delacorte Press, New York, NY, 1968.

Cleland, David I., and King, William R., Management: A Systems Approach, McGraw-Hill, New York, NY, 1972.

Cleland, David I., and King, William R., Systems Analysis and Project Management, McGraw-Hill, New York, NY, 1983.

Datapro Research Corporation, "Strategic Planning in Telecommunication," Datapro Management of Telecommunications, Delran, NJ, March 1986.

Gable, Robert A., Telecommunications Department Management, Artech House, Norwood, MA, 1999.

Goldman, James E., Applied Data Communications: A Business-Oriented Approach, Wiley, New York, 1995.

Gordon, Geoffrey, System Simulation, 2nd ed., Prentice-Hall, Englewood Cliffs, NJ, 1978.

Green, J. H., The Irwin Handbook of Telecommunications Management, 3rd ed., McGraw-Hill, New York, NY, 2001.

Haines, Stephen G., The Systems Thinking Approach to Strategic Planning and Management, St. Lucie Press, Boca Raton, FL, 2000.

Hickson, David J., editor, Top Decisions: Strategic Decision-Making in Organizations, Jossey-Bass Publishers, San Francisco, CA, 1986.

Jantsch, Eric, "Forecasting and Systems Approach: A Frame of Reference," Management Science, Vol. 19, No. 12, August 1973.

Jackson, Michael C., Systems Thinking: Creative Holism for Managers, John Wiley & Sons, Chichester, UK, 2003.

Klir, George J., Facets of Systems Science, 2nd ed., Springer, New York, 2001.

Schoening, H. M., Business Management of Telecommunications, Pearson Prentice Hall, Upper Saddle River, NJ, 2005.

Smol, G., Hamer, M. P. R., and Hills, M. T., Telecommunications: A Systems Approach, George Allen & Unwin, London, UK, 1976.

Wagner, Harvey M., Principles of Operations Research, 2nd ed., Prentice-Hall, Englewood Cliffs, NJ, 1975.

Wheelen, Thomas L. and Hunger, J. David, Concepts in Strategic Management and Business Policy, 8th ed., Prentice Hall, Upper Saddle River, NJ, 2002.

Acronym Guide

2B1Q	two binary, one quaternary	ARM	asynchronous response mode
2B1Q	two binary, one quaternary	ARMS	authentication, rating, mediation, and settlement
2G	second generation	ARP	address resolution protocol
3DES	triple data encryption standard	ARPANET	Advanced Research Projects Agency Network
3G	third generation		
3GPP	third-generation partnership project	ARPU	average revenue per customer
3R	regeneration, reshaping, and retraining	A-Rx	analog receiver
4B3T	four binary, three ternary	AS	application server OR autonomous system
4F/BDPR	four-fiber bidirectional dedicated protection ring	ASAM	ATM subscriber access multiplexer
4F/BSPR	four-fiber bidirectional shared protection ring	ASC	Accredited Standards Committee
		ASCII	American Standard Code for Information Interchange
4G	fourth generation		
AAA	authentication, authorization, and accounting	ASE	amplified spontaneous emission
		ASIC	application-specific integrated circuit
AAL–[x]	ATM adaptation layer–x	ASIP	application-specific instruction processor
ABC	activity-based costing	ASON	automatically switched optical network
ABR	available bit rate	ASP	application service provider
AC	alternating current OR authentication code	ASR	access service request OR answer-seizure rate OR automatic service request OR automatic speech recognition
ACD	automatic call distributor		
ACF	admission confirmation		
ACH	automated clearinghouse	ASSP	application-specific standard part
ACL	access control list	ASTN	automatically switched transport network OR analog switched telephone network
ACLEP	adaptive code excited linear prediction		
ACM	address complete message		
ACR	alternate carrier routing OR anonymous call rejection	ATC	automatic temperature control
		ATIS	Alliance for Telecommunications Industry Solutions
ADM	add/drop multiplexer OR asymmetric digital multiplexer		
		ATM	asynchronous transfer mode OR automated teller machine
ADPCM	adaptive differential pulse code modulation		
		ATMF	ATM Forum
ADS	add/drop switch	ATP	analog twisted pair
ADSI	analog display services interface	ATU-C	ADSL transmission unit-CO
ADSL	asymmetric digital subscriber line	ATU-R	ADSL transmission unit-remote
AES	advanced encryption standard	A-Tx	analog transceiver
AFE	analog front end	AUI	attachment unit interface
AGW	agent gateway	AVI	audio video interleaved
AIM	advanced intelligent messaging	AWG	American Wire Gauge OR arrayed waveguide grating
AIN	advanced intelligent network		
ALI	automatic location identification	AYUTOS	as-yet-unthought-of services
AM	amplitude modulation	B2B	business-to-business
AMA	automatic messaging account	B2C	business-to-consumer
AMI	alternate mark inversion	BCSM	basic call state model
AMPS	advanced mobile phone service	BDCS	broadband digital cross-connect system
AN	access network	BDPR	bidirectional dedicated protection ring
ANI	automatic number identification	BE	border element
ANM	answer message	BER	bit-error rate
ANSI	American National Standards Institute	BERT	bit error–rate test
AOL	America Online	BGP	border gateway protocol
AON	all-optical network	BH	busy hour
AP	access point OR access provider	BHCA	busy hour call attempt
APC	automatic power control	BI	bit rate independent
API	application programming interface	BICC	bearer independent call control
APON	ATM passive optical network	BID	bit rate identification
APS	automatic protection switching	BIP	bit interactive parity
ARCNET	attached resource computer network	B–ISDN	broadband ISDN
ARI	assist request instruction	BLEC	broadband local-exchange carrier OR building local-exchange carrier

BLES	broadband loop emulation services	CDMS	configuration and data management server
BLSR	bidirectional line-switched ring	CDN	control directory number
BML	business management layer	CDPD	cellular digital packet data
BOC	Bell operating company	CDR	call detail record OR clock and data recovery
BOF	business operations framework		
BOND	back-office network development	CD–ROM	compact disc–read-only memory
BOSS	broadband operating system software	CWDM	coarse wavelength division multiplexing
BPON	broadband passive optical network	CE	customer edge
BPSK	binary phase shift keying	CEI	comparable efficient interface
B–RAS	broadband–remote access server	CEO	chief executive officer
BRI	basic rate interface	CER	customer edge router
BSA	business services architecture	CERT	computer emergency response team
BSPR	bidirectional shared protection ring	CES	circuit emulation service
BSS	base-station system OR business support system	CES	circuit emulation service
		CESID	caller emergency service identification
BTS	base transceiver station	CEV	controlled environment vault
BVR	best-value routing	CFB/NA	call forward busy/not available
BW	bandwidth	CFO	chief financial officer
CA	call agent	CGI	common gateway interface
CAC	call admission control OR carrier access code OR connection admission control	CHN	centralized hierarchical network
		C–HTML	compressed HTML
		CIC	circuit identification code
CAD	computer-aided design	CID	caller identification
CAGR	compound annual growth rate	CIM	common information model
CALEA	Communications Assistance for Law Enforcement Act	CIMD2	computer interface message distribution 2
		CIO	chief information officer
CAM	computer-aided manufacture	CIP	classical IP over ATM
CAMEL	customized application of mobile enhanced logic	CIR	committed information rate
		CIT	computer integrated telephone
CAP	competitive access provider OR carrierless amplitude and phase modulation OR CAMEL application part	CLASS	custom local-area signaling services
		CLE	customer-located equipment
		CLEC	competitive local-exchange carrier
CAPEX	capital expenditures/expenses	CLI	command-line interface OR call-line identifier
CAR	committed access rate		
CARE	customer account record exchange	CLID	calling-line identification
CAS	channel-associated signaling OR communications applications specification	CLLI	common language location identifier
		CLR	circuit layout record
		CM	cable modem
CAT	conditional access table OR computer-aided telephony	CM&B	customer management and billing
		CMIP	common management information protocol
CATV	cable television		
C-band	conventional band	CMISE	common management information service element
CBDS	connectionless broadband data service		
CBR	constant bit rate	CMOS	complementary metal oxide semiconductor
CBT	core-based tree		
CC	control component	CMRS	commercial mobile radio service
CCB	customer care and billing	CMTS	cable modem termination system
CCF	call-control function	CNAM	calling name (also defined as "caller identification with name" and simply "caller identification")
CCI	call clarity index		
CCITT	Consultative Committee on International Telegraphy and Telephony		
		CNAP	CNAM presentation
CCK	complementary code keying	CNS	customer negotiation system
CCR	call-completion ratio	CO	central office
CCS	common channel signaling	CODEC	coder-decoder OR compression/decompression
CD	chromatic dispersion OR compact disc		
cDCF	conventional dispersion compensation fiber	COI	community of interest
		COO	chief operations officer
CDD	content delivery and distribution	COPS	common open policy service
CDDI	copper-distributed data interface	CORBA	common object request broker architecture
CDMA	code division multiple access		
CDMP	cellular digital messaging protocol	CoS	class of service

COT	central office terminal	DBS	direct broadcast satellite
COTS	commercial off-the-shelf	DC	direct current
COW	cell site on wheels	DCC	data communications channel
CP	connection point	DCF	discounted cash flow OR dispersion compensation fiber
CPAS	cellular priority access service		
CPC	calling-party category (also calling-party control OR calling-party connected)	DCLEC	data competitive local-exchange carrier
		DCM	dispersion compensation module
CPE	customer-premises equipment	DCN	data communications network
CPI	continual process improvement	DCOM	distributed component object model
CPL	call-processing language	DCS	digital cross-connect system OR distributed call signaling
CPLD	complex programmable logic device		
CPN	calling-party number	DCT	discrete cosine transform
CPU	central processing unit	DDN	defense data network
CR	constraint-based routing	DDS	dataphone digital service
CRC	cyclic redundancy check OR cyclic redundancy code	DECT	Digital European Cordless Telecommunication
CRIS	customer records information system	demarc	demarcation point
CR–LDP	constraint-based routed–label distribution protocol	DEMS	digital electronic messaging service
		DES	data encryption standard
CRM	customer-relationship management	DFB	distributed feedback
CRTP	compressed real-time transport protocol	DFC	dedicated fiber/coax
CRV	call reference value	DGD	differentiated group delay
CS	client signal	DGFF	dynamic gain flattening filter
CS–[x]	capability set [x]	DHCP	dynamic host configuration protocol
CSA	carrier serving area	DiffServ	differentiated services
CSCE	converged service-creation and execution	DIN	digital information network
CSCF	call-state control function	DIS	distributed interactive simulation
CSE	CAMEL service environment	DITF	Disaster Information Task Force
CS–IWF	control signal interworking function	DLC	digital loop carrier
CSM	customer-service manager	DLCI	data-link connection identifier
CSMA/CA	carrier sense multiple access with collision avoidance	DLE	digital loop electronics
		DLEC	data local-exchange carrier
CSMA/CD	carrier sense multiple access with collision detection	DLR	design layout report
		DM	dense mode
CSN	circuit-switched network	DMD	dispersion management device
CSP	communications service provider OR content service provider	DMS	digital multiplex system
		DMT	discrete multitone
CSR	customer-service representative	DN	distinguished name
CSU	channel service unit	DNS	domain name server OR domain naming system
CSV	circuit-switched voice		
CT	computer telephony	DOC	department of communications
CT–2	cordless telephony generation 2	DOCSIS	data over cable service interface specifications
CTI	computer telephony integration		
CTIA	Cellular Telecommunications & Internet Association	DOD	Department of Defense
		DOJ	Department of Justice
CTO	chief technology officer	DoS	denial of service
CWD	centralized wavelength distribution	DOS	disk operating system
CWDM	coarse wavelength division multiplexing	DOSA	distributed open signaling architecture
		DOT	Department of Transportation
CWIX	cable and wireless Internet exchange	DP	detection point
DAC	digital access carrier	DPC	destination point code
DACS	digital access cross-connect system	DPE	distributed processing environment
DAM	DECT authentication module	DPT	dial pulse terminate
DAMA	demand assigned multiple access	DQoS	dynamic quality of service
DAML	digital added main line	D-Rx	digital receiver
DARPA	Defense Advanced Research Projects Agency	DS–[x]	digital signal [level x]
		DSAA	DECT standard authentication algorithm
DAVIC	Digital Audio Video Council	DSC	DECT standard cipher
DB	database	DSCP	DiffServ code point
dB	decibel(s)	DSF	dispersion-shifted fiber
DBMS	database management system	DSL	digital subscriber line [also xDSL]
dBrn	decibels above reference noise	DSLAM	digital subscriber line access multiplexer

DSLAS	DSL–ATM switch
DSP	digital signal processor OR digital service provider
DSS	decision support system
DSSS	direct sequence spread spectrum
DSU	data service unit OR digital service unit
DTH	direct-to-home
DTMF	dual-tone multifrequency
DTV	digital television
D-Tx	digital transceiver
DVB	digital video broadcast
DVC	dynamic virtual circuit
DVD	digital video disc
DVMRP	distance vector multicast routing protocol
DVoD	digital video on demand
DVR	digital video recording
DWDM	dense wavelength division multiplexing
DXC	digital cross-connect
E911	enhanced 911
EAI	enterprise application integration
EAP	extensible authentication protocol
EBITDA	earnings before interest, taxes, depreciation, and amortization
EC	electronic commerce
ECD	echo-cancelled full-duplex
ECRM	echo canceller resource module
ECTF	Enterprise Computer Telephony Forum
EDA	electronic design automation
EDF	electronic distribution frame OR erbium-doped fiber
EDFA	erbium-doped fiber amplifier
EDGE	enhanced data rates for GSM evolution
EDI	electronic data interchange
EDSX	electronic digital signal cross-connect
EFM	Ethernet in the first mile
EFT	electronic funds transfer
EJB	enterprise Java beans
ELAN	emulated local-area network
ELEC	enterprise local-exchange carrier
EM	element manager
EMI	electromagnetic interference
EML	element-management layer
EMS	element-management system OR enterprise messaging server
E–NNI	external network-to-network interface
ENUM	telephone number mapping
E–O	electrical-to-optical
EO	end office
EoA	Ethernet over ATM
EOC	embedded operations channel
EoVDSL	Ethernet over VDSL
EPD	early packet discard
EPON	Ethernet PON
EPROM	erasable programmable read-only memory
ERP	enterprise resource planning
ESCON	enterprise systems connectivity
ESS	electronic switching system
ETC	establish temporary connection
EtherLEC	Ethernet local-exchange carrier
ETL	extraction, transformation, and load
eTOM	enhanced telecom operations map

ETSI	European Telecommunications Standards Institute
EU	European Union
EURESCOM	European Institute for Research and Strategic Studies in Telecommunications
EXC	electrical cross-connect
FAB	fulfillment, assurance, and billing
FAQ	frequently asked question
FBG	fiber Bragg grating
FCAPS	fault, configuration, accounting, performance, and security
FCC	Federal Communications Commission
FCI	furnish charging information
FCIF	flexible computer-information format
FDA	Food and Drug Administration
FDD	frequency division duplex
FDDI	fiber distributed data interface
FDF	fiber distribution frame
FDM	frequency division multiplexing
FDMA	frequency division multiple access
FDS–1	fractional DS–1
FE	extended framing
FEC	forward error correction
FEPS	facility and equipment planning system
FEXT	far-end crosstalk
FHSS	frequency hopping spread spectrum
FICON	fiber connection
FITL	fiber-in-the-loop
FM	fault management OR frequency modulation
FOC	firm order confirmation
FOT	fiber-optic terminal
FOTS	fiber-optic transmission system
FP	Fabry-Perot [laser]
FPB	flex parameter block
FPGA	field programmable gate array
FPLMTS	future public land mobile telephone system
FPP	fast-packet processor
FR	frame relay
FRAD	frame-relay access device
FSAN	full-service access network
FSC	framework services component
FSN	full-service network
FT	fixed-radio termination
FT1	fractional T1
FTC	Federal Trade Commission
FTE	full-time equivalent
FTP	file transfer protocol
FTP3	file transfer protocol 3
FTTB	fiber-to-the-building
FTTC	fiber to the curb
FTTCab	fiber-to-the-cabinet
FTTEx	fiber-to-the-exchange
FTTH	fiber-to-the-home
FTTN	fiber-to-the-neighborhood
FTTS	fiber-to-the-subscriber
FTTx	fiber-to-the-x
FWM	four-wave mixing
FX	foreign exchange
GA	genetic algorithm

Gb	gigabit	HSIA	high-speed Internet access
GbE	gigabit Ethernet [also GE]	HSP	hosting service provider
GBIC	gigabit interface converter	HTML	hypertext markup language
Gbps	gigabits per second	HTTP	hypertext transfer protocol
GCRA	generic cell rate algorithm	HVAC	heating, ventilating, and air-conditioning
GDIN	global disaster information network	HW	hardware
GDMO	guidelines for the definition of managed objects	IAD	integrated access device
		IAM	initial address message
GE	[see GbE]	IAS	integrated access service OR Internet access server
GEO	geosynchronous Earth orbit		
GETS	government emergency telecommunications service	IAST	integrated access, switching, and transport
		IAT	inter-arrival time
GFF	gain flattening filter	IBC	integrated broadband communications
GFR	guaranteed frame rate	IC	integrated circuit
Ghz	gigahertz	ICD	Internet call diversion
GIF	graphics interface format	ICDR	Internet call detail record
GIS	geographic information services	ICL	intercell linking
GKMP	group key management protocol	ICMP	Internet control message protocol
GMII	gigabit media independent interface	ICP	integrated communications provider OR intelligent communications platform
GMLC	gateway mobile location center		
GMPCS	global mobile personal communications services		
		ICS	integrated communications system
GMPLS	generalized MPLS	ICW	Internet call waiting
GNP	gross national product	IDC	Internet data center OR International Data Corporation
GOCC	ground operations control center		
GPIB	general-purpose interface bus	IDE	integrated development environment
GPRS	general packet radio service	IDES	Internet data exchange system
GPS	global positioning system	IDF	intermediate distribution frame
GR	generic requirement	IDL	interface definition language
GRASP	greedy randomized adaptive search procedure	IDLC	integrated digital loop carrier
		IDS	intrusion detection system
GSA	Global Mobile Suppliers Association	IDSL	integrated services digital network DSL
GSM	Global System for Mobile Communications	IEC	International Electrotechnical Commission OR International Engineering Consortium
GSMP	generic switch management protocol		
GSR	gigabit switch router		
GTT	global title translation	IEEE	Institute of Electrical and Electronics Engineers
GUI	graphical user interface		
GVD	group velocity dispersion	I-ERP	integrated enterprise resource planning
GW	gateway	IETF	Internet Engineering Task Force
HCC	host call control	IFITL	integrated [services over] fiber-in-the-loop
HD	home domain	IFMA	International Facility Managers Association
HDLC	high-level data-link control		
HDML	handheld device markup language	IFMP	Ipsilon flow management protocol
HDSL	high-bit-rate DSL	IGMP	Internet group management protocol
HDT	host digital terminal	IGP	interior gateway protocol
HDTV	high-definition television	IGRP	interior gateway routing protocol
HDVMRP	hierarchical distance vector multicast routing protocol	IGSP	independent gateway service provider
		IHL	Internet header length
HEC	head error control OR header error check	IIOP	Internet inter–ORB protocol
HEPA	high-efficiency particulate arresting	IIS	Internet Information Server
HFC	hybrid fiber/coax	IKE	Internet key exchange
HIDS	host intrusion detection system	ILA	in-line amplifier
HLR	home location register	ILEC	incumbent local-exchange carrier
HN	home network	ILMI	interim link management interface
HOM	high-order mode	IM	instant messaging
HomePNA	Home Phoneline Networking Alliance [also HomePNA2]	IMA	inverse multiplexing over ATM
		IMAP	Internet message access protocol
HomeRF	Home Radio Frequency Working Group	IMRP	Internet multicast routing protocol
HQ	headquarters	IMSI	International Mobile Subscriber Identification
HSCSD	high-speed circuit-switched data	IMT	intermachine trunk OR International Mobile Telecommunications
HSD	high-speed data		

IMTC	International Multimedia Teleconferencing Consortium	ITU–T	ITU–Telecommunication Standardization Sector
IN	intelligent network	ITV	Internet television
INAP AU	INAP adaptation unit	IVR	interactive voice response
INAP	intelligent network application part	IVRU	interactive voice-response unit
INE	intelligent network element	IWF	interworking function
InfoCom	information communication	IWG	interworking gateway
INM	integrated network management	IWU	interworking unit
INMD	in-service, nonintrusive measurement device	IXC	interexchange carrier
I–NNI	internal network-to-network interface	J2EE	Java Enterprise Edition
INT	[point-to-point] interrupt	J2ME	Java Micro Edition
InterNIC	Internet Network Information Center	J2SE	Java Standard Edition
IntServ	integrated services	JAIN	Java APIs for integrated networks
IOF	interoffice facility	JCAT	Java coordination and transactions
IOS	intelligent optical switch	JCC	JAIN call control
IP	Internet protocol	JDBC	Java database connectivity
IPBX	Internet protocol private branch exchange	JDMK	Java dynamic management kit
IPcoms	Internet protocol communications	JMAPI	Java management application programming interface
IPDC	Internet protocol device control	JMX	Java management extension
IPDR	Internet protocol data record	JPEG	Joint Photographic Experts Group
IPe	intelligent peripheral	JSCE	JAIN service-creation environment
IPG	intelligent premises gateway	JSIP	Java session initiation protocol
IPO	initial public offering OR Internet protocol over optical	JSLEE	JAIN service logic execution environment
IPoA	Internet protocol over ATM	JTAPI	Java telephony application programming interface
IPQoS	Internet protocol quality of service	JVM	Java virtual machine
IPSec	Internet protocol security	kbps	kilobits per second
IPTel	IP telephony	kHz	kilohertz
IPv6	Internet protocol version 6	km	kilometer
IPX	Internet package exchange	L2F	Layer-2 forwarding
IR	infrared	L2TP	Layer-2 tunneling protocol
IRU	indefeasible right to user	LAC	L2TP access concentrator
IS	information service OR interim standard	LAI	location-area identity
IS-IS	intermediate system to intermediate system	LAN	local-area network
		LANE	local-area network emulation
ISA	industry standard architecture	LATA	local access and transport area
ISAPI	Internet server application programmer interface	LB311	location-based 311
		L-band	long band
ISC	integrated service carrier OR International Softswitch Consortium	LBS	location-based services
		LC	local convergence
ISDF	integrated service development framework	LCD	liquid crystal display
		LCP	link control protocol
ISDN	integrated services digital network	LD	laser diode OR long distance
ISDN–BA	ISDN basic access	LDAP	lightweight directory access protocol
ISDN–PRA	ISDN primary rate access	LD–CELP	low delay–code excited linear prediction
ISEP	intelligent signaling endpoint	LDP	label distribution protocol
ISM	industrial, scientific, and medical OR integrated service manager	LDS	local digital service
		LE	line equipment OR local exchange
ISO	International Organization for Standardization	LEAF®	large-effective-area fiber
		LEC	local-exchange carrier
ISOS	integrated software on silicon	LED	light-emitting diode
ISP	Internet service provider	LEO	low Earth orbit
ISUP	ISDN user part	LEOS	low Earth-orbiting satellite
ISV	independent software vendor	LER	label edge router
IT	information technology OR Internet telephony	LES	loop emulation service
		LIDB	line information database
ITSP	Internet telephony service provider	LL	long line
ITTP	information technology infrastructure library	LLC	logical link control
		LMDS	local multipoint distribution system
ITU	International Telecommunication Union	LMN	local network management

LMOS	loop maintenance operation system
LMP	link management protocol
LMS	loop-management system OR loop-monitoring system OR link-monitoring system
LNNI	LANE network-to-network interface
LNP	local number portability
LNS	L2TP network server
LOL	loss of lock
LOS	line of sight OR loss of signal
LPF	low-pass filter
LQ	listening quality
LRN	local routing number
LRQ	location request
LSA	label switch assignment OR link state advertisement
LSB	location-sensitive billing
LSMS	local service management system
LSO	local service office
LSP	label-switched path
LSR	label-switched router OR leaf setup request OR local service request
LT	line terminator OR logical terminal
LTE	lite terminating equipment
LUNI	LANE user network interface
LX	local exchange
M2PA	message transfer protocol 2 peer-to-peer adaptation
M2UA	message transfer protocol 2–user adaptation layer
M3UA	message transfer protocol 3–user adaptation layer
MAC	media access control
MADU	multiwave add/drop unit
MAN	metropolitan-area network
MAP	mobile applications part
MAS	multiple-application selection
Mb	megabit
MB	megabyte
MBAC	measurement-based admission control
MBGP	multicast border gateway protocol
MBone	multicast backbone
Mbps	megabits per second
MC	multipoint controller
MCC	mobile country code
MCU	multipoint control unit
MDF	main distribution frame
MDSL	multiple DSL
MDTP	media device transport protocol
MDU	multiple-dwelling unit
MEGACO	media gateway control
MEMS	micro-electromechanical system
MExE	mobile execution environment
MF	multifrequency
MFJ	modified final judgment
MG	media gateway
MGC	media gateway controller
MGCF	media gateway control function
MGCP	media gateway control protocol
MHz	megahertz
MIB	management information base

MII	media independent interface
MIME	multipurpose Internet mail extensions
MIMO	multiple inputs, multiple outputs
MIN	mobile identification number
MIPS	millions of instructions per second
MIS	management information system
MITI	Ministry of International Trade and Industry (in Japan)
MLT	mechanized loop testing
MM	mobility management
MMDS	multichannel multipoint distribution system
MMPP	Markov-Modulated Poisson Process
MMS	multimedia message service
MMUSIC	Multiparty Multimedia Session Control [working group]
MNC	mobile network code
MOM	message-oriented middleware
MON	metropolitan optical network
MOP	method of procedure
MOS	mean opinion score
MOSFP	multicast open shortest path first
MOU	minutes of use OR memorandum of understanding
MPC	mobile positioning center
MPEG	Moving Pictures Experts Group
MPI	message passing interface
MPLambdaS	multiprotocol lambda switching
MPLS	multiprotocol label switching
MPOA	multiprotocol over ATM
MPoE	multiple point of entry
MPoP	metropolitan point of presence
MPP	massively parallel processor
MPx	MPEG–Layer x
MRC	monthly recurring charge
MRS	menu routing system
MRSP	mobile radio service provider
ms	millisecond
MSC	mobile switching center
MSF	Multiservice Switch Forum
MSIN	mobile station identification number
MSNAP	multiple services network access point
MSO	multiple-system operator
MSP	management service provider
MSPP	multiservice provisioning platform
MSS	multiple-services switching system
MSSP	mobile satellite service provider
MTA	message transfer agent
MTBF	mean time between failures
MTP [x]	message transfer part [x]
MTTR	mean time to repair
MTU	multiple-tenant unit
MVL	multiple virtual line
MWIF	Mobile Wireless Internet Forum
MZI	Mach-Zender Interferometer
N11	(refers to FCC–managed dialable service codes such as 311, 411, and 911)
NA	network adapter
NAFTA	North America Free Trade Agreement
NANC	North American Numbering Council
NANP	North American Numbering Plan

NAP	network access point	NSAP	network service access point
NARUC	National Association of Regulatory Utility Commissioners	NSAPI	Netscape server application programming interface
NAS	network access server	NSCC	network surveillance and control center
NASA	National Aeronautics and Space Administration	NSDB	network and services database
NAT	network address translation	NSP	network service provider OR network and service performance
NATA	North American Telecommunications Association	NSTAC	National Security Telecommunications Advisory Committee
NBN	node-based network	NT	network termination OR new technology
NCP	network control protocol	NTN	network terminal number
NCS	national communications system OR network connected server	NTSC	National Television Standards Committee
NDA	national directory assistance	NVP	network voice protocol
NDM–U	network data management–usage	NZ–DSF	nonzero dispersion-shifted fiber
NDSF	non-dispersion-shifted fiber	O&M	operations and maintenance
NE	network element	OA&M	operations, administration, and maintenance
NEAP	non-emergency answering point	OADM	optical add/drop multiplexer
NEBS	network-equipment building standards	OAM&P	operations, administration, maintenance, and provisioning
NEL	network-element layer		
NEXT	near-end crosstalk	OBF	Ordering and Billing Forum
NFS	network file system	OBLSR	optical bidirectional line-switched ring
NG	next generation	OC–[x]	optical carrier–[level x]
NGCN	next-generation converged network	OCBT	ordered core-based protocol
NGDLC	next-generation digital loop carrier	OCD	optical concentration device
NGF	next-generation fiber	OCh	optical channel
NGN	next-generation network	OCR	optical character recognition
NGOSS	next-generation operations system and software OR next-generation OSS	OCS	original call screening
		OCU	office channel unit
NHRP	next-hop resolution protocol	OCX	open compact exchange
NI	network interface	OD	origin-destination
NIC	network interface card	ODBC	open database connectivity
NID	network interface device	ODSI	optical domain services interface
NIDS	network intrusion detection system	O–E	optical-to-electrical
NIIF	Network Interconnection Interoperability Forum	O–EC	optical–electrical converter
		OECD	Organization for Economic Cooperation and Development
NIS	network information service		
NIU	network interface unit	OEM	original equipment manufacturer
nm	nanometer	O–E–O	optical-to-electrical-to-optical
NML	network-management layer	OEXC	opto-electrical cross-connect
NMS	network-management system	OFDM	orthogonal frequency division multiplexing
NND	name and number delivery		
NNI	network-to-network interface	OIF	Optical Internetworking Forum
NNTP	network news transport protocol	OLA	optical line amplifier
NOC	network operations center	OLAP	on-line analytical processing
NOMAD	national ownership, mobile access, and disaster communications	OLI	optical link interface
		OLT	optical line termination OR optical line terminal
NP	number portability		
NPA	numbering plan area	OLTP	on-line transaction processing
NPAC	Number Portability Administration Center	OMC	Operations and Maintenance Center
		OMG	Object Management Group
NPN	new public network	OMS SW	optical multiplex section switch
NP–REQ	number-portable request query	OMS	optical multiplex section
NPV	net present value	OMSSPRING	optical multiplex section shared protection ring
NRC	Network Reliability Council OR nonrecurring charge		
		ONA	open network architecture
NRIC	Network Reliability and Interoperability Council	ONE	optical network element
		ONI	optical network interface
NRSC	Network Reliability Steering Committee	ONMS	optical network-management system
NRZ	non–return to zero	ONT	optical network termination
NS/EP	national security and emergency preparedness	ONTAS	optical network test access system
		ONU	optical network unit

OP	optical path	PE	provider edge
OPEX	operational expenditures/expenses	PER	packed encoding rules
OPS	operator provisioning station	PERL	practical extraction and report language
OPTIS	overlapped PAM transmission with interlocking spectra	PESQ	perceptual evolution of speech quality
OPXC	optical path cross-connect	PFD	phase-frequency detector
ORB	object request broker	PHB	per-hop behavior
ORT	operational readiness test	PHY	physical layer
OS	operating system	PIC	point-in-call OR predesignated interexchange carrier OR primary interexchange carrier
OSA	open service architecture		
OSC	optical supervisory panel		
OSD	on-screen display	PICS	plug-in inventory control system
OSGI	open services gateway initiative	PIM	personal information manager OR protocol-independent multicast
OSI	open systems interconnection		
OSMINE	operations systems modification of intelligent network elements	PIN	personal identification number
		PINT	PSTN and Internet Networking [IETF working group]
OSN	optical-service network		
OSNR	optical signal-to-noise ratio	PINTG	PINT gateway
OSP	outside plant OR open settlement protocol	PKI	public key infrastructure
		PLA	performance-level agreement
OSPF	open shortest path first	PLC	planar lightwave circuit OR product life cycle
OSS	operations support system		
OSS/J	OSS through Java	PLCP	physical layer convergence protocol
OSU	optical subscriber unit	PLL	phase locked loop
OTM	optical terminal multiplexer	PLMN	public land mobile network
OTN	optical transport network	PLOA	protocol layers over ATM
OUI	optical user interface	PM	performance monitoring
O-UNI	optical user-to-network interface	PMD	physical-medium dependent OR polarization mode dispersion
OUSP	optical utility services platform		
OVPN	optical virtual petabits network OR optical virtual private network	PMDC	polarization mode dispersion compensator
		PMO	present method of operation
OWSR	optical wavelength switching router	PMP	point-to-multipoint
OXC	optical cross-connect	PN	personal number
P&L	profit and loss	PNNI	private network-to-network interface
PABX	private automatic branch exchange	PnP	plug and play
PACA	priority access channel assignment	PO	purchase order
PACS	picture archiving communications system	PODP	public office dialing plan
		POET	partially overlapped echo-cancelled transmission
PAL	phase alternate line		
PAM	Presence and Availability Management [Forum] OR pulse amplitude modulation	POF	plastic optic fiber
		POH	path overhead
		POIS	packet optical interworking system
PAMS	perceptual analysis measurement system	PON	passive optical network
PAN	personal access network	PoP	point of presence
PBCC	packet binary convolutional codes	POP3	post office protocol 3
PBN	point-to-point–based network OR policy-based networking	POS	packet over SONET OR point of service
		PosReq	position request
PBX	private branch exchange	POT	point of termination
PC	personal computer	POTS	plain old telephone service
PCF	physical control field	PP	point-to-point
PCI	peripheral component interconnect	PPD	partial packet discard
PCM	pulse code modulation	PPP	point-to-point protocol
PCN	personal communications network	PPPoA	point-to-point protocol over ATM
PCR	peak cell rate	PPPoE	point-to-point protocol over Ethernet
PCS	personal communications service	PPTP	point-to-point tunneling protocol
PDA	personal digital assistant	PP–WDM	point-to-point–wavelength division multiplexing
PDC	personal digital cellular		
PDD	post-dial delay	PQ	priority queuing
PDE	position determination equipment	PRI	primary rate interface
PDH	plesiochronous digital hierarchy	ps	picosecond
PDN	public data network	PSAP	public safety answering point
PDP	policy decision point	PSC	Public Service Commission
PDSN	packet data serving node		
PDU	protocol data unit		

PSD	power spectral density	RPR	resilient packet ring
PSDN	public switched data network	RPRA	Resilient Packet Ring Alliance
PSID	private system identifier	RPT	resilient packet transport
PSN	public switched network	RQMS	requirements and quality measurement system
PSPDN	packet-switched public data network		
PSQM	perceptual speech quality measure	RRQ	round-robin queuing or registration request
PSTN	public switched telephone network		
PTE	path terminating equipment	RSU	remote service unit
PTN	personal telecommunications number service	RSVP	resource reservation protocol
		RSVP–TE	resource reservation protocol–traffic engineering
PTP	point-to-point		
PTT	Post Telephone and Telegraph Administration	RT	remote terminal
		RTCP	real-time conferencing protocol
PUC	public utility commission	RTOS	real-time operating system
PVC	permanent virtual circuit	RTP	real-time transport protocol
PVM	parallel virtual machine	RTSP	real-time streaming protocol
PVN	private virtual network	RTU	remote test unit
PWS	planning workstation	RxTx	receiver/transmitter
PXC	photonic cross-connect	RZ	return to zero
QAM	quadrature amplitude modulation	SAM	service access multiplexer
QoE	quality of experience	SAN	storage-area network
QoS	quality of service	SAP	service access point OR session announcement protocol
QPSK	quaternary phase shift keying		
QSDG	QoS Development Group	SAR	segmentation and reassembly
RAD	rapid application development	S-band	short band
RADIUS	remote authentication dial-in user service	SBS	stimulated Brillouin scattering
		SCAN	switched-circuit automatic network
RADSL	rate-adaptive DSL	SCCP	signaling connection control part
RAM	remote access multiplexer	SCCS	switching control center system
RAN	regional-area network	SCE	service-creation environment
RAP	resource allocation protocol	SCF	service control function
RAS	remote access server	SCL	service control language
RBOC	regional Bell operating company	SCM	service combination manager OR station class mark OR subscriber carrier mark
RCP	remote call procedure		
RCU	remote control unit		
RDBMS	relational database management system		
RDC	regional distribution center	SCN	service circuit node OR switched-circuit network
RDSLAM	remote DSLAM		
REL	release	SCP	service control point
RF	radio frequency	SCR	sustainable cell rate
RFC	request for comment	SCSI	small computer system interface
RFI	request for information	SCSP	server cache synchronization protocol
RFP	request for proposal	SCTP	simple computer telephony protocol OR simple control transport protocol OR stream control transmission protocol
RFPON	radio frequency optical network		
RFQ	request for quotation		
RGU	revenue-generating unit		
RGW	residential gateway	SD	selective discard
RHC	regional holding company	SD&O	service development and operations
RIAC	remote instrumentation and control	SDA	separate data affiliate
RIP	routing information protocol	SDB	service design bureau
RISC	reduced instruction set computing	SDC	service design center
RJ	registered jack	SDF	service data function
RLL	radio in the loop	SDH	synchronous digital hierarchy
RM	resource management	SDM	service-delivery management OR shared data model
RMA	request for manual assistance		
RMI	remote method invocation	SDN	software-defined network
RMON	remote monitoring	SDP	session description protocol
ROADM	reconfigurable optical add/drop multiplexer	SDRP	source demand routing protocol
		SDSL	symmetric DSL
ROBO	remote office/branch office	SDTV	synchronous digital hierarchy
ROI	return on investment	SDV	switched digital video
RPC	remote procedure call	SE	service element
		SEC	Securities and Exchange Commission
RPF	reverse path forwarding	SEE	service-execution environment

SEP	signaling endpoint
ServReq	service request
SET	secure electronic transaction
SFA	sales force automation
SFD	start frame delimiter
SFF	small form-factor
SFGF	supplier-funded generic element
SG	signaling gateway
SG&A	selling, goods, and administration OR sales, goods, and administration
SGCP	simple gateway control protocol
SGSN	serving GPRS support node
SHDSL	single-pair high-bit-rate DSL
SHLR	standalone home location register
SHV	shareholder value
SI	systems integrator
SIBB	service-independent building block
SIC	service initiation charge
SICL	standard interface control library
SID	silence indicator description
SIF	SONET Interoperability Forum
sigtran	Signaling Transport [working group]
SIM	subscriber identity module OR service interaction manager
SIP CPL	SIP call processing language
SIP	session initiation protocol
SIP–T	session initiation protocol for telephony
SISO	single input, single output
SIU	service interface unit
SIVR	speaker-independent voice recognition
SKU	stock-keeping unit
SL	service logic
SLA	service-level agreement
SLC	subscriber line carrier
SLEE	service logic execution environment
SLIC	subscriber line interface circuit
SLO	service-level objective
SM	sparse mode
SMC	service management center
SMDI	simplified message desk interface
SMDS	switched multimegabit data service
SME	small-to-medium enterprise
SMF	single-mode fiber
SML	service management layer
SMP	service management point
SMPP	short message peer-to-peer protocol
SMS	service-management system OR short message service
SMSC	short messaging service center
SMTP	simple mail transfer protocol
SN	service node
SNA	service node architecture OR service network architecture
SNAP	subnetwork access protocol
SNMP	simple network-management protocol
SNPP	simple network paging protocol
SNR	signal-to-noise ratio
SO	service objective
SOA	service order activation
SOAC	service order analysis and control
SOAP	simple object access protocol
SOCC	satellite operations control center

SOE	standard operating environment
SOHO	small office/home office
SON	service order number
SONET	synchronous optical network
SOP	service order processor
SP	service provider OR signaling point
SPC	stored program control
SPE	synchronous payload envelope
SPF	shortest path first
SPIRITS	Service in the PSTN/IN Requesting Internet Service [working group]
SPIRITSG	SPIRITS gateway
SPM	self-phase modulation OR subscriber private meter
SPoP	service point of presence
SPX	sequence packet exchange
SQL	structured query language
SQM	service quality management
SRF	special resource function
SRP	source routing protocol
SRS	stimulated Raman scattering
srTCM	single-rate tri-color marker
SS	softswitch
SS7	signaling system 7
SSE	service subscriber element
SSF	service switching function
SSG	service selection gateway
SSL	secure sockets layer
SSM	service and sales management
SSMF	standard single-mode fiber
SSP	service switching point
STE	section terminating equipment
STM	synchronous transfer mode
STN	service transport node
STP	shielded twisted pair OR signal transfer point OR spanning tree protocol
STR	signal-to-resource
STS	synchronous transport signal
SUA	SCCP user adaptation
SVC	switched virtual circuit
SW	software
SWAN	storage wide-area network
SWAP	shared wireless access protocol
SWOT	strengths, weaknesses, opportunities, and threats
SYN	IN synchronous transmission
TALI	transport adapter layer interface
TAPI	telephony application programming interface
TAT	terminating access trigger OR termination attempt trigger OR transatlantic telephone cable
Tb	terabit
TBD	to be determined
Tbps	terabits per second
TC	tandem connect
TCAP	transactional capabilities application part
TCB	transfer control block
TCIF	Telecommunications Industry Forum
TCL	tool command language
TCM	time compression multiplexing
TCO	total cost of ownership

TCP	transmission control protocol	TV	television
TCP/IP	transmission control protocol/Internet protocol	UA	user agent
		UADSL	universal ADSL
TC–PAM	trellis coded–pulse amplitude modulation	UAK	user-authentication key
TDD	time division duplex	UAWG	Universal ADSL Working Group
TDM	time division multiplex	UBR	unspecified bit rate
TDMA	time division multiple access	UBT	ubiquitous bus technology
TDMDSL	time division multiplex digital subscriber line	UCP	universal computer protocol
		UCS	uniform communication standard
TDR	time domain reflectometer OR transaction detail record	UDDI	universal description, discovery, and integration
TE	traffic engineering	UDP	user datagram protocol
TEAM	transport element activation manager	UDR	usage detail record
TED	traffic engineering database	UI	user interface
TEM	telecommunications equipment manufacturer	ULH	ultra-long-haul
		UM	unified messaging
TFD	toll-free dialing	UML	unified modeling language
THz	terahertz	UMTS	Universal Mobile Telecommunications System
TIA	Telecommunications Industry Association		
TIMS	transmission impairment measurement set	UN	United Nations
TINA	Telecommunications Information Networking Architecture	UNE	unbundled network element
		UNI	user network interface
TINA-C	Telecommunications Information Networking Architecture Consortium	UOL	unbundled optical loop
		UPC	usage parameter control
TIPHON	Telecommunications and Internet Protocol Harmonization over Networks	UPI	user personal identification
		UPS	uninterruptible power supply
TIWF	trunk interworking function	UPSR	unidirectional path-switched ring
TKIP	temporal key integrity protocol	URI	uniform resource identifier
TL1	transaction language 1	URL	universal resource locator
TLDN	temporary local directory number	USB	universal serial bus
TLS	transparent LAN service OR transport-layer security	USTA	United States Telecom Association
		UTOPIA	Universal Test and Operations Interface for ATM
TLV	tag length value		
TMF	TeleManagement Forum	UTS	universal telephone service
TMN	telecommunications management network	UWB	ultra wideband
TMO	trans-metro optical	UWDM	ultra-dense WDM
TN	telephone number	V&H	vertical and horizontal
TNO	telecommunications network operator	VAD	voice activity detection
TO&E	table of organization and equipment	VAN	value-added network
TOM	telecom operations map	VAR	value-added reseller
ToS	type of service	VAS	value-added service
TP	twisted pair	VASP	value-added service provider
TPM	transaction processing monitor	VBNS	very–high-speed backbone network service
TPS–TC	transmission control specific–transmission convergence		
		VBR	variable bit rate
TR	technical requirement OR tip and ring	VBR–nrt	variable bit rate–non–real-time
TRA	technology readiness assessment	VBR–rt	variable bit rate–real time
TRIP	telephony routing over Internet protocol	VC	virtual circuit OR virtual channel
trTCM	two-rate tri-color marker	VCC	virtual channel connection
TSB	telecommunication system bulletin	VCI	virtual channel identifier
TSC	terminating call screening	VCLEC	voice CLEC
TSI	time slot interchange	VCO	voltage-controlled oscillator
TSP	telecommunications service provider	VCR	videocassette recorder
TSS	Telecommunications Standardization Section	VCSEL	vertical cavity surface emitting laser
		VD	visited domain
TTC	Telecommunications Technology Committee	VDM	value delivery model
		VDSL	very-high–data-rate DSL
TTCP	test TCP	VeDSL	voice-enabled DSL
TTL	transistor-transistor logic	VGW	voice gateway
TTS	text-to-speech OR TIRKS® table system	VHE	virtual home environment
		VHS	video home system
TUI	telephone user interface	VITA	virtual integrated transport and access
TUP	telephone user part		

VLAN	virtual local-area network OR voice local-area network	WB DCS	wideband DCS
VLR	visitor location register	WCDMA	wideband CDMA
VLSI	very-large-scale integration	WCT	wavelength converting transponder
VM	virtual machine	WDCS	wideband digital cross-connect
VMS	voice-mail system	WDM	wavelength division multiplexing
VoADSL	voice over ADSL	WECA	wireless Ethernet compatibility alliance
VoATM	voice over ATM	WEP	wired equivalent privacy
VoB	voice over broadband	WFA	work and force administration
VoD	video on demand	WFQ	weighted fair queuing
VoDSL	voice over DSL	Wi-Fi	wireless fidelity
VoFR	voice over frame relay	WIM	wireless instant messaging
VoIP	voice over IP	WiMAX	worldwide interoperability for microwave access
VON	voice on the Net	WIN	wireless intelligent network
VoP	voice over packet	WLAN	wireless local-area network
VOQ	virtual output queuing	WLL	wireless local loop
VoT1	voice over T1	WMAP	wireless messaging application programming interface
VP	virtual path	WML	wireless markup language
VPDN	virtual private dial network	WNP	wireless local number portability
VPI	virtual path identifier	WRED	weighted random early discard
VPIM	voice protocol for Internet messaging	WS	work station
VPN	virtual private network	WSP	wireless session protocol
VPR	virtual path ring	WTA	wireless telephony application
VPRN	virtual private routed network	WUI	Web user interface
VRU	voice response unit	WVPN	wireless VPN
VSAT	very-small–aperture terminal	WWCUG	wireless/wireline closed user group
VSI	virtual switch interface	WWW	World Wide Web
VSM	virtual services management	XA	transaction management protocol
VSN	virtual service network	XC	cross-connect
VSR	very short reach	XD	extended distance
VT	virtual tributary	xDSL	[see DSL]
VTN	virtual transport network	XML	extensible markup language
VToA	voice traffic over ATM	XPM	cross-phase modulation
VVPN	voice virtual private network	XPS	cross-point switch
VXML	voice extensible markup language	xSP	specialized service provider
W3C	World Wide Web Consortium	XT	crosstalk
WAN	wide-area network	XTP	express transport protocol
WAP	wireless application protocol	Y2K	year 2000
WATS	wide-area telecommunications service		